# ADVANCES IN IMAGING AND ELECTRON PHYSICS

## VOLUME 119

Aspects of Image Processing and Compression

EDITOR-IN-CHIEF

## PETER W. HAWKES
*CEMES—Centre National de la Recherche Scientifique*
*Toulouse, France*


ASSOCIATE EDITORS

## BENJAMIN KAZAN
*Xerox Corporation*
*Palo Alto Research Center*
*Palo Alto, California*

## TOM MULVEY
*Department of Electronic Engineering and Applied Physics*
*Aston University*
*Birmingham, United Kingdom*

# Advances in Imaging and Electron Physics

Aspects of Image Processing and Compression

EDITED BY
PETER W. HAWKES

*CEMES—Centre National de la Recherche Scientifique*
*Toulouse, France*

VOLUME 119

## ACADEMIC PRESS
A Harcourt Science and Technology Company

San Diego  San Francisco  New York  Boston
London  Sydney  Tokyo

This book is printed on acid-free paper.

Copyright © 2001 by ACADEMIC PRESS

All Rights Reserved.
No part of this publication may be reproduced or transmitted in any form or by any means, electronic or mechanical, including photocopy, recording, or any information storage and retrieval system, without permission in writing from the Publisher.

The appearance of the code at the bottom of the first page of a chapter in this book indicates the Publisher's consent that copies of the chapter may be made for personal or internal use of specific clients. This consent is given on the condition, however, that the copier pay the stated per copy fee through the Copyright Clearance Center, Inc. (222 Rosewood Drive, Danvers, Massachusetts 01923), for copying beyond that permitted by Sections 107 or 108 of the U.S. Copyright Law. This consent does not extend to other kinds of copying, such as copying for general distribution, for advertising or promotional purposes, for creating new collective works, or for resale. Copy fees for pre-2001 chapters are as shown on the title pages. If no fee code appears on the title page, the copy fee is the same as for current chapters.
1076-5670/01 $35.00

Explicit permission from Academic Press is not required to reproduce a maximum of two figures or tables from an Academic Press chapter in another scientific or research publication provided that the material has not been credited to another source and that full credit to the Academic Press chapter is given.

Academic Press
*A Harcourt Science and Technology Company*
525 B Street, Suite 1900, San Diego, California 92101-4495, USA
http://www.academicpress.com

Academic Press
Harcourt Place, 32 Jamestown Road, London NW1 7BY, UK
http://www.academicpress.com

International Standard Serial Number: 1076-5670
International Standard Book Number: 0-12-014761-0

PRINTED IN THE UNITED STATES OF AMERICA
01  02  03  04  SB  9  8  7  6  5  4  3  2  1

# CONTENTS

CONTRIBUTORS . . . . . . . . . . . . . . . . . . . . . . . . . . . vii
PREFACE . . . . . . . . . . . . . . . . . . . . . . . . . . . . . . . ix
FUTURE CONTRIBUTIONS . . . . . . . . . . . . . . . . . . . . . xi

### Binary, Gray-Scale, and Vector Soft Mathematical Morphology: Extensions, Algorithms, and Implementations
M. I. VARDAVOULIA, A. GASTERATOS, AND I. ANDREADIS

I. Introduction . . . . . . . . . . . . . . . . . . . . . . . . . . . 1
II. Standard Mathematical Morphology . . . . . . . . . . . . . . 3
III. Soft Mathematical Morphology . . . . . . . . . . . . . . . . 14
IV. Soft Morphological Structuring Element Decomposition . . . 23
V. Fuzzy Soft Mathematical Morphology . . . . . . . . . . . . . 27
VI. Implementations . . . . . . . . . . . . . . . . . . . . . . . . 37
VII. Conclusions . . . . . . . . . . . . . . . . . . . . . . . . . . 52
References . . . . . . . . . . . . . . . . . . . . . . . . . . . 52

### Still Image Compression with Lattice Quantization in Wavelet Domain
MIKHAIL SHNAIDER AND ANDREW P. PAPLIŃSKI

I. Introduction . . . . . . . . . . . . . . . . . . . . . . . . . . . 56
II. Quantization of Wavelet Coefficients . . . . . . . . . . . . . 57
III. Lattice Quantization Fundamentals . . . . . . . . . . . . . . 70
IV. Lattices . . . . . . . . . . . . . . . . . . . . . . . . . . . . . 75
V. Quantization Algorithms for Selected Lattices . . . . . . . . 85
VI. Counting the Lattice Points . . . . . . . . . . . . . . . . . . 89
VII. Scaling Algorithm . . . . . . . . . . . . . . . . . . . . . . . 97
VIII. Selecting a Lattice for Quantization . . . . . . . . . . . . . . 100
IX. Entropy Coding of Lattice Vectors . . . . . . . . . . . . . . . 105
X. Experimental Results . . . . . . . . . . . . . . . . . . . . . 111
XI. Conclusions . . . . . . . . . . . . . . . . . . . . . . . . . . 116
XII. Cartan Matrices of Some Root Systems . . . . . . . . . . . . 117
References . . . . . . . . . . . . . . . . . . . . . . . . . . . 119

### Morphological Scale-Spaces
PAUL T. JACKWAY

I. Introduction . . . . . . . . . . . . . . . . . . . . . . . . . . . 124
II. Multiscale Morphology . . . . . . . . . . . . . . . . . . . . 131

III. Multiscale Dilation-Erosion Scale-Space . . . . . . . . . . . . . 139
IV. Multiscale Closing-Opening Scale-Space. . . . . . . . . . . . . 146
V. Fingerprints in Morphological Scale-Space . . . . . . . . . . . . 153
VI. Structuring Functions for Scale-Space . . . . . . . . . . . . . . 161
VII. A Scale-Space for Regions. . . . . . . . . . . . . . . . . . . . . 170
VIII. Summary, Limitations, and Future Work . . . . . . . . . . . . . 179
IX. Appendix . . . . . . . . . . . . . . . . . . . . . . . . . . . . . . 181
References . . . . . . . . . . . . . . . . . . . . . . . . . . . . 186

### The Processing of Hexagonally Sampled Images
RICHARD C. STAUNTON

I. Introduction . . . . . . . . . . . . . . . . . . . . . . . . . . . . . 192
II. Image Sampling on a Hexagonal Grid . . . . . . . . . . . . . . . 196
III. Processor Architecture . . . . . . . . . . . . . . . . . . . . . . 218
IV. Binary Image Processing . . . . . . . . . . . . . . . . . . . . . 237
V. Monochrome Image Processing. . . . . . . . . . . . . . . . . . 247
VI. Conclusions . . . . . . . . . . . . . . . . . . . . . . . . . . . . 256
References . . . . . . . . . . . . . . . . . . . . . . . . . . . . 260

### Space-Variant Two-Dimensional Filtering of Noisy Images
ALBERTO DE SANTIS, ALFREDO GERMANI, AND LEOPOLDO JETTO

I. Introduction . . . . . . . . . . . . . . . . . . . . . . . . . . . . . 268
II. Kalman Filtering . . . . . . . . . . . . . . . . . . . . . . . . . . 271
III. The Image Model . . . . . . . . . . . . . . . . . . . . . . . . . 276
IV. Image Restoration . . . . . . . . . . . . . . . . . . . . . . . . . 285
V. New Research Developments. . . . . . . . . . . . . . . . . . . 294
VI. Numerical Results . . . . . . . . . . . . . . . . . . . . . . . . . 303
VII. Conclusions . . . . . . . . . . . . . . . . . . . . . . . . . . . . 310
VIII. Appendices . . . . . . . . . . . . . . . . . . . . . . . . . . . . 310
References . . . . . . . . . . . . . . . . . . . . . . . . . . . . 316

INDEX . . . . . . . . . . . . . . . . . . . . . . . . . . . . . . . . . . 319

# CONTRIBUTORS

Numbers in parentheses indicate the pages on which the authors' contributions begin.

I. ANDREADIS (1), Department of Electrical and Computer Engineering, Democritus University of Thrace, GR-67100 Xanthi, Greece

ALBERTO DE SANTIS (267), Dipartimento di Informatica e Sistemistica, Università degli Studi "La Sapienza" di Roma, 00184 Rome, Italy

A. GASTERATOS (1), Department of Electrical and Computer Engineering, Democritus University of Thrace, GR-67100 Xanthi, Greece

ALFREDO GERMANI (267), Dipartimento di Ingegneria Elettrica, Università dell'Aquila, 67100 Monteluco (L'Aquila), Italy and Istituto di Analisi dei Sistemi ed Informatica del CNR–Viale Manzoni 30, 00185, Rome, Italy

PAUL T. JACKWAY (123), Cooperative Research Centre for Sensor Signal and Information Processing, School of Computer Science and Electrical Engineering, The University of Queensland, Brisbane, Queensland 4072, Australia

LEOPOLDO JETTO (267), Dipartimento di Elettronica e Automatica, Università di Ancona, 60131 Ancona, Italy

ANDREW P. PAPLIŃSKI (55), School of Computer Science and Software Engineering, Monash University, Clayton, Victoria 3168, Australia

MIKHAIL SHNAIDER (55), Telstra Research Laboratories, Ormond, Victoria 3204, Australia

RICHARD C. STAUNTON (191), School of Engineering, University of Warwick, Coventry CV4 7AL, United Kingdom

M. I. VARDAVOULIA (1), Department of Electrical and Computer Engineering, Democritus University of Thrace, GR-67100 Xanthi, Greece

# PREFACE

The latest volume in these Advances is a new venture in the series. Regular readers will be aware that almost every volume contains several contributions, which fall within the very broad catchment area of AIEP but which frequently have little in common. For many years, successive editors at Academic Press have urged me to submit "thematic volumes" but I have resisted this, partly because it seems to me preferable that every reader can hope to find something for him in every volume. And also, long experience of authors (among whom I include myself) has taught me that deadlines are regarded as boundlessly elastic: thematic volumes would inevitably be delayed to the justified annoyance of those who did send in their chapters on time. A former Academic Press editorial director even went to the trouble of publishing an article on "The pitfalls of publishing," the title of which is eloquent (A. Watkinson, *Nature* **300**, 1982, 111): "All publishers meet people who will never contribute a chapter again [to a multi-author work] because the experience has been too horrific. The edited book is where the publisher–author interface is raw and bleeding. What happens and what can go wrong?"

Nevertheless, there is much to be said for gathering related materials between the same two covers, and the present volume is the first of two thematic volumes of AIEP in which a group of chapters from earlier volumes is reprinted after such updating and revision as the authors wish to make. Here, the theme is Image Processing and Compression, with chapters on mathematical morphology, image compression by means of wavelets, hexagonal sampling, and space-variant filtering. These all seemed to me important enough to warrant reissue in this form, though the choice was exceedingly difficult and inevitably invidious. I am most grateful to the contributors to this volume for agreeing to republication and for the work of revision. Their chapters first appeared in vol. 110 (M. I. Vardavoulia, A. Gasteratos, and I. Andreadis), vol. 109 (M. Shnaider and A. P. Paplinski, vol. 99 (P. T. Jackway and A. de Santis, A. Germani and L. Jetto), and vol. 107 (R. C. Staunton).

A further thematic volume is planned (vol. 121) on Electron Microscopy and Electron Holography, after which we resume publication of original surveys as usual. A glance at the following list of future contributions shows how abundant and wide-ranging is the material in prospect.

Peter Hawkes

# FUTURE CONTRIBUTIONS

**T. Aach**
Lapped transforms

**G. Abbate**
New developments in liquid-crystal-based photonic devices

**S. Ando**
Gradient operators and edge and corner detection

**D. Antzoulatos**
Use of the hypermatrix

**A. Arnéodo, N. Decoster, P. Kestener and S. Roux**
A wavelet-based method for multifractal image analysis

**M. Barnabei and L. Montefusco**
Algebraic aspects of signal and image processing

**L. Bedini, E. Salerno and A. Tonazzini (vol. 120)**
Discontinuities and image restoration

**C. Beeli**
Structure and microscopy of quasicrystals

**I. Bloch**
Fuzzy distance measures in image processing

**R. D. Bonetto (vol. 120)**
Characterization of texture in scanning electron microscope images

**G. Borgefors**
Distance transforms

**A. Carini, G. L. Sicuranza and E. Mumolo**
V-vector algebra and Volterra filters

**Y. Cho**
Scanning nonlinear dielectric microscopy

**E. R. Davies**
Mean, median and mode filters

**H. Delingette**
Surface reconstruction based on simplex meshes

**A. Diaspro**
Two-photon excitation in microscopy

**R. G. Forbes**
Liquid metal ion sources

**E. Förster and F. N. Chukhovsky**
X-ray optics

**A. Fox**
The critical-voltage effect

**L. Frank and I. Müllerová**
Scanning low-energy electron microscopy

**A. Garcia**
Sampling theory

**L. Godo & V. Torra**
Aggregation operators

**P. Hartel, D. Preikszas, R. Spehr, H. Mueller and H. Rose (vol. 120)**
Design of a mirror corrector for low-voltage electron microscopes

**P. W. Hawkes**
Electron optics and electron microscopy: conference proceedings and abstracts as source material

**M. I. Herrera**
The development of electron microscopy in Spain

**J. S. Hesthaven**
Higher-order accuracy computational methods for time-domain electromagnetics

**K. Ishizuka**
Contrast transfer and crystal images

**I. P. Jones**
ALCHEMI

**W. S. Kerwin and J. Prince**
The kriging update model

**B. Kessler**
Orthogonal multiwavelets

**G. Kögel**
Positron microscopy

**W. Krakow**
Sideband imaging

**N. Krueger**
The application of statistical and deterministic regularities in biological and artificial vision systems

**B. Lahme**
Karhunen–Loeve decomposition

**J. Marti (vol. 120)**
Image segmentation

**C. L. Matson**
Back-propagation through turbid media

**S. Mikoshiba and F. L. Curzon**
Plasma displays

**M. A. O'Keefe**
Electron image simulation

**N. Papamarkos and A. Kesidis**
The inverse Hough transform

**M. G. A. Paris and G. d'Ariano**
Quantum tomography

**C. Passow**
Geometric methods of treating energy transport phenomena

**E. Petajan**
HDTV

**F. A. Ponce**
Nitride semiconductors for high-brightness blue and green light emission

**T.-C. Poon**
Scanning optical holography

**H. de Raedt, K. F. L. Michielsen and J. Th. M. Hosson**
Aspects of mathematical morphology

**H. Rauch**
The wave-particle dualism

**D. Saad, R. Vicente and A. Kabashima**
Error-correcting codes

**O. Scherzer**
Regularization techniques

**G. Schmahl**
X-ray microscopy

**S. Shirai**
CRT gun design methods

**T. Soma**
Focus-deflection systems and their applications

**I. Talmon**
Study of complex fluids by transmission electron microscopy

**M. Tonouchi**
Terahertz radiation imaging

**N. M. Towghi**
$I_p$ norm optimal filters

**T. Tsutsui and Z. Dechun**
Organic electroluminescence, materials and devices

**Y. Uchikawa**
Electron gun optics

**D. van Dyck**
Very high resolution electron microscopy

**J. S. Walker**
Tree-adapted wavelet shrinkage

**C. D. Wright and E. W. Hill**
Magnetic force microscopy

**F. Yang and M. Paindavoine**
Pre-filtering for pattern recognition using wavelet transforms and neural networks

**M. Yeadon**
Instrumentation for surface studies

**S. Zaefferer**
Computer-aided crystallographic analysis in TEM

# Binary, Gray-Scale, and Vector Soft Mathematical Morphology: Extensions, Algorithms, and Implementations

## M. I. VARDAVOULIA, A. GASTERATOS, AND I. ANDREADIS

*Department of Electrical and Computer Engineering*
*Democritus University of Thrace, GR-67100 Xanthi, Greece*

I. Introduction . . . . . . . . . . . . . . . . . . . . . . . . . . . . 1
II. Standard Mathematical Morphology . . . . . . . . . . . . . . . . . 3
   A. Binary Morphology . . . . . . . . . . . . . . . . . . . . . . . 3
   B. Basic Algebraic Properties . . . . . . . . . . . . . . . . . . . . 5
   C. Gray-scale Morphology with Flat Structuring Elements . . . . . . . . . 6
   D. Gray-scale Morphology with Gray-scale Structuring Elements . . . . . . . 7
   E. Vector Morphology for Color Image Processing . . . . . . . . . . . . 9
   F. Fuzzy Morphology . . . . . . . . . . . . . . . . . . . . . . . 13
III. Soft Mathematical Morphology . . . . . . . . . . . . . . . . . . . 14
   A. Binary Soft Morphology . . . . . . . . . . . . . . . . . . . . . 14
   B. Gray-scale Soft Morphology with Flat Structuring Elements . . . . . . . 15
   C. Gray-scale Soft Morphology with Gray-scale Structuring Elements . . . . . 16
   D. Vector Soft Morphology for Color Image Processing . . . . . . . . . . 16
IV. Soft Morphological Structuring Element Decomposition . . . . . . . . . . 23
V. Fuzzy Soft Mathematical Morphology . . . . . . . . . . . . . . . . . 27
   A. Definitions . . . . . . . . . . . . . . . . . . . . . . . . . . 27
   B. Compatibility with Soft Mathematical Morphology . . . . . . . . . . . 34
   C. Algebraic Properties of Fuzzy Soft Mathematical Morphology . . . . . . . 35
VI. Implementations . . . . . . . . . . . . . . . . . . . . . . . . . 37
   A. Threshold Decomposition . . . . . . . . . . . . . . . . . . . . 38
   B. Majority Gate . . . . . . . . . . . . . . . . . . . . . . . . . 39
      1. Algorithm Description . . . . . . . . . . . . . . . . . . . . 39
      2. Systolic Array Implementation for Soft Morphological Filtering . . . . . 40
      3. Architecture for Decomposition of Soft Morphological Structuring Elements 44
   C. Histogram Technique . . . . . . . . . . . . . . . . . . . . . . 47
   D. Vector Standard Morphological Operation Implementation . . . . . . . . 50
VII. Conclusions . . . . . . . . . . . . . . . . . . . . . . . . . . . 52
   References . . . . . . . . . . . . . . . . . . . . . . . . . . . 52

## I. INTRODUCTION

Mathematical morphology is an active and growing area of image processing and analysis. It is based on set theory and topology (Matheron, 1975; Serra, 1982; Haralick *et al.*, 1986; Giardina and Dougherty, 1988). Mathematical morphology studies the geometric structure inherent within the image. For this reason it uses a predetermined geometric shape known as the structuring element. Erosion, which is the basic morphological operation, quantifies the way in which the structuring element fits into the image. Mathematical morphology has provided solutions to many tasks, where image processing can be applied, such as in remote sensing, optical character recognition, radar image sequence recognition, medical imaging, etc. Soft mathematical morphology was introduced by Koskinen *et al.* (1991). In this approach the definitions of the standard morphological operations were slightly relaxed in such a way that a degree of robustness was achieved while most of their desirable properties were maintained. Soft morphological filters are less sensitive to additive noise and to small variations in object shape than standard morphological filters. They have found applications mainly in noise removal, in areas such as medical imaging and digital TV (Harvey, 1998).

The extension of concepts of mathematical morphology to color image processing is not straightforward, because there is not an obvious and unambiguous method of fully ordering vectors (Barnett, 1976). Componentwise morphological techniques, which are based on marginal subordering, do not take into consideration the correlation among color components; thus, they are not vector preserving. Transformation techniques have been used to decorrelate color components and then apply componentwise gray-scale techniques (Goutsias *et al.*, 1995). Morphological techniques that are based on reduced or partial subordering imply the existence of multiple suprema (infima); Thus, they could introduce ambiguity in the resultant data (Comer and Delp, 1998). The crucial point in developing a vector morphology theory for color image processing is the definition of vector-preserving infimum and supremum operators with unique outcome in a properly selected color space.

Another relatively new approach to mathematical morphology is fuzzy mathematical morphology. A fuzzy morphological framework was introduced by Sinha and Dougherty (1992). In this framework the images are treated not as crisp binary sets but as fuzzy sets. Set union and intersection have been replaced by fuzzy bold union and bold intersection, respectively, in order to formulate fuzzy erosion and dilation, respectively. This attempt to adapt mathematical morphology into fuzzy set theory is not unique. Several other attempts have been developed independently by researchers, and they are all described and discussed by Bloch and Maitre (1995). Several fuzzy mathematical morphologies are grouped and compared, and their properties are studied. A general framework unifying all these approaches is also demonstrated.

In this paper recent trends in soft mathematical morphology are presented. The rest of the paper is organized as follows. Binary, gray-scale, and vector standard morphological operations, their algebraic properties, and fuzzy morphology are discussed in Section II. Soft mathematical morphology is described in Section III. The definitions of vector soft morphological operations, their basic properties, and their use in color impulse noise attenuation are also presented in this section. A soft morphological structuring element-decomposition technique is introduced in Section IV. The definitions of fuzzy soft morphological operations and their algebraic properties are provided in Section V. Several implementations of soft morphological filters and an implementation of vector morphological filters are analyzed in Section VI. Concluding remarks are made in Section VII.

## II. Standard Mathematical Morphology

The considerations for the structuring element used by Haralick *et al.* (1987) have been adopted for the basic morphological operations. Also, the notations of the extensions of the basic morphological operations (soft morphology, fuzzy morphology and fuzzy soft morphology) are based on the same consideration. Moreover, throughout the paper the discrete case is considered, i.e., all sets belong to the Cartesian grid $Z^2$.

### A. Binary Morphology

Let the set $A$ denote the image under process and the set $B$ denote the structuring element. Binary erosion and dilation are defined:

$$A \ominus B = \bigcap_{x \in B} (A)_{-x} \tag{1}$$

and

$$A \oplus B = \bigcup_{x \in B} (A)_x, \tag{2}$$

respectively, where $A$, $B$ are sets of $Z^2$ and $(A)_x$ is the translation of $A$ by $x$, which is defined as follows:

$$(A)_x = \{c \in Z^2 \mid c = a + x \text{ for some } a \in A\} \tag{3}$$

A case of binary erosion and dilation is illustrated in Examples II.1 and II.2, respectively.

***Example II.1*** $A \ominus B$ results from $A_{(0,0)} \cap A_{(0,-1)}$, according to Eq. (1). The adopted coordinate system is (row, column). The arrows denote the origin of the coordinate system and its direction.

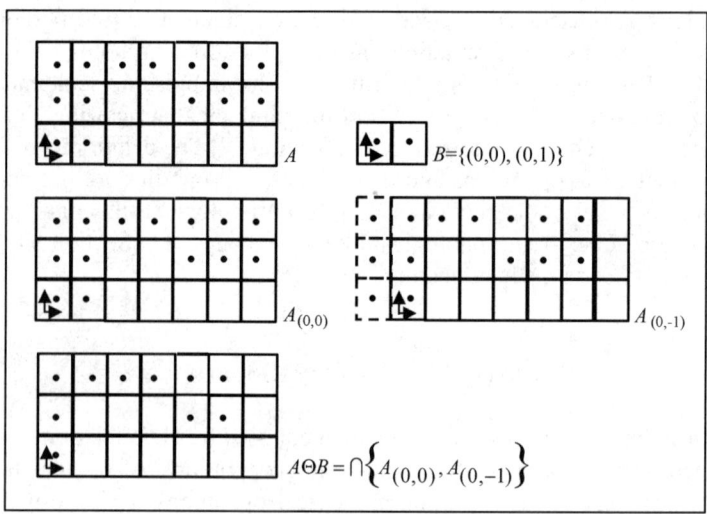

***Example II.2***  $A \oplus B$ results from $A_{(0,0)} \cup A_{(0,1)}$, according to Eq. (2).

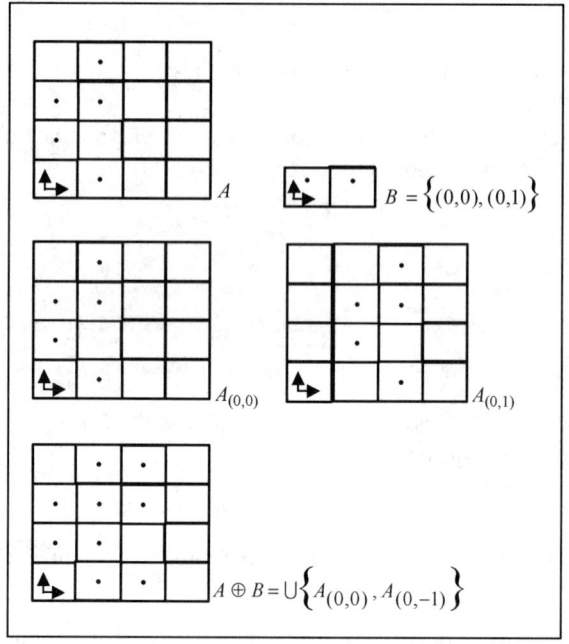

The definitions of binary opening and closing are

$$A \circ B = (A \ominus B) \oplus B \qquad (4)$$

and
$$A \bullet B = (A \oplus B) \ominus B, \tag{5}$$
respectively.

## B. Basic Algebraic Properties

The basic algebraic properties of the morphological operations are provided in this section.

**Theorem II.1** *Duality Theorem*
*Erosion and dilation are dual operations:*
$$(A \ominus B)^C = A^C \oplus B^S, \tag{6}$$
*where $A^C$ is the complement of A, defined as*
$$A^C = \{x \in Z^2 \mid x \notin A\}, \tag{7}$$
*and $B^S$ is the reflection of B, defined as*
$$B^S = \{x \mid \text{for some } b \in B, x = -b\}. \tag{8}$$
*Opening and closing are also dual operations:*
$$(A \bullet B)^C = A^C \circ B^S \tag{9}$$

**Theorem II.2** *Translation Invariance*
*Both erosion and dilation are translation invariant operations:*
$$(A)_x \oplus B = (A \oplus B)_x \tag{10}$$
and
$$(A)_x \ominus B = (A \ominus B)_x, \tag{11}$$
respectively.

**Theorem II.3** *Increasing Operations*
*Both erosion and dilation are increasing operations:*
$$A \subseteq B \Rightarrow A \ominus C \subseteq B \ominus C, \tag{12}$$
$$A \subseteq B \Rightarrow A \oplus D \subseteq B \oplus D. \tag{13}$$

**Theorem II.4** *Distributivity*
*Erosion distributes over set intersection and dilation distributes over set union:*
$$(A \cap B) \ominus C = (A \ominus C) \cap (B \ominus C) \tag{14}$$
and
$$(A \cup B) \oplus C = (A \oplus C) \cup (B \oplus C), \tag{15}$$
respectively.

**Theorem II.5** *Antiextensivity–Extensivity*
Erosion is an antiextensive operation, provided that the origin belongs to the structuring element:

$$0 \in B \Rightarrow A \ominus B \subseteq A. \tag{16}$$

Similarly, dilation is extensive if the origin belongs to the structuring element:

$$0 \in B \Rightarrow A \subseteq A \oplus B. \tag{17}$$

**Theorem II.6** *Idempotency*
Opening and closing are idempotent, i.e., their successive applications do not further change the previously transformed result:

$$A \circ B = (A \circ B) \circ B \tag{18}$$

and

$$A \bullet B = (A \bullet B) \bullet B \tag{19}$$

### C. Gray-scale Morphology with Flat Structuring Elements

The definitions of morphological erosion and dilation of a function $f: F \to Z$ by a flat structuring element (set) $B$ are

$$(f \ominus B)(x) = \min\{f(y) \mid y \in (B)_x\} \tag{20}$$

and

$$(f \oplus B)(x) = \max\{f(y) \mid y \in (B^S)_x\}, \tag{21}$$

respectively, where $x, y \in Z^2$ are the spatial coordinates and $F \subseteq Z^2$ is the domain of the gray-scale image (function).

Examples II.3 and II.4 demonstrate how we can use Eqs. (20) and (21) to perform erosion and dilation, respectively, of a function by a flat structuring element.

**Example II.3** $f(1, 2) = 3$, $x = (1, 2)$, $B = \{(0, 0), (0, 1), (-1, 0), (-1, 1)\}$. According to Eq. (20):

$$\begin{aligned} f \ominus g(1, 2) &= \min\{f(0+1, 0+2), f(0+1, 1+2), f(-1+1, 0+2), \\ &\quad f(-1+1, 1+2)\} \\ &= \min\{f(1, 2), f(1, 3), f(0, 2), f(0, 3)\} = \min\{3, 2, 7, 5\} = 2 \end{aligned}$$

# EXTENSIONS, ALGORITHMS, AND IMPLEMENTATIONS 7

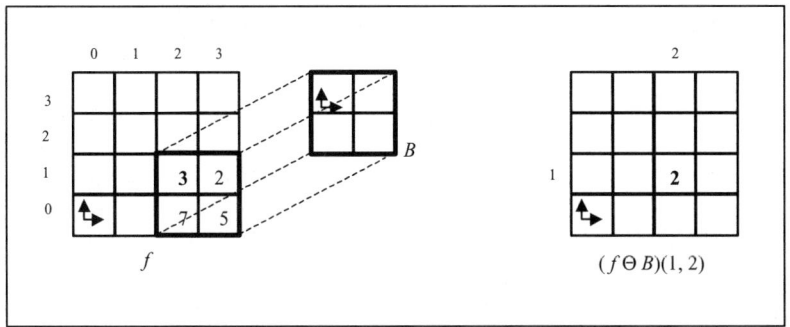

**Example II.4** $f(1, 2) = 3$, $x = (1, 2)$, $B = \{(0, 0), (0, 1), (-1, 0), (-1, 1)\}$ and, consequently, $B^S = \{(0, 0), (0, -1), (1, 0), (1, -1)\}$. According to Eq. (21):

$$(f \oplus B)(1, 2) = \max\{f(0+1, 0+2), f(0+1, -1+2), f(1+1, 0+2), f(1+1, -1+2)\}$$
$$= \max\{f(1, 2), f(1, 1), f(2, 2), f(2, 1)\}$$
$$= \max\{3, 6, 8, 5\} = 8.$$

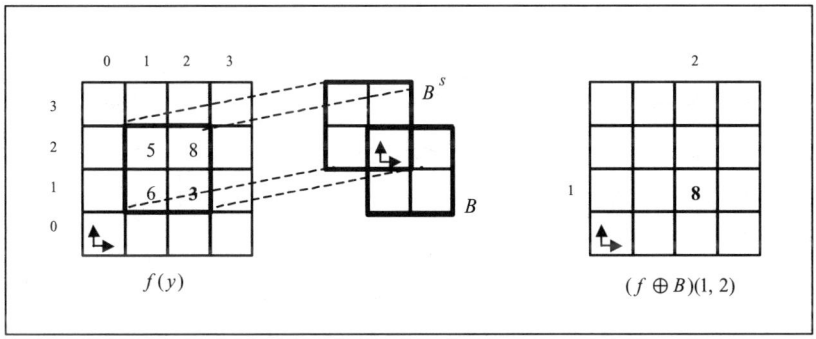

## D. Gray-scale Morphology with Gray-scale Structuring Elements

The definitions of erosion and dilation of a function $f: F \to Z$ by a gray-scale structuring element $g: G \to Z$ are

$$(f \ominus g)(x) = \min_{y \in G}\{f(x + y) - g(y)\} \tag{22}$$

and

$$(f \oplus g)(x) = \max_{\substack{y \in G \\ x-y \in F}}\{f(x - y) + g(y)\}, \tag{23}$$

respectively, where $x, y \in Z^2$ are the spatial coordinates and $F, G \subseteq Z^2$ are the domains of the gray-scale image (function) and gray-scale structuring element, respectively.

An application of Eqs. (22) and (23) is illustrated in Examples 5 and 6, respectively.

**Example II.5**  $f(1, 2) = 3, x = (1, 2), G = \{(0, 0), (0, 1), (-1, 0), (-1, 1)\}$. According to Eq. (22):

$$\begin{aligned} f \ominus g(1, 2) &= \min\{f(1+0, 2+0) - g(0, 0), f(1+0, 2+1) - g(0, 1), \\ &\quad f(1-1, 2+0) - g(-1, 0), f(1-1, 2+1) - g(-1, 1)\} \\ &= \min\{f(1, 2) - g(0, 0), f(1, 3) - g(0, 1), f(0, 2) - g(-1, 0), \\ &\quad f(0, 3) - g(-1, 1)\} \\ &= \min\{3 - 2, 2 - 1, 7 - 4, 5 - 3)\} \\ &= \min\{1, 1, 3, 2\} = 1. \end{aligned}$$

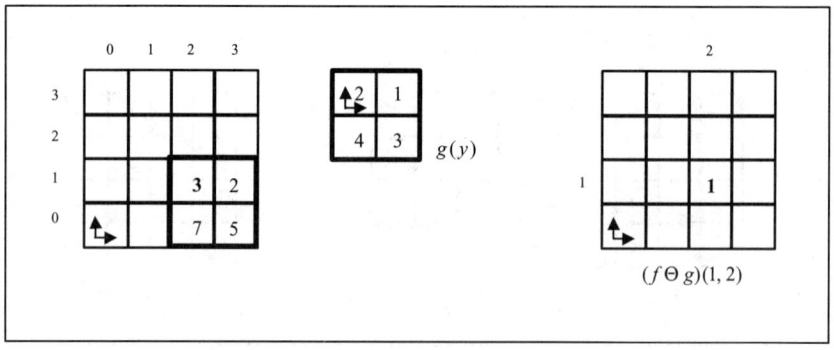

**Example II.6**  $f(1, 2) = 3, x = (1, 2), G = \{(0, 0), (0, 1), (-1, 0), (-1, 1)\}$. According to Eq. (23):

$$\begin{aligned} (f \oplus g)(1, 2) &= \max\{f(1-0, 2-0) + g(0, 0), f(1-0, 2-1) + g(0, 1), \\ &\quad f(1+1, 2-0) + g(-1, 0), f(1+1, 2-1) + g(-1, 1)\} \\ &= \max\{f(1, 2) + g(0, 0), f(1, 1) + g(0, 1), f(2, 2) + g(-1, 0), \\ &\quad f(2, 1), +g(-1, 1)\} \\ &= \max\{3 + 2, 6 + 1, 8 + 4, 5 + 3) \\ &= \max\{5, 7, 12, 8\} = 12\} \end{aligned}$$

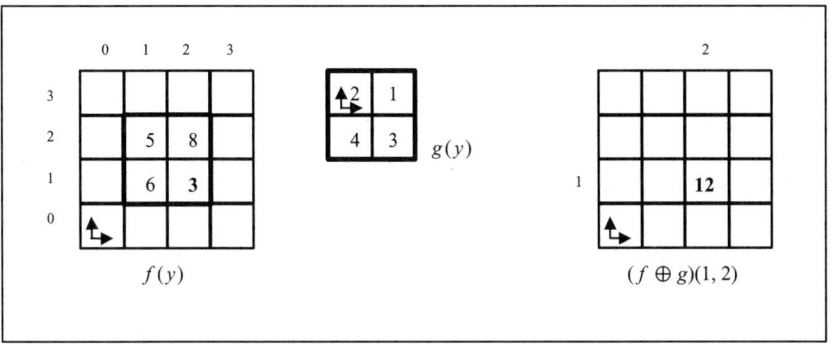

Gray-scale erosion and dilation possess the properties of binary erosion and dilation, respectively.

### E. Vector Morphology for Color Image Processing

Morphological operations suitable for color image processing are defined taking into consideration the following: (1) They should treat colors as vectors (i.e., they should not be componentwise operations), so they can utilize the correlation between color components (Goutsias *et al.*, 1995). (2) They should be vector preserving, so that they do not introduce new vectors (colors) not present in the original data (Talbot *et al.*, 1998). (3) They should produce unique results in all cases, so that they do not introduce ambiguity in the resultant data (Comer and Delp, 1998). (4) They should have the same basic properties with their gray-scale counterparts. (5) They should reduce to their gray-scale counterparts when the vector dimension is 1.

Thus, the definitions for vector morphological operations are extracted by means of vector-preserving supremum and infimum operators, properly defined in a selected color space—i.e., the HSV color space that is user oriented; it depicts colors in a way that approaches human perception. In this space a color is a vector with the components hue ($h \in [0, 360]$), saturation ($s \in [0, 1]$), and value ($v \in [0, 1]$). In the following sections such a color will be denoted by $c(h, s, v)$.

Consider that the HSV space is equipped with the $<_c$ conditional suborder relationship (Barnett, 1976), so that

$$c_1(h_1, s_1, v_1) <_c c_2(h_2, s_2, v_2) \Leftrightarrow \begin{cases} v_1 < v_2 \\ \text{or} \\ v_1 = v_2 \quad \text{and} \quad s_1 > s_2 \\ \text{or} \\ v_1 = v_2 \quad \text{and} \quad s_1 = s_2 \quad \text{and} \quad h_1 < h_2 \end{cases}$$
(24)

and

$$c_1(h_1, s_1, v_1) \underset{c}{=} c_2(h_2, s_2, v_2) \Leftrightarrow v_1 = v_2 \text{ and } s_1 = s_2 \text{ and } h_1 = h_2 \quad (25)$$

Then, if $SB_n$ is an arbitrary subset of the HSV space, which includes $n$ vectors $c_1(h_1, s_1, v_1), c_2(h_2, s_2, v_2), \ldots, c_n(h_n, s_n, v_n)$, the $\underset{c}{\wedge}$ infimum operator in $SB_n$ is defined as follows:

$$\underset{c}{\wedge} SB_n = \underset{c}{\wedge} \{c_1(h_1, s_1, v_1), c_2(h_2, s_2, v_2), \ldots, c_n(h_n, s_n, v_n)\}$$

$$= c_k(h_k, s_k, v_k): \begin{cases} v_k = \min\{v_1, v_2, \ldots, v_n\} \\ \text{if } \nexists i \neq j : v_i = v_j = \min\{v_1, v_2, \ldots, v_n\} \\ \quad 1 \leq i,j \leq n \\ \text{or} \\ v_k = v_i = v_j = \min\{v_1, v_2, \ldots, v_n\} \text{ and} \\ s_k = \max\{s_i, s_j\} \\ \text{if } \exists i \neq j : v_i = v_j = \min\{v_1, v_2, \ldots, v_n\} \text{ and } s_i \neq s_j \\ \quad 1 \leq i,j \leq n \\ \text{or} \\ v_k = v_i = v_j = \min\{v_1, v_2, \ldots, v_n\} \text{ and} \\ s_k = \max\{s_i, s_j\} \text{ and } h_k = \min\{h_i, h_j\} \\ \text{if } \exists i \neq j : v_i = v_j = \min\{v_1, v_2, \ldots, v_n\} \text{ and } s_i = s_j \\ \quad 1 \leq i,j \leq n \end{cases}$$

$$(26)$$

The $\underset{c}{\vee}$ supremum operator in $SB_n$ is defined in a similar way.

From previous definition it is obvious that the $\underset{c}{\wedge}$ and $\underset{c}{\vee}$ operators are vector preserving, because they always produce as a result one of the input vectors included in $SB_n$. These operators are used to define vector morphological operations that are vector preserving, as well.

Let us consider two functions $f, g : R^n \rightarrow \text{HSV}$, i.e., two $n$-dimensional color images, where $f$ is the image under process (the input image) and $g$ is the structuring element. If $f(x) = c(h_{xf}, s_{xf}, v_{xf}), x \in R^n$, then the definitions of vector erosion and dilation, respectively, are

$$(f \ominus g)(x) = \underset{c}{\wedge} \underset{y \in G}{\{f(x+y) - g(y)\}} \quad (27)$$

$$(f \oplus g)(x) = \underset{c}{\vee} \underset{\substack{y \in G \\ x-y \in F}}{\{f(x-y) + g(y)\}} \quad (28)$$

where $x, y \in R^n$ are the spatial coordinates and $F, G \subseteq R^n$ are the domains of the color input image (function) and the color structuring element, respectively. Moreover, $\ominus$, $\oplus$ denote vector subtraction and addition, respectively, which are defined as follows:

$$f(k) - g(k) = c(h_{kf} - h_{kg}, s_{kf} - s_{kg}, v_{kf} - v_{kg}), \quad k \in R^n \quad (29)$$

with

$$h_{kf} - h_{kg} = 0 \quad \text{if } h_{kf} - h_{kg} < 0$$
$$s_{kf} - s_{kg} = 0 \quad \text{if } s_{kf} - s_{kg} < 0$$
$$v_{kf} - v_{kg} = 0 \quad \text{if } v_{kf} - v_{kg} < 0$$

and

$$f(k) + g(k) = c(h_{kf} + h_{kg}, s_{kf} + s_{kg}, v_{kf} + v_{kg}) \tag{30}$$

with

$$h_{kf} + h_{kg} = 360 \quad \text{if } h_{kf} + h_{kg} > 360$$
$$s_{kf} + s_{kg} = 1 \quad \text{if } s_{kf} + s_{kg} > 1$$
$$v_{kf} - v_{kg} = 1 \quad \text{if } v_{kf} + v_{kg} > 1$$

Vector opening and closing are defined similarly to their gray-scale counterparts.

It has been proven that the defined vector morphological operations possess the same basic properties with their gray-scale counterparts: extensivity or antiextensivity, increasing or decreasing monotony, translation invariance, and duality. In addition, they are identical to their gray-scale counterparts when the vector dimension is 1. Consequently, the proposed vector morphology is compatible to gray-scale morphology.

Examples II.7 and II.8 demonstrate the cases of vector erosion and dilation, respectively.

**Example II.7** $f(1, 2) = c(0, 0.8, 0.9)$, $x = (1, 2)$, $G = \{(0, 0), (0, 1), (-1, 0), (-1, 1)\}$. According to Eq. (27):

$$\begin{aligned}
f \ominus g(1, 2) &= \bigwedge_c \{f(1+0, 2+0) - g(0, 0), f(1+0, 2+1) - g(0, 1), \\
&\quad f(1-1, 2+0) - g(-1, 0), f(1-1, 2+1) - g(-1, 1)\} \\
&= \bigwedge_c \{f(1, 2) - g(0, 0), f(1, 3) - g(0, 1), f(0, 2) - g(-1, 0), \\
&\quad f(0, 3) - g(-1, 1)\} \\
&= \bigwedge_c \{c(0, 0.8, 0.9) - c(10, 0.4, 0.6), c(30, 0.5, 0.7) \\
&\quad - c(10, 0.4, 0.3), c(60, 0.5, 0.8) \\
&\quad - c(55, 0.4, 0.3), c(80, 0.9, 0.6) - c(75, 0.4, 0.3)\} \\
&= \bigwedge_c \{c(0, 0.4, 0.3), c(20, 0.1, 0.4), c(5, 0.1, 0.5), c(5, 0.5, 0.3)\} \\
&= c(5, 0.5, 0.3)
\end{aligned}$$

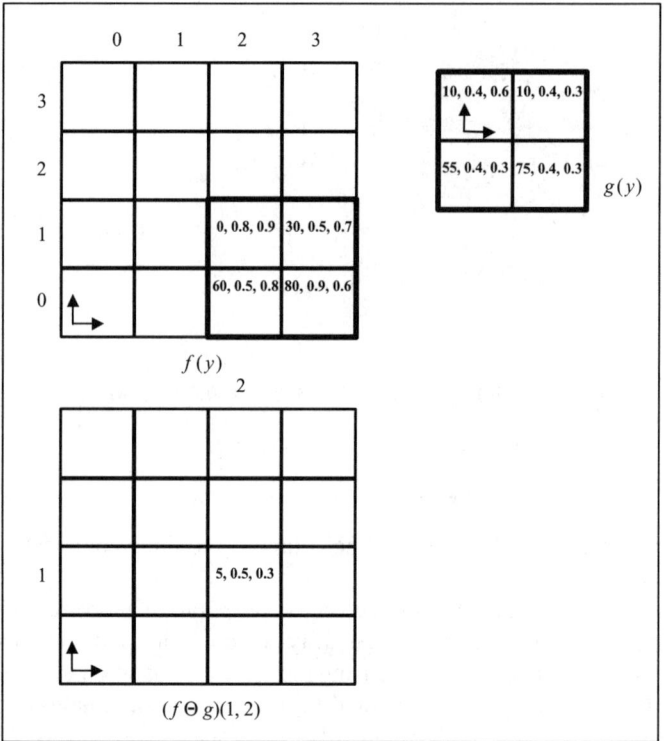

**Example II.8** $f(1, 2) = c(80, 0.6, 0.6)$, $x = (1, 2)$, $G = \{(0, 0), (0, 1), (-1, 0), (-1, 1)\}$. According to Eq. (28):

$$(f \oplus g)(1, 2) = \bigvee_c \{f(1 - 0, 2 - 0) + g(0, 0), f(1 - 0, 2 - 1) + g(0, 1),$$

$$f(1 + 1, 2 - 0) + g(-1, 0), f(1 + 1, 2 - 1) + g(-1, 1)\}$$

$$= \bigvee_c \{f(1, 2) + g(0, 0), f(1, 1) + g(0, 1), f(2, 2) + g(-1, 0),$$

$$f(2, 1) + g(-1, 1)\}$$

$$= \bigvee_c \{c(80, 0.6, 0.6) + c(30, 0.2, 0.2), c(40, 0.5, 0.6)$$

$$+ c(0, 0.2, 0.2), c(60, 0.4, 0.4)$$

$$+ c(20, 0.2, 0.2), c(0, 0.4, 0.7) + c(50, 0.2, 0.2)\}$$

$$= \bigvee_c \{c(110, 0.8, 0.8), c(40, 0.7, 0.8), c(80, 0.6, 0.6),$$

$$c(50, 0.6, 0.9)\} = c(50, 0.6, 0.9)$$

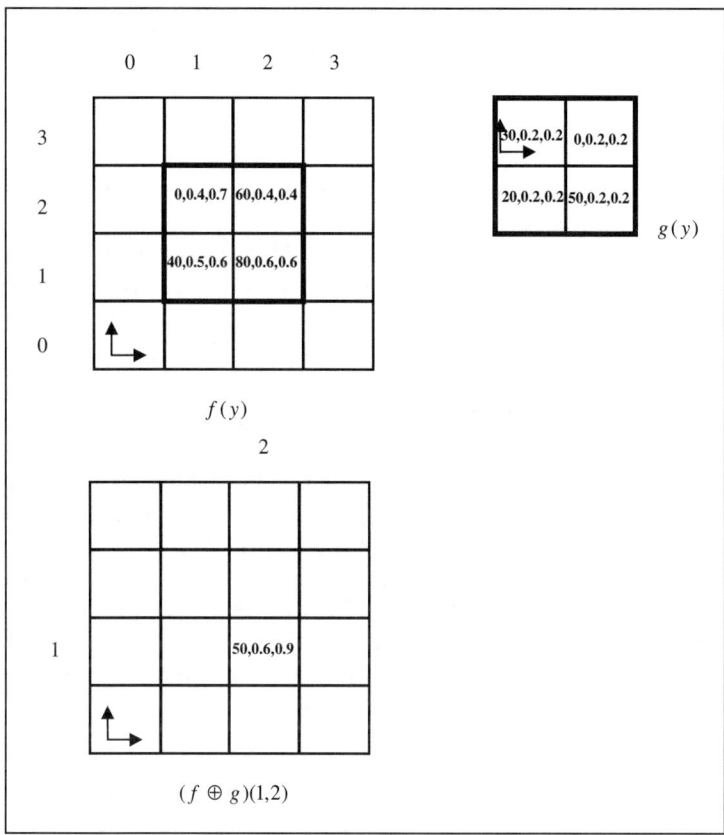

### F. Fuzzy Morphology

In this paper the definitions introduced by Sinha and Dougherty (1992) are used. These are a special case of the framework presented by Bloch and Maitre (1995). In this approach, fuzzy mathematical morphology is studied in terms of fuzzy fitting. The fuzziness is introduced by the degree to which the structuring element fits into the image. The operations of erosion and dilation of a fuzzy image by a fuzzy structuring element having a bounded support are defined in terms of their membership functions:

$$\mu_{A\ominus B}(x) = \min_{y\in B}[\min[1, 1 + \mu_A(x+y) - \mu_B(y)]]$$
$$= \min\left[1, \min_{y\in B}[1, +\mu_A(x+y) - \mu_B(y)]\right] \quad (31)$$

and

$$\mu_{A \oplus B}(x) = \max_{y \in B}[\max[0, \mu_A(x-y) + \mu_B(y) - 1]]$$
$$= \max\left[0, \max_{y \in B}[\mu_A(x-y) + \mu_B(y) - 1]\right] \qquad (32)$$

where $x, y \in Z^2$ are the spatial coordinates and $\mu_A$, $\mu_B$ are the membership functions of the image and the structuring element, respectively.

It is obvious from Eqs. (31) and (32) that the results of both fuzzy erosion and dilation have membership functions whose values are within the interval [0, 1].

### III. Soft Mathematical Morphology

In soft morphological operations, the maximum or the minimum operations used in standard gray-scale morphology are replaced by weighted order statistics. A weighted order statistic is a certain element of a list, the members of which have been ordered. Some of the members from the original unsorted list, participate with a weight greater than 1, i.e., they are repeated more than once, before sorting (David, 1981; Pitas and Venetsanolpoulos, 1990). Furthermore, in soft mathematical morphology the structuring element B is divided into two subsets; the core $B_1$ and the soft boundary $B_2$.

#### A. Binary Soft Morphology

The basic definitions of the binary soft erosion and dilation are (Pu and Shih, 1995):

$$(A \ominus [B_1, B_2, k])(x) = \{x \in A \mid (k \times \mathrm{Card}[A \cap (B_1)_x] + \mathrm{Card}[A \cap (B_2)_x]) \\ \geq k\mathrm{Card}[B_1] + \mathrm{Card}[B_2] - k + 1\} \qquad (33)$$

and

$$(A \oplus [B_1, B_2, k])(x) = \{x \in A \mid (k \times \mathrm{Card}[A \cap (B_1^S)_x]) \\ + \mathrm{Card}[A \cap (B_2^S)_x]) \geq k\} \qquad (34)$$

respectively, where $k$ is called the "order index," which determines the number of times that the elements of core participate into the result, and Card $[X]$ denotes the cardinality of set $X$, i.e., the number of the elements of $X$.

In the extreme case, where the order index $k=1$ or, alternatively, $B = B_1(B_2 = \emptyset)$, soft morphological operations are reduced to standard morphological operations.

***Example III.1*** The following example demonstrates a case of soft binary dilation and erosion.

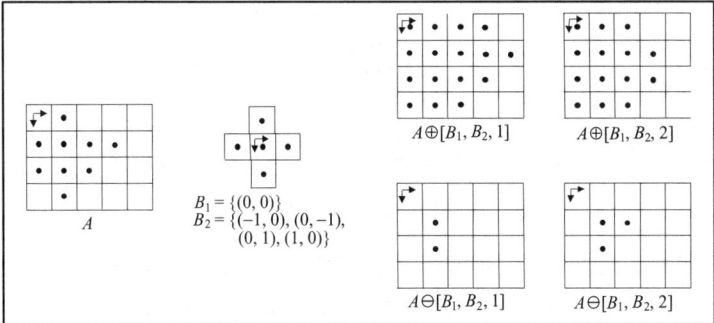

If $k > \text{Card}[B_2]$, soft morphological operations are affected only by the core $B_1$, i.e., using $B_1$ as the structuring element. Therefore, in this case the nature of soft morphological operations is not preserved (Kuosmanen and Astola, 1995; Pu and Shih, 1995). For this reason the constraint $k \leq \min\{\text{Card}(B)/2, \text{Card}(B_2)\}$ is used. In the preceding example $\min(\text{Card}(B)/2, \text{Card}(B_2)) = 2.5$; therefore, only the cases $k=1$ and $k=2$ are considered. For $k=1$ the results of both dilation and erosion are the same as those that would have been obtained by applying Eqs. (2) and (1), respectively.

### B. Gray-scale Soft Morphology with Flat Structuring Elements

The definitions of soft morphology were first introduced by Koskinen et al. (1991) as transforms of a function by a set. In the definition of soft dilation, the reflection of the structuring element is used, so that in the case of $k=1$ the definitions comply with (Haralick et al., 1986).

$$(f \ominus [B_1, B_2, k])(x) = \min{}^{(k)}(\{k \diamond f(y) \mid y \in (B_1)_x\} \cup (\{f(z) \mid z \in (B_2)_x\}) \tag{35}$$

and

$$(f \oplus [B_1, B_2, k])(x) = \max{}^{(k)}(\{k \diamond (y) \mid y \in (B_1^S)_x\} \cup (\{f(z) \mid z \in (B_2^S)_x\}) \tag{36}$$

respectively, where $\min^{(k)}$ and $\max^{(k)}$ are the $k$th smallest and the $k$th largest element of the multiset, respectively; a multiset is a collection of objects, where the repetition of objects is allowed and the symbol $\diamond$ denotes the repetition, i.e., $\{k \diamond f(x)\} = \{f(x), f(x), \ldots, f(x)\}$ ($k$ times).

### C. Gray-scale Soft Morphology with Gray-scale Structuring Elements

Soft morphological erosion of a gray-scale image $f: F \to Z$ by a soft gray-scale structuring element $[\alpha, \beta, k]: B \to Z$ is (Pu and Shih, 1995):

$$f \ominus [\alpha, \beta, k](x) = \min_{\substack{y \in B_1 \\ z \in B_2}}^{(k)} (\{k \diamond (f(x+y) - \alpha(y))\} \cup \{f(x+z) - \beta(z)\}) \tag{37}$$

Soft morphological dilation of $f$ by $[\alpha, \beta, k]$ is

$$f \oplus [\alpha, \beta, k](x) = \max_{\substack{(x-y),(x-z) \in F \\ z \in B_1 \\ z \in B_2}}^{(k)} (\{k \diamond (f(x-y) + \alpha(y))\} \cup \{f(x-z) - \beta(z)\}) \tag{38}$$

where $x, y, z \in Z^2$ are the spatial coordinates, $\alpha: B_1 \to Z$ is the core of the gray-scale structuring element, $\beta: B_2 \to Z$ is the soft boundary of the gray-scale structuring element, and $F, B_1, B_2 \subseteq Z^2$ are the domains of the gray-scale image, the core of the gray-scale structuring element, and the soft boundary of the gray-scale structuring element, respectively.

Figure 1 demonstrates one-dimensional soft morphological operations and the effect of the order index $k$. The same structuring element is used for both operations. It is a one-dimensional structuring element with five discrete values. The central value corresponds to its core and it is equal to 30. Additionally, it denotes the origin. The four remaining values belong to its soft boundary and they are equal to 20. From both Figures 1a and b it is obvious that the greater the value of the order index, the better the fitting.

### D. Vector Soft Morphology for Color Image Processing

In this section an approach to soft color image mathematical morphology is presented. This extends the vector standard morphology theory discussed in Section II.E in the same way that gray-scale soft morphology extends the gray-scale standard morphology theory. Vector soft morphology, like gray-scale soft morphology, aims at improving the behavior of vector standard morphological

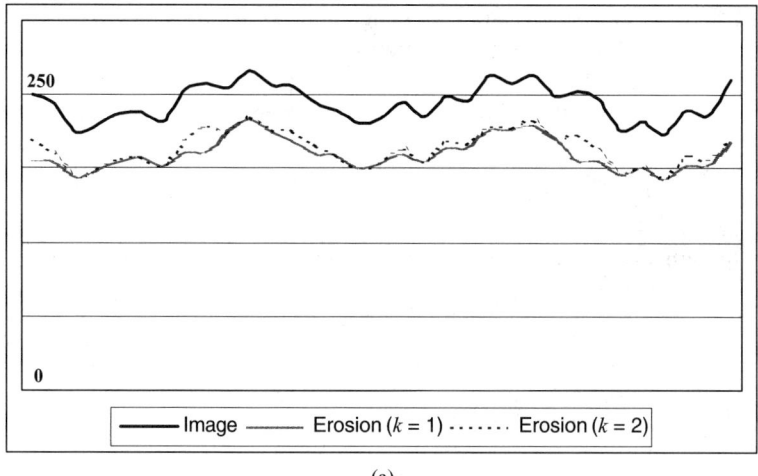

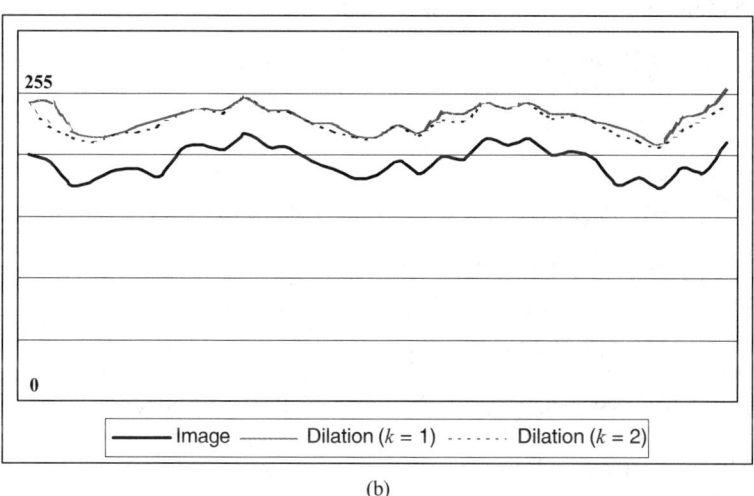

FIGURE 1. Illustration of one-dimensional soft morphological operations and the effect of the order index $k$; (a) soft erosion and (b) soft dilation.

filters in noisy environments. It retains the concept of splitting the structuring element in two parts: the core and the soft boundary. It also preserves the concept of the order index $k$, which implies that the core "weights" more than the soft boundary in the calculation of the result. Furthermore, it uses the relational operator $<_c$ (Eq. (24)) in order to rank the vector values included in a multiset of HSV space vectors. Here again the $k$th-order statistic is the result of the vector soft morphological operation.

Let $SB_n$ be an arbitrary subset of the HSV space, which includes $n$ vectors $c_1, c_2, \ldots, c_n$, and $SB_{n(\text{ord})}$ be the set of the ordered values $c_{(1)}, c_{(2)}, \ldots, c_{(n)}$ i.e.,

$$SB_{n(\text{ord})} = \{c_{(1)}, c_{(2)}, \ldots, c_{(n)}\}, \quad c_{(1)} \underset{c}{\leq} c_{(2)} \underset{c}{\leq} \cdots \underset{c}{\leq} c_{(n)}.$$

Then the $k$th smallest and the $k$th largest vector, respectively, in $SB_n$ are

$$\min{}_c^{(k)}(SB_n) = c_{(k)}, \quad 1 \leq k \leq n \tag{39}$$

and

$$\max{}_c^{(k)}(SB_n) = c_{(n-k+1)}, \quad 1 \leq k \leq n \tag{40}$$

Therefore, vector soft erosion and dilation of a color image $f$ by a color structuring element $g(f, g : R^n \to \text{HSV})$ are defined as follows.

$$(f \ominus [\beta, \alpha, k])(x) = \min{}_c^{(k)}(\{k \diamond (f(x+y) - a(y))\} \cup \{f(x+z) - \beta(z)\})$$
$$\scriptstyle y \in B_1 \atop z \in B_2$$
$$\tag{41}$$

$$(f \oplus [\beta, \alpha, k])(x) = \max{}_c^{(k)}(\{k \diamond (f(x+y) + a(y))\} \cup \{f(x-z) + \beta(z)\})$$
$$\scriptstyle (x-y), (x-z) \in F \atop z \in B_1 \atop z \in B_2$$
$$\tag{42}$$

where $x, y, z \in R^n$ are the spatial coordinates, $a: B_1 \to R^n$ is the core of the color structuring element, $\beta: B_2 \to R^n$ is the soft boundary of the color structuring element, and $F, B_1, B_2 \subseteq R^n$ are the domains of the color image, the core of the color structuring element, and the soft boundary of the color structuring element, respectively. In addition, $(-)$ and $(+)$ are the vector subtraction and addition operations defined in Eqs. (29) and (30), respectively. Vector soft opening and closing are defined similarly to their gray-scale counterparts.

Vector soft morphology is compatible to gray-scale soft morphology. In vector soft morphology, as in gray-scale soft morphology, the restriction $k \leq \min\{\text{Card}(B)/2, \text{Card}(B_2)\}$ for the order index $k$ ensures that the nature of soft morphological operations is preserved. Moreover, primary and secondary operations of vector soft morphology are reduced to their gray-scale counterparts when they are applied to gray-scale images. In addition, vector soft erosion,

FIGURE 2. (a) Original color image "Veta," (b) image after vector standard erosion by $g = [\beta, \alpha, 1]$, (c) image after vector soft erosion by $[\beta, \alpha, 2]$, and (d) image after vector soft erosion by $[\beta, \alpha, 4]$.

FIGURE 3. (a) Original color image "Veta's birthday," (b) image corrupted by 6% positive and negative HSV impulse noise with $p_1 = p_2 = p_3 = 0$ and $p_S = 1$, (c) resulting image after vector standard opening by $g = [\beta, \alpha, 1]$, (d) resulting image after vector soft opening by $[\beta, \alpha, 2]$, and (e) resulting image after vector soft opening by $[\beta, \alpha, 4]$.

dilation opening, and dilation closing possess the same basic properties with gray-scale soft erosion, dilation, opening, and closing, respectively.

**Theorem III.1** *Duality Theorem*
*Vector soft erosion and dilation are dual operations:*

$$-(f \oplus [\beta, \alpha, k])(x) = (-f \ominus [\beta^s, \alpha^s, k])(x) \qquad (43)$$

*Vector soft opening and closing are also dual operations:*

$$-(f \circ [\beta, \alpha, k]) = -f \bullet [\beta^s, \alpha^s, k] \qquad (44)$$

**Theorem III.2** *Translation Invariance*
*Vector soft erosion and dilation are translation invariant:*

$$(f \ominus [\beta, \alpha, k])_y + j = (f_y + j) \ominus [\beta, \alpha, k], \qquad y \in R^n, j \in \text{HSV} \quad (45)$$
$$(f \oplus [\beta, \alpha, k])_y + j = (f_y + j) \oplus [\beta, \alpha, k], \qquad y \in R^n, j \in \text{HSV} \quad (46)$$

*Vector soft opening and closing are translation invariant, as well:*

$$(f \circ [\beta, \alpha, k])_y + j = (f_y + j) \circ [\beta, \alpha, k], \qquad y \in R^n, j \in \text{HSV} \quad (47)$$
$$(f \bullet [\beta, \alpha, k])_y + j = (f_y + j) \bullet [\beta, \alpha, k], \qquad y \in R^n, j \in \text{HSV} \quad (48)$$

**Theorem III.3** *Increasing Operations*
*Vector soft erosion and dilation are monotonically increasing operations:*

$$f_1 \ll f_2 \Rightarrow \begin{cases} f_1 \ominus [\beta, \alpha, k] \ll f_2 \ominus [\beta, \alpha, k] \\ f_1 \oplus [\beta, \alpha, k] \ll f_2 \oplus [\beta, \alpha, k] \end{cases} \qquad (49)$$

where $g \ll f \Leftrightarrow G \subseteq F$ and $g(x) \underset{c}{\leq} f(x) \, \forall x \in G$.
*Vector opening and closing are monotonically increasing operations, as well:*

$$f_1 \ll f_2 \Rightarrow \begin{cases} f_1 \circ [\beta, \alpha, k] \ll f_2 \circ [\beta, \alpha, k] \\ f_1 \bullet [\beta, \alpha, k] \ll f_2 \bullet [\beta, \alpha, k] \end{cases} \qquad (50)$$

**Theorem III.4** *Antiextensivity–Extensivity*
*If the origin lies inside the core of the structuring element, vector soft erosion is antiextensive and vector soft dilation is extensive:*

$$0 \in \alpha \Rightarrow \begin{cases} (f \ominus [\beta, \alpha, k])(x) \underset{c}{\leq} f(x) \\ (f \oplus [\beta, \alpha, k])(x) \underset{c}{\geq} f(x) \end{cases} \qquad (51)$$

*On the contrary, vector soft opening is not in general antiextensive and vector soft closing is not in general extensive:*

$$\exists x \in R^n : \begin{cases} (f \circ [\beta, \alpha, k])(x) \underset{c}{\geq} f(x) \\ (f \bullet [\beta, \alpha, k])(x) \underset{c}{\geq} f(x) \end{cases} \quad (52)$$

**Theorem III.5** *Idempotency*
*Like their gray-scale counterparts, vector soft opening and closing are not, in general, idempotent.*

$$\exists f, g : R^n \to \text{HSV} \ \textit{such that} \begin{cases} f \circ [\beta, \alpha, k] \neq (f \circ [\beta, \alpha, k]) \circ [\beta, \alpha, k] \\ f \bullet [\beta, \alpha, k] \neq (f \bullet [\beta, \alpha, k]) \bullet [\beta, \alpha, k] \end{cases} \quad (53)$$

The main characteristic of morphological methods is that they take into consideration the geometrical shape of the objects to be analyzed. However, standard morphological operations are highly sensitive to noise. In some applications this sensitivity may cause problems: prefiltering to remove noise is necessary; if this prefiltering is not done very carefully, it may result in corruption of the shape of objects to be studied, thus degrading the overall performance of the system. Gray-scale soft mathematical morphology was introduced by Koskinen et al. (1991), as an extension of gray-scale standard mathematical morphology, in order to improve the behavior of gray-scale standard morphological filters in noisy environments: gray-scale soft morphological operations are less sensitive to impulse noise and to small variations in object shape (Koskinen and Astola, 1994) compared to the corresponding gray-scale standard morphological operations. Experimental results show that vector soft morphological operations act in a similar way with their gray-scale counterparts: they are advantageous regarding small detail preservation and impulse noise attenuation in comparison to the corresponding vector standard morphological operations. This is illustrated in Figures 2 and 3.

More specifically, Figure 2 demonstrates the effect of the order index $k$ in object shape and small detail preservation. From Figures 2b–d it is obvious that, as in gray-scale soft morphology, the greater the value of the order index, the better the detail preservation. Comparing Figures 2b–d, it can be also observed that the smaller the value of $k$, the closer the behavior of a vector soft morphological transform is to that of the corresponding vector standard morphological transform, just as in the case of soft gray-scale morphology (Kuosmanen and Astola, 1995). This is one more similarity of vector soft and gray-scale soft

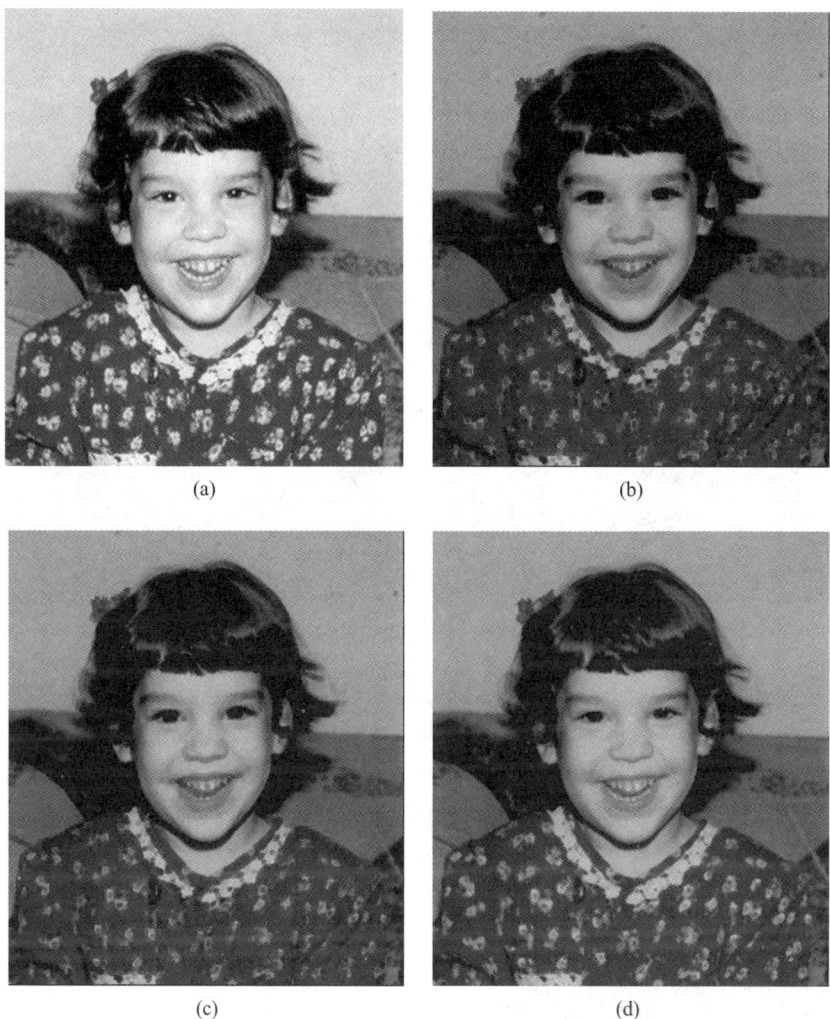

FIGURE 2. (a) Original color image "Veta," (b) image after vector standard erosion by $g = [\beta, \alpha, 1]$, (c) image after vector soft erosion by $[\beta, \alpha, 2]$, and (d) image after vector soft erosion by $[\beta, \alpha, 4]$.

morphological transforms. For instance, in Figure 2 it can be seen that, for soft vector erosion, $f \ominus g \leq_c f \ominus [\beta, \alpha, k] \leq_c f \ominus [\beta, \alpha, k+1] \leq_c f$, which holds for gray-scale soft erosion as well.

Figure 3 illustrates that vector soft morphological transforms are advantageous in color impulse noise elimination, in comparison to the corresponding

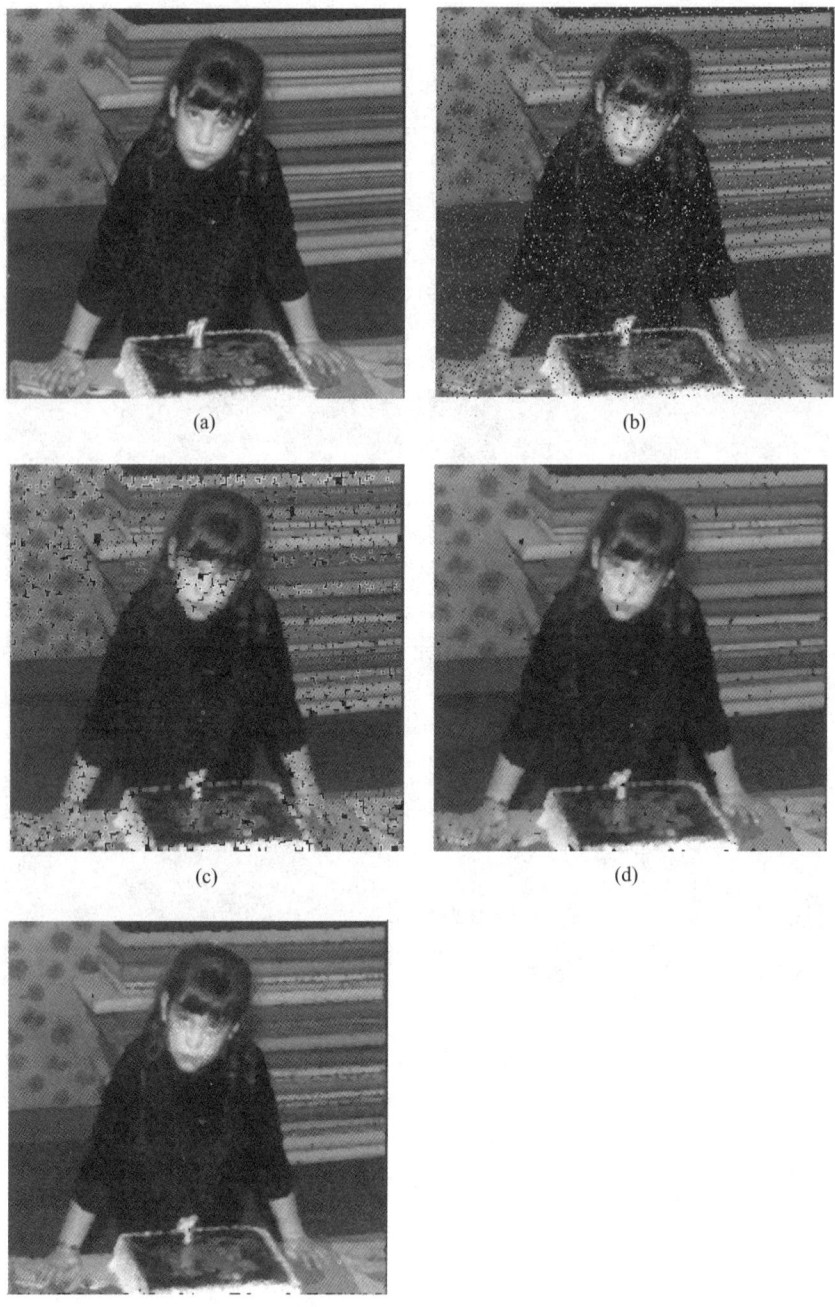

vector standard morphological transforms. It can be seen that vector soft opening (closing) removes both positive and negative color impulse noise, just as in the gray-scale case. It can also be observed that the increase of the order index $k$ increases the noise-removal capability and the detail-preservation capability, as well.

At this point it must be mentioned that a significant problem in the study of impulse noise removal from color images is the lack of a generally accepted multivariate impulse noise model. Recently various such models have been proposed. The following color impulse noise model (Plataniotis et al., 1999) has been used in our experiments:

$$c_n = \begin{cases} c_0 = c(h, s, v) & \text{with probability } (1 - p) \\ c(d, s, v) & \text{with probability } p_1 p \\ c(h, d, v) & \text{with probability } p_2 p \\ c(h, s, d) & \text{with probability } p_3 p \\ c(d, d, d) & \text{with probability } p_S p \end{cases} \quad (54)$$

where $c_0$ is the original vector (the noncontaminated color), $c_n$ is the noisy vector, and $d$ is the impulse value. Furthermore, $P$ is the degree of impulse noise distortion, $p_S = 1 - (p_1 + p_2 + p_3)$, and $p_1 + p_2 + p_3 \leq 1$. The positive or negative impulse value $d$ is properly placed in the range of each vector component.

## IV. SOFT MORPHOLOGICAL STRUCTURING ELEMENT DECOMPOSITION

A soft morphological structuring element decomposition technique is described in this section (Gasteratos et al., 1998c). According to this technique, the domain $B$ of the structuring element is divided into smaller, nonoverlapping subdomains B1, B2, ..., B$n$. Also, $B1 \cup B2 \cup \cdots \cup Bn = B$. The soft morphological structuring elements obtain values from these domains, and they are denoted by $[\lambda_1, \mu_1, k], [\lambda_2, \mu_2, k], \ldots, [\lambda_n, \mu_n, k]$, respectively. These have a common origin, which is the origin of the original structuring element. Additionally, the points of B that belong to its core are also points of the cores of B1, B2, ..., B$n$ and the points of B that belong to the soft boundary are also points

---

FIGURE 3. (a) Original color image "Veta's birthday," (b) image corrupted by 6% positive and negative HSV impulse noise with $p_1 = p_2 = p_3 = 0$ and $p_S = 1$, (c) resulting image after vector standard opening by $g = [\beta, \alpha, 1]$, (d) resulting image after vector soft opening by $[\beta, \alpha, 2]$, and (e) resulting image after vector soft opening by $[\beta, \alpha, 4]$.

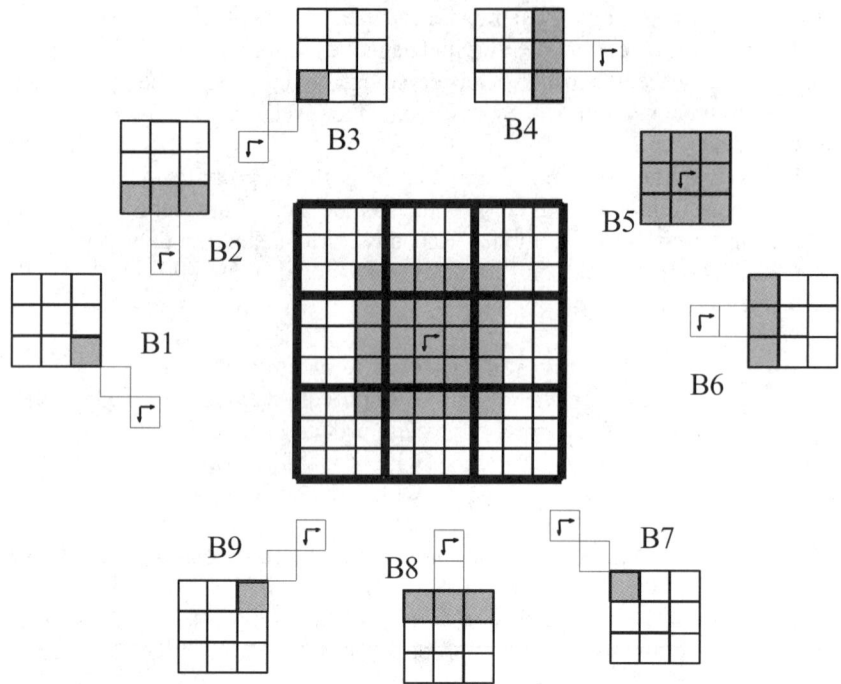

FIGURE 4. Example of a 4 × 4 soft morphological structuring element decomposition.

of the soft boundaries of B1, B2, ..., Bn. This process is graphically illustrated in Figure 4. In this figure the core of the structuring element is denoted by the shaded area.

Soft dilation and erosion are computed as follows:

$$f \oplus [\alpha, \beta, k](x) = \max_{i=1}^{n}{}^{(k)}\left[\max_{\substack{(x-y)\in B_1 \\ (x-z)\in B_2 \\ j=1}}^{k}{}^{(j)} (\{k \diamond (f(x-y) + \lambda_i(y))\} \cup \{f(x-z) + \mu_i(z)\})\right] \qquad (55)$$

$$f \ominus [\alpha, \beta, k](x) = \max_{i=1}^{n}{}^{(k)}\left[\max_{\substack{(x-y)\in B_1 \\ (x-z)\in B_2 \\ j=1}}^{k}{}^{(j)} (\{k \diamond (f(x+y) - \lambda_i(y))\} \cup \{f(x+z) - \mu_i(z)\})\right] \qquad (56)$$

respectively, where $B_1$ and $B_2$ are the domain of the core and the soft boundary of the large structuring element $[\alpha, \beta, k]: B \rightarrow Z$.

*Proof.*

$$\forall y \in B_1 : \alpha(y) = \bigcup_{i=1}^{n} \lambda_i(y)$$

$$\Rightarrow f(x-y) + \alpha(y) = \bigcup_{i=1}^{n}[f(x-y) + \lambda_i(y)], \quad (x-y) \in B_1$$

$$\Rightarrow k \diamond (f(x-y) + \alpha(y)) = k \diamond \left(\bigcup_{i=1}^{n}[f(x-y) + \lambda_i(y)]\right)$$

$$= k \diamond (f(x-y) + \lambda_1(y), f(x-y) + \lambda_2(y), \ldots, f(x-y) + \lambda_n(y)), (x-y) \in B_1 \quad (57)$$

Also,

$$\forall z \in B_2 : \beta(z) = \bigcup_{i=1}^{n} \mu_i(z)$$

$$\Rightarrow f(x-z) + \beta(z) = \bigcup_{i=1}^{n}[f(x-z) + \mu_i(z)]$$

$$= f(x-z) + \mu_1(z), f(x-z) + \mu_2(z), \ldots, f(x-z) + \mu_n(z), \quad (x-z) \in B_2 \quad (58)$$

Through Eqs. (38), (57), and (58) we obtain:

$$f \oplus [\alpha, \beta, k](x)$$

$$= \max_{\substack{(x-y) \in B_1 \\ (x-z) \in B_2}}^{(k)} \begin{pmatrix} \{k \diamond (f(x-y) + \lambda_1(y), f(x-y) \\ + \lambda_2(y), \ldots, f(x-y) + \lambda_n(y))\} \cup \\ \{f(x-z) + \mu_1(z), f(x-z) + \mu_2(z), \ldots, \\ f(x-z) + \mu_n(z)\} \end{pmatrix}$$

$$= \max_{\substack{(x-y) \in B_1 \\ (x-z) \in B_2}}^{(k)} \begin{pmatrix} \{k \diamond (f(x-y) + \lambda_1(y))\} \cup \{f(x-z) + \mu_1(z)\}, \\ \{k \diamond (f(x-y) + \lambda_2(y))\} \cup \{f(x-z) + \mu_2(z)\}, \\ \ldots \\ \{k \diamond (f(x-y) + \lambda_n(y))\} \cup \{f(x-z) + \mu_n(z)\} \end{pmatrix}$$

$$= \max_{\substack{(x-y) \in B_1 \\ (x-z) \in B_2 \\ i=1}}^{n} {}^{(k)} [\{k \diamond (f(x-y) + \lambda_i(y))\} \cup \{f(x-z) + \mu_i(z)\}]$$

This equation can be expressed in terms of order statistics of the multiset as follows:

$$f \oplus [\alpha, \beta, k](x)$$
$$= \max_{i=1}^{n\,(k)} [\max_{\substack{(x-y)\in B_1 \\ (x-z)\in B_2}}^{(N)} (\{k \diamond (f(x-y) + \lambda_i(y))\} \cup \{f(x-z) + \mu_i(z)\}),$$

$$\max_{\substack{(x-y)\in B_1 \\ (x-z)\in B_2}}^{(N-1)} (\{k \diamond (f(x-y) + \lambda_i(y))\} \cup \{f(x-z) + \mu_i(z)\}),$$

$$\vdots$$

$$\max_{\substack{(x-y)\in B_1 \\ (x-z)\in B_2}} (\{k \diamond (f(x-y) + \lambda_i(y))\} \cup \{f(x-z) + \mu_i(z)\})]$$

where $N$ is the number of the elements of the multiset.

However, if an element is not greater than the local $(N-k)$th-order statistic, then it cannot be greater than the global $(N-k)$th-order statistic. Therefore, the terms $\max^{(N)}, \ldots, \max^{(k+1)}$ can be omitted:

$$f \oplus [\alpha, \beta, k](x)$$
$$= \max_{i=1}^{n\,(k)} [\max_{\substack{(x-y)\in B_1 \\ (x-z)\in B_2}}^{(k)} (\{k \diamond (f(x-y) + \lambda_i(y))\} \cup \{f(x-z) + \mu_i(z)\}),$$

$$\max_{\substack{(x-y)\in B_1 \\ (x-z)\in B_2}}^{(k-1)} (\{k \diamond (f(x-y) + \lambda_i(y))\} \cup \{f(x-z) + \mu_i(z)\}),$$

$$\vdots$$

$$\max_{\substack{(x-y)\in B_1 \\ (x-z)\in B_2}} (\{k \diamond (f(x-y) + \lambda_i(y))\} \cup \{f(x-z) + \mu_i(z)\})]$$

$$= \max_{i=1}^{n\,(k)} [\max_{\substack{(x-y)\in B_1 \\ (x-z)\in B_2 \\ j-1}}^{k\;(j)} (\{k \diamond (f(x-y) + \lambda_i(y))\} \cup \{f(x-z) + \mu_i(z)\})]$$

Equation (56) can be proved similarly. ∎

**Example IV.1** Consider the following image $f$ and soft structuring element $[\alpha, \beta]$:

EXTENSIONS, ALGORITHMS, AND IMPLEMENTATIONS 27

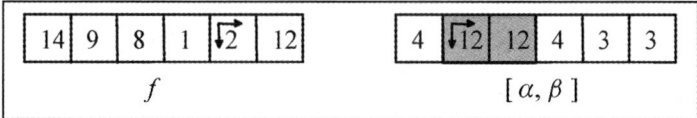

Soft dilation at point (0, 0) for $k = 2$, according to Eq. (38), is

$$f \oplus [\alpha, \beta, 2](0, 0) = \max^{(2)}(\{2 \diamond (14, 13)\} \cup \{16, 12, 12, 17\})$$
$$= \max^{(2)}(14, 14, 13, 13, 16, 12, 12, 17) = 16$$

According to the proposed technique, the structuring is divided into three structuring elements:

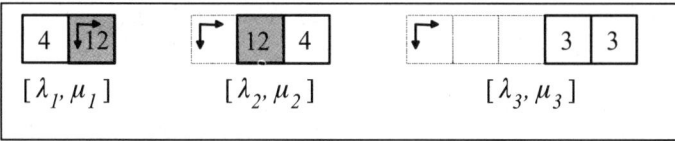

The following multisets are obtained from the preceding structuring elements: $\{2 \diamond (14), 16\}$, $\{2 \diamond (13), 12\}$ and $\{12, 17\}$, for the first, the second, and the third structuring elements, respectively. From these multisets the max and $\max^{(2)}$ elements are retained: $(\{16, 14\}, \{13, 13\}$ and $\{17, 12\})$. The $\max^{(2)}$ of the union of these multisets, i.e., 16, is the result of soft dilation at point (0, 0). It should be noted that although 16 is the max of the first multiset, it is also the $\max^{(2)}$ of the global multiset.

## V. FUZZY SOFT MATHEMATICAL MORPHOLOGY

### A. Definitions

Fuzzy soft mathematical morphology operations are defined taking into consideration that in soft mathematical morphology the structuring element is divided into two subsets, i.e., the core and the soft boundary, from which the core "weights" more than the soft boundary in the formation of the final result. Also, depending on $k$, the $k$th-order statistic provides the result of the operation. Fuzzy soft morphological operation should also preserve the notion of fuzzy fitting (Sinha and Dougherty, 1992). Thus, the definitions for fuzzy soft erosion and fuzzy soft dilation are (Gasteratos et al., 1998a):

$$\mu_{A \ominus [B_1, B_2, k]}(x) = \min[1, \min_{\substack{y \in B_1 \\ z \in B_2}}^{(k)} (\{k \diamond (\mu_A(x + y) - \mu_{B_1}(y) + 1)\} \cup$$
$$\{\mu_A(x + z) - \mu_{B_2}(z) + 1\})] \quad (59)$$

and

$$\mu_{A\oplus[B_1,B_2,k]}(x) = \max[0, \max_{\substack{(x-y)\in B_1 \\ (x-z)\in B_2}}^{(k)}(\{k \diamond (\mu_A(x-y) + \mu_{B_1}(y) - 1)\} \cup$$
$$\{\mu_A(x-z) + \mu_{B_2}(z) - 1\})] \tag{60}$$

respectively, where $x, y, z, \in Z^2$, are the spatial coordinates and $\mu_A, \mu_{B_1}, \mu_{B_2}$ are the membership functions of the image, the core of the structuring element, and the soft boundary of the structuring element. Additionally, for the fuzzy structuring element $B \subset Z^2$: $B = B_1 \cup B_2$ and $B_1 \cap B_2 = \emptyset$.

It is obvious that for $k = 1$, Eqs. (59) and (60) are reformed to Eqs. (31) and (32), respectively, i.e., standard fuzzy morphology.

*Example V.1* Let us consider the image $A$ and the structuring element $B$. Fuzzy soft erosion and fuzzy soft dilation are computed for cases $k = 1$ and $k = 2$.

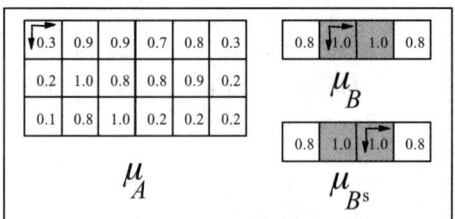

In order to preserve the nature of soft morphological operations, the constraint $k \leq \min\{\text{Card}(B)/2, \text{Card}(B_2)\}$ is adopted in fuzzy soft mathematical morphology as well as in soft mathematical morphology. In this example only the cases of $k = 1$ and $k = 2$ are considered, in order to comply with this constraint.

*Case 1* ($k = 1$): The fuzzy soft erosion of the image is calculated as follows:

$$\mu_E(0, 0) = \mu_{A\ominus[B_1,B_2,1]}(0, 0) = \min[1, \min[0.3 - 1 + 1, 0.9 - 1 + 1, 0.9 - 0.8 + 1]] = 0.3$$

$$\mu_E(0, 1) = \min[1, \min[0.3 - 0.8 + 1, 0.9 - 1 + 1, 0.9 - 1 + 1, 0.7 - 0.8 + 1]] = 0.5$$

$$\vdots$$

$$\mu_E(5, 2) = \min[1, \min[0.2 - 0.8 + 1, 0.2 - 1 + 1]] = 0.2$$

Therefore, the eroded image is:

| 0.3 | 0.5 | 0.7 | 0.5 | 0.3 | 0.3 |
|---|---|---|---|---|---|
| 0.2 | 0.4 | 0.8 | 0.4 | 0.2 | 0.2 |
| 0.1 | 0.3 | 0.2 | 0.2 | 0.2 | 0.2 |

$\mu_{A \ominus [B_1, B_2, 1]}$

The values of the eroded image at points (0, 2) and (1, 2) are higher than the rest values of the image. This agrees with the notion of fuzzy fitting, because the structuring element fits better only at these points than at the rest points of the image. Fuzzy erosion quantifies the degree of structuring element fitting. The larger the number of pixels of the structuring element, the more difficult the fitting. Furthermore, fuzzy soft erosion shrinks the image. If fuzzy image $A$ is considered as a noisy version of a binary image (Sinha and Dougherty, 1992), then the object of interest consists of points (0, 1), (0, 2), (0, 3), (0, 4), (1, 1), (1, 2), (1, 3), (1, 4), (2, 1) and (2, 2), and the rest is the background. By eroding the image with a 4-pixel horizontal structuring element, it would be expected that the eroded image would comprise points (0, 2) and (1, 2). This is exactly what has been obtained.

Similarly, the dilation of the image is calculated as follows:

$$\mu_D(0, 0) = \mu_{A \oplus [B_1, B_2, 1]}(0, 0)$$
$$= \max[0, \max[0.3 + 1 - 1, 0.9 + 0.8 - 1]] = 0.7$$
$$\mu_D(0, 1) = \max[0, \max[0.3 + 1 - 1, 0.9 + 1 - 1, 0.9 + 0.8 - 1]] = 0.9$$
$$\vdots$$
$$\mu_D(5, 2) = \max[0, \max[0.2 + 0.8 - 1, 0.2 + 1 - 1, 0.2 + 1 - 1]] = 0.2$$

Therefore, the dilated image is:

| 0.7 | 0.9 | 0.9 | 0.9 | 0.8 | 0.8 |
|---|---|---|---|---|---|
| 0.8 | 1.0 | 1.0 | 0.8 | 0.9 | 0.9 |
| 0.6 | 0.8 | 1.0 | 1.0 | 0.8 | 0.2 |

$\mu_{A \oplus [B_1, B_2, 1]}$

As it can be seen, fuzzy soft dilation expands the image. In other words the dilated image includes the points of the original image and also the points (0, 0), (0, 5), (1, 0), (1, 5), (2, 0), (2, 3) and (2, 4).

*Case 2* ($k=2$): The erosion of the image is calculated as follows:

$$\mu_E(0,0) = \mu_{A\ominus[B_1,B_2,2]}(0,0) = \min\left[1, \min^{(2)}[0.3, 0.3, 0.9, 0.9, 1.1]\right] = 0.3$$
$$\mu_E(0,1) = \min\left[1, \min^{(2)}[0.5, 0.9, 0.9, 0.9, 0.9, 0.9]\right] = 0.9$$
$$\vdots$$
$$\mu_E(5,2) = \min\left[1, \min^{(2)}[0.4, 0.2, 0.2]\right] = 0.2$$

The eroded image for $k=2$ is:

| 0.3 | 0.9 | 0.7 | 0.7 | 0.3 | 0.3 |
|-----|-----|-----|-----|-----|-----|
| 0.2 | 0.8 | 0.8 | 0.8 | 0.2 | 0.2 |
| 0.1 | 0.4 | 0.2 | 0.2 | 0.2 | 0.2 |

$$\mu_{A\ominus[B_1,B_2,2]}$$

In this case the values of the eroded image at points (0, 1), (0, 2), (0, 3), (1, 1), (1, 2) and (1, 3) are higher than the rest values of the image. This is in agreement with the notion of fuzzy soft fitting. At these points the $k$ repeated "high-value" pixels, which are combined with the core of the structuring element, and the pixels that are combined with the soft boundary of the structuring element are greater than or equal to $k\,\text{Card}[B_1] + \text{Card}[B_2] - k + 1$.

Similarly, the dilation of the image is calculated:

$$\mu_D(0,0) = \mu_{A\oplus[B_1,B_2,2]}(0,0) = \max\left[0, \max^{(2)}[0.3, 0.3, 0.7]\right] = 0.3$$
$$\mu_D(0,1) = \max\left[0, \max^{(2)}[0.3, 0.3, 0.9, 0.9, 0.7]\right] = 0.9$$
$$\vdots$$
$$\mu_D(5,2) = \max\left[0, \max^{(2)}[0.4, 0.2, 0.2, 0.2, 0.2]\right] = 0.2$$

Therefore, the dilated image for $k=2$ is:

| 0.3 | 0.9 | 0.9 | 0.9 | 0.8 | 0.8 |
|-----|-----|-----|-----|-----|-----|
| 0.2 | 1.0 | 1.0 | 0.8 | 0.9 | 0.9 |
| 0.1 | 0.8 | 1.0 | 1.0 | 0.2 | 0.2 |

$$\mu_{A\oplus[B_1,B_2,2]}$$

Here again fuzzy soft dilation expands the image, but more softly than when $k=1$. This means that certain points $((0, 0), (1, 0), (2, 0),$ and $(2, 4))$, which were considered image points (when $k=1$, now $k=2$), belong to the background. The greater the value of $k$, the less the effect of dilation.

Finally, fuzzy soft opening and closing are defined as

$$\mu_{A \circ [B_1,B_2,k]}(x) = \mu_{(A \ominus [B_1,B_2,k]) \oplus [B_1,B_2,k]}(x) \tag{61}$$

and

$$\mu_{A \bullet [B_1,B_2,k]}(x) = \mu_{(A \oplus [B_1,B_2,k]) \ominus [B_1,B_2,k]}(x) \tag{62}$$

respectively.

Illustration of the basic fuzzy soft morphological operations is given through one-dimensional and two-dimensional signals. Figure 5 depicts fuzzy soft morphological erosion and dilation in one-dimensional space. More specifically, Figure 5a shows the initial one-dimensional signal and fuzzy soft erosion for $k=1$ and for $k=2$. Figure 5b shows the initial one-dimensional signal and fuzzy soft dilation for $k=1$ and for $k=2$. Figure 5c shows the structuring element. The core of the structuring element is the shaded area, and the rest area of the structuring element is its soft boundary. From Figures 5a and 5b it becomes clear that the action of the structuring element becomes more effective when $k=1$, i.e., the results of both fuzzy soft erosion and dilation are more visible in the case of $k=1$ than in the case of $k=2$. Moreover, both erosion and dilation preserve the details of the original image better in the case of $k=2$ than in the case of $k=1$.

Figure 6 presents the result of fuzzy soft morphological erosion and dilation on a two-dimensional image. More specifically, Figures 6a and 6b present the initial image and the structuring element, respectively. The image in Figure 6b has been considered as an array of fuzzy singletons (Goetcharian, 1980). The results of fuzzy soft erosion ($k=1$) after the first and second iterations are presented in Figures 6c and 6d, respectively. The white area is reduced after each iteration. The white area of the eroded image (Figure 6c) is the area of the initial image, where the structuring element fits better. Similarly, Figures 6e and 6f present the results of fuzzy soft erosion ($k=3$) after the first and second iteration, respectively. Comparing Figures 6c and 6e, it becomes clear that the greater $k$ is, the less visible are the results of fuzzy soft erosion. Figures 6g and 6h depict the results of fuzzy soft dilation ($k=1$) after the first and second iterations, respectively. In the case of fuzzy soft dilation, the white area increases. Similarly, Figures 6i and 6j show the results of fuzzy soft dilation ($k=3$) after the first and the second iteration, respectively. Again, the greater $k$ is, the less visible are the results of fuzzy soft dilation.

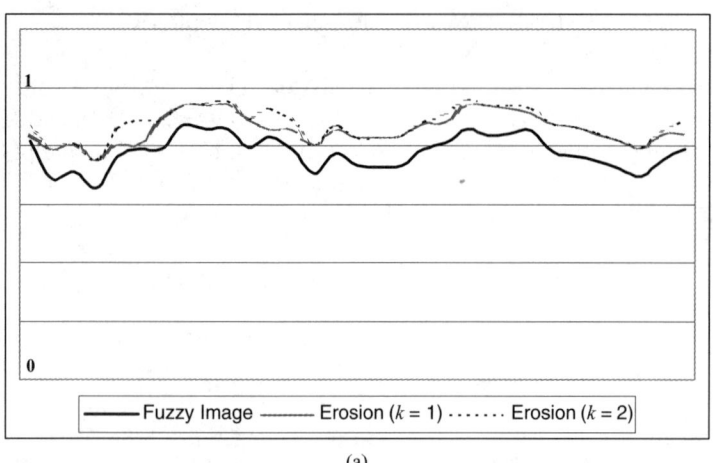

(a)

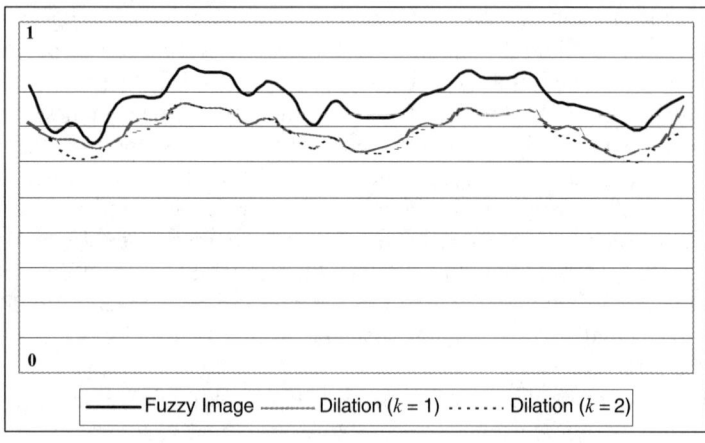

(b)

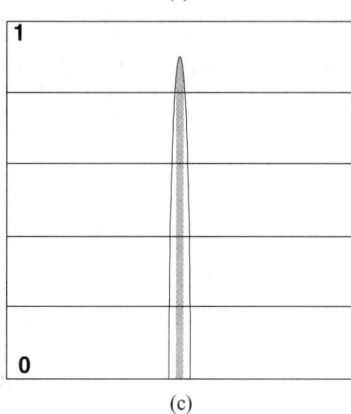

(c)

FIGURE 5. Illustration of one-dimensional fuzzy soft morphological operations and the effect of the order index $k$: (a) fuzzy soft erosion, (b) a fuzzy soft dilation, and (c) the structuring element.

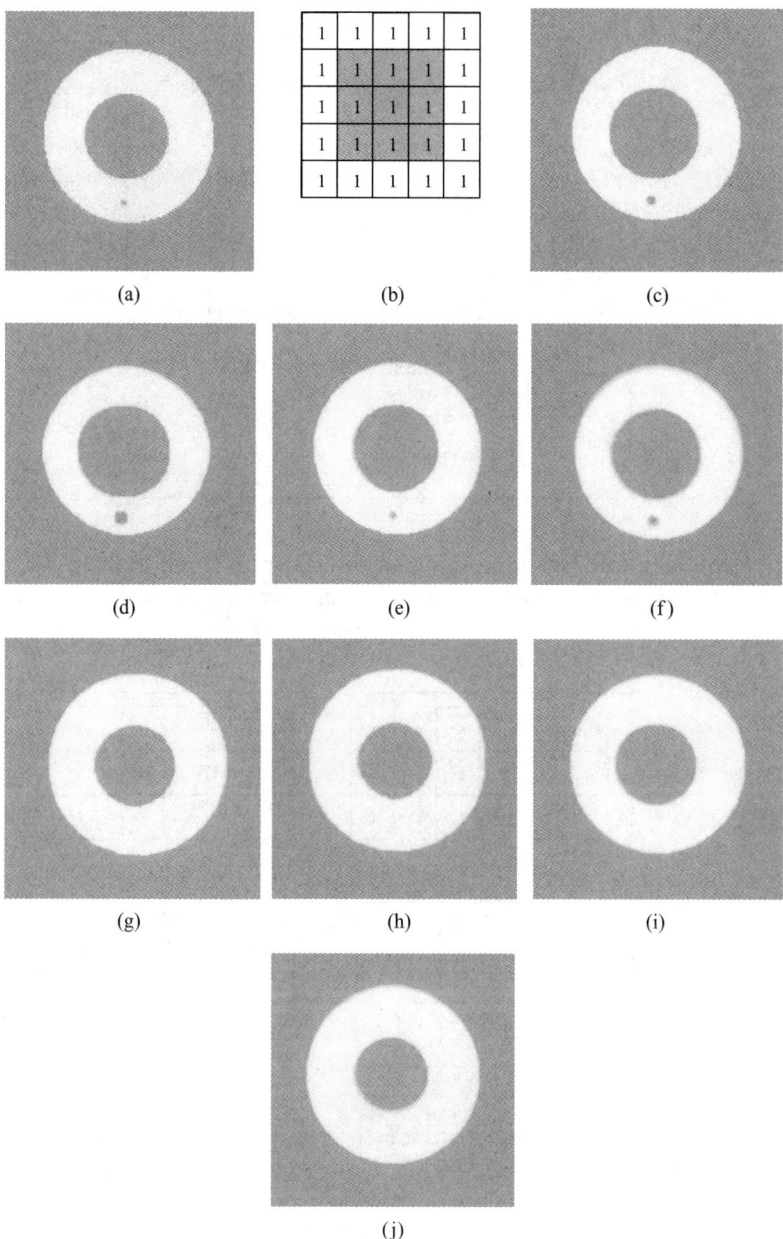

FIGURE 6. (a) Image, (b) structuring element, (c) fuzzy soft erosion ($k=1$) after the first iteration, (d) fuzzy soft erosion ($k=1$) after the second iteration, (e) fuzzy soft erosion ($k=3$) after the first iteration, (f) fuzzy soft erosion ($k=3$) after the second iteration, (g) fuzzy soft dilation ($k=1$) after the first iteration, (h) fuzzy soft dilation ($k=1$) after the second iteration, (i) fuzzy soft dilation ($k=3$) after the first iteration, and (j) fuzzy soft dilation ($k=3$) after the second iteration.

## B. Compatibility with Soft Mathematical Morphology

Let us consider Example V.1. By thresholding image $A$ and structuring element $B$ (using a threshold equal to 0.5), the following binary image and binary structuring element are obtained:

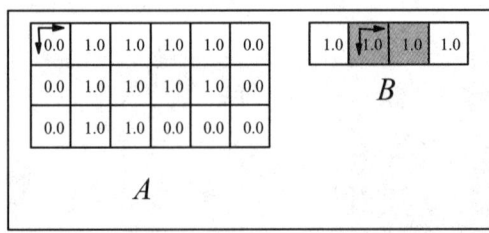

By applying soft binary erosion and soft binary dilation to image $A$ with structuring element $B$, the following images are obtained for $k=1$ and $k=2$:

$k=1$:

| ↓0.0 | 0.0 | 1.0 | 0.0 | 0.0 | 0.0 |
|---|---|---|---|---|---|
| 0.0 | 0.0 | 1.0 | 0.0 | 0.0 | 0.0 |
| 0.0 | 0.0 | 0.0 | 0.0 | 0.0 | 0.0 |

| ↓1.0 | 1.0 | 1.0 | 1.0 | 1.0 | 1.0 |
|---|---|---|---|---|---|
| 1.0 | 1.0 | 1.0 | 1.0 | 1.0 | 1.0 |
| 1.0 | 1.0 | 1.0 | 1.0 | 1.0 | 0.0 |

$k=2$:

| ↓0.0 | 1.0 | 1.0 | 1.0 | 0.0 | 0.0 |
|---|---|---|---|---|---|
| 0.0 | 1.0 | 1.0 | 1.0 | 0.0 | 0.0 |
| 0.0 | 0.0 | 0.0 | 0.0 | 0.0 | 0.0 |

| ↓0.0 | 1.0 | 1.0 | 1.0 | 1.0 | 1.0 |
|---|---|---|---|---|---|
| 0.0 | 1.0 | 1.0 | 1.0 | 1.0 | 1.0 |
| 0.0 | 1.0 | 1.0 | 1.0 | 0.0 | 0.0 |

It is obvious that these results are identical to those of Example V.1, when the same threshold value is used. This was expected, because binary soft morphology quantifies the soft fitting in a crisp way, whereas fuzzy soft erosion quantifies the soft fitting in a fuzzy way. The same results are obtained using a threshold equal to 0.55. However, when fuzzy soft morphology and thresholding with a threshold equal to or greater than 0.6 on the one hand and thresholding with the same threshold and soft morphology on the other hand are applied, different results will be obtained. This means that, in general, the operations do not commute.

## C. Algebraic Properties of Fuzzy Soft Mathematical Morphology

**Theorem V.1** *Duality Theorem*
Fuzzy soft erosion and dilation are dual operations:

$$\mu_{A^c \oplus [B_1, -B_2, k]}(x) = \mu_{(A \ominus [B_1, B_2, k])^c}(x) \tag{63}$$

Opening and closing are also dual operations:

$$\mu_{(A \bullet [B_1, B_2, k])^c}(x) = \mu_{A^c \circ [-B_1, -B_2, k]}(x) \tag{64}$$

**Theorem V.2** *Translation Invariance*
Fuzzy soft erosion and dilation are translation invariant:

$$\mu_{(A)_u \ominus [B_1, B_2, k]}(x) = \left(\mu_{A \ominus [B_1, B_2, k]}(x)\right)_u \tag{65}$$

where $u \in Z^2$.

**Theorem V.3** *Increasing Operations*
Both fuzzy soft erosion and dilation are increasing operations:

$$\mu_A < \mu_{A'} \Rightarrow \begin{cases} \mu_{A \ominus [B_1, B_2, k]}(x) < \mu_{A' \ominus [B_1, B_2, k]}(x) \\ \mu_{A \oplus [B_1, B_2, k]}(x) < \mu_{A' \oplus [B_1, B_2, k]}(x) \end{cases} \tag{66}$$

where $A$ and $A'$, are two images with membership functions $\mu_A$ and $\mu_{A'}$, respectively and $\mu_A(x) < \mu_{A'}(x), \forall x \in Z^2$.

**Theorem V.4** *Distributivity*
Fuzzy soft erosion is not distributive over intersection, as it is in standard morphology:

$$\exists x \in Z^2 \quad \text{and} \quad \exists A1, A2, B \subseteq Z^2 | \mu_{(A1 \cap A2) \ominus [B_1, B_2, k]}(x)$$
$$\neq \mu_{(A1 \ominus [B_1, B_2, k]) \cap (A2 \ominus [B_1, B_2, k])}(x) \tag{67}$$

***Example V.2*** Consider the following image $A$ and structuring element $B$, where image $A$ is the intersection of images $A1$ and $A2$.

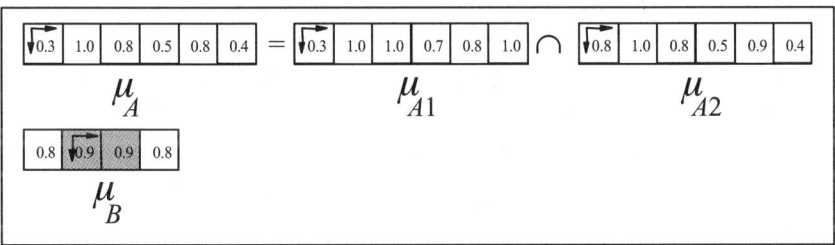

The fuzzy soft erosion for $k=2$ of $A, A1, A2$ and the intersection of the eroded $A1$ and the eroded $A2$ are:

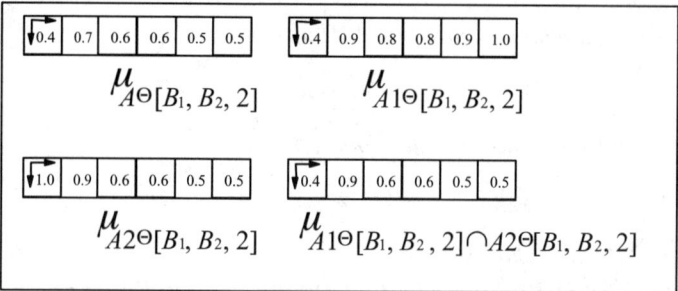

In general, fuzzy soft dilation does not distribute over union:
$$\exists x \in Z^2 \quad \text{and} \quad \exists A1, A2, B \subseteq Z^2 \,|\, \mu_{(A1 \cup A2) \oplus [B_1, B_2, k]}(x)$$
$$\neq \mu_{(A1 \oplus [B_1, B_2, k]) \cup (A2 \oplus [B_1, B_2, k])}(x) \quad (68)$$

**Theorem V.5** *Antiextensivity–Extensivity*
*Fuzzy soft opening is not antiextensive. If it were antiextensive, then $\mu_{A \circ [B_1, B_2, k]}(x) \leq \mu_A(x), \forall x \in Z^2$. The following example shows that $\exists x \in Z^2 \,|\, \mu_{A \circ [B_1, B_2, k]}(x) > \mu_A(x)$.*

**Example V.3** Consider the image $A$ and the structuring element $B$ for $k=2$. In this example

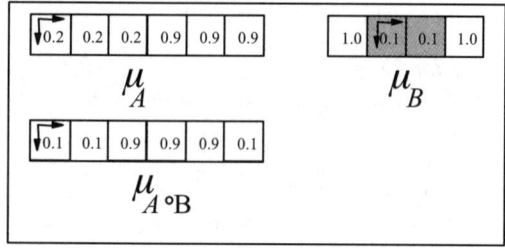

$$\mu_{A \circ [B_1, B_2, k]}(0, 2) = 0.9 > \mu_A(0, 2) = 0.2$$

which means that fuzzy soft opening is not antiextensive.

Similarly, it can be shown that, in general, fuzzy soft closing is not extensive: $\exists x \in Z^2$ and $A, B \subset Z^2 | \mu_{A \bullet [B_1, B_2, k]}(x) < \mu_A(x)$.

**Theorem V.6** *Idempotency*
*In general, fuzzy soft opening is not idempotent:*
$$\exists x \in Z^2 \quad \text{and} \quad \exists A, B \subseteq Z^2 | \mu_{A \circ [B_1, B_2, k]}(x) \neq \mu_{(A \circ [B_1, B_2, k]) \circ [B_1, B_2, k]}(x) \quad (69)$$

This is illustrated by the following example.

*Example V.4*  Consider the image $A$ and the structuring element $B$ for $k = 1$.

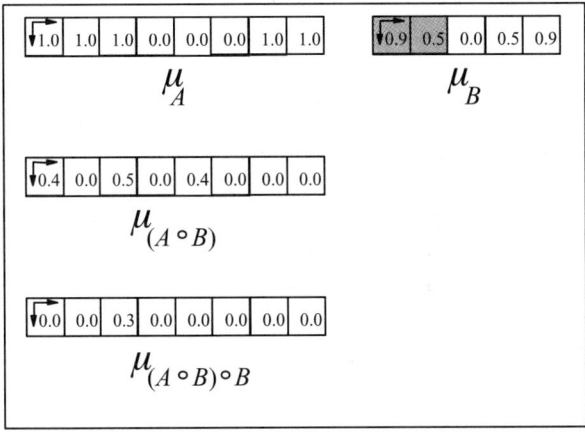

From this example it is obvious that fuzzy soft opening is not idempotent.

By the duality theorem [Eq. (64)], it can be proved that, in general, fuzzy soft closing is also not idempotent:

$$\exists x \in Z^2 \quad \text{and} \quad \exists A, B \subseteq Z^2 \mid \mu_{A \bullet [B_1, B_2, k]}(x) \neq \mu_{(A \bullet [B_1, B_2, k]) \bullet [B_1, B_2, k]}(x) \quad (70)$$

## VI. Implementations

Soft morphological operations are based on weighted order statistics and, therefore, algorithms such as mergesort and quicksort, which were developed for the computation of weighted order statistics, can be used for the computation of soft morphological filters (Kuosmanen and Astola, 1995). The average complexity of the quicksort algorithm is $O(N \log N)$, where $N$ is the number of elements to be sorted (Pitas and Venetsanopoulos, 1990). Therefore, the average complexity for a soft morphological operation utilizing a soft structuring element $[\alpha, \beta, k]: B \rightarrow Z$ is $O((k \, \text{Card}[B_1] + \text{Card}[B_2]) \log(k \, \text{Card}[B_1] + \text{Card}[B_2]))$.

Hardware implementations of soft morphological operations include the threshold decomposition and the majority gate techniques. These structures, along with an algorithm based on a local histogram, are described in some detail in this section.

## A. Threshold Decomposition

The threshold decomposition (Wendt et al., 1985) is a well-known technique for hardware implementation of nonlinear filters. The implementation of soft morphological filters in hardware, using the threshold decomposition technique, has been described in Shih and Pu (1995) and Pu and Shih (1995). According to this approach, both the gray-scale image and the gray-scale structuring element are decomposed into $2^b$ binary images $f_i$ and $2^b$ structuring elements $\beta_i$, respectively. Binary soft morphological operations are performed on the binary images by the binary soft structuring elements and then a maximum or a minimum selection at each position is performed, depending whether the operation is soft dilation or soft erosion, respectively. Finally, the addition of the corresponding binary pixels is performed. Figure 7 demonstrates this technique for soft dilation.

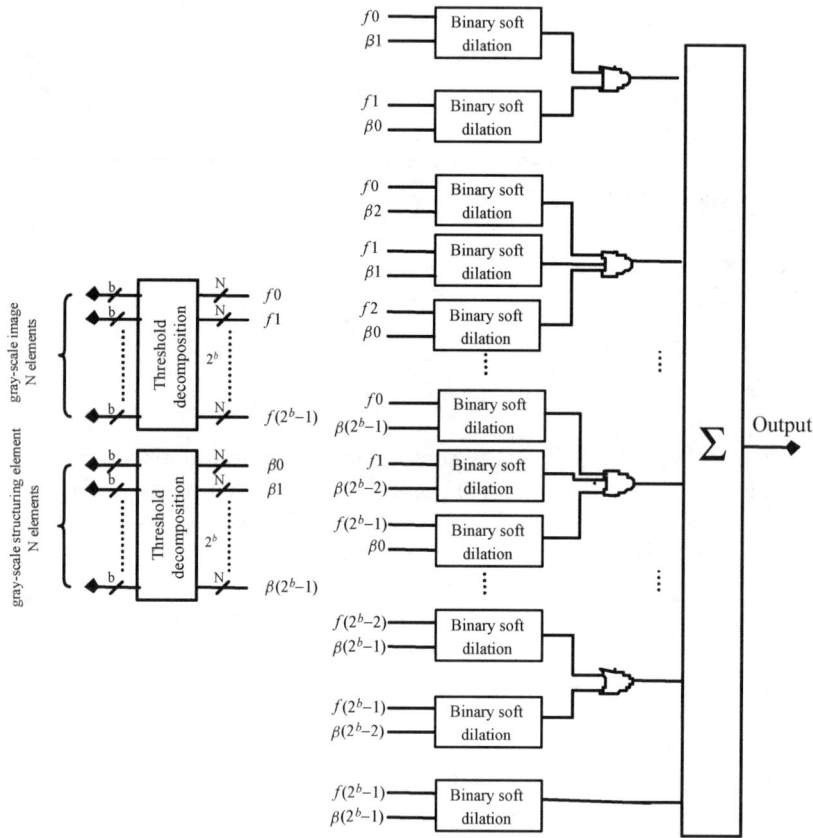

FIGURE 7. Illustration of the threshold decomposition technique for soft dilation.

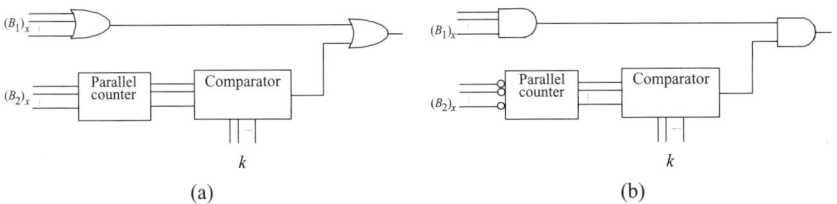

FIGURE 8. Implementation of binary (a) soft morphological dilation and (b) soft morphological erosion.

The logic-gate implementation of binary soft morphological dilation and erosion are shown in Figures 8a and 8b, respectively. The parallel counter counts the number of 1s of the input signal and the comparator compares them to the order index $k$ and outputs 1 when this number is greater than or equal to $k$.

It is obvious that this technique, although it can achieve high-speed computation times, is hardware demanding because it is realized using simple binary structures. Its hardware complexity grows exponentially both with the structuring element size and the resolution of the pixels, i.e., its hardware complexity is $O(2^N 2^b)$.

## B. Majority Gate

### 1. Algorithm Description

The majority gate algorithm is an efficient bit serial algorithm suitable for the computation of the median filter (Lee and Jen, 1992). According to this algorithm, the MSBs of the numbers within the data window are first processed. The other bits are then processed sequentially until the less significant bits (LSBs) are reached. Initially, a set of signals (named the rejecting flag signals) are set to 1. These signals indicate which numbers are candidates to be the median value. If the majority of the MSBs are found to be 1s, then the MSB of the output is 1; otherwise, it is 0. The majority is computed through a CMOS programmable device, shown in Figure 9. In the following stage the bits of the numbers whose MSBs have been rejected by means of the rejecting flag signals are not taken into account. The majority selection procedure continues in the next stages until the median value is found.

Gasteratos et al. (1997a) have proposed an improvement of this algorithm for the implementation of any rank filter using a single hardware structure. This is based on the concept that by having a method to compute the median value of $4N + 1$ numbers and by being able to control $2N$ of these numbers,

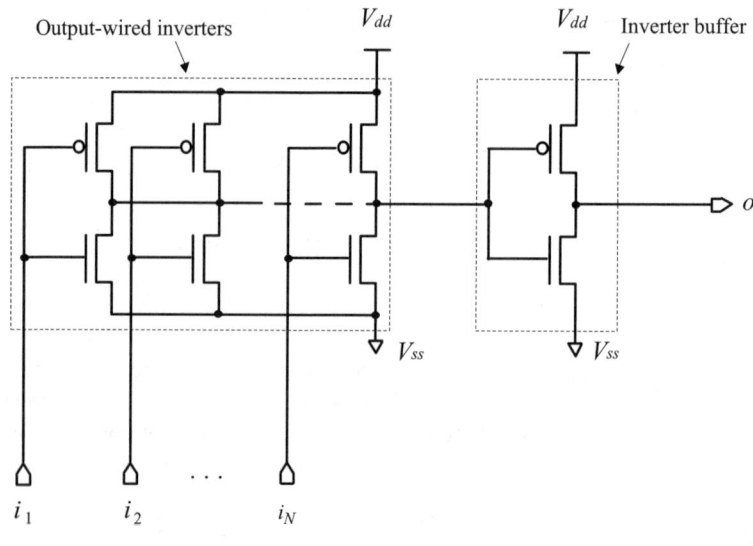

FIGURE 9. Programmable CMOS majority gate.

any order statistic of the rest $2N + 1$ numbers can be determined. Suppose that there are $W = 2N + 1$ numbers $x_i$, the $r$th-order statistic of which is required. The $2N + 1$ inputs are the numbers $x_i$, whereas the rest are dummy inputs $d_i$ ($0 < 1 \leq 2N$). The binary values of the dummy inputs can be either $00 \ldots 0$ or $11 \ldots 1$. This implies that when the $W'$ numbers are ordered in ascending sequence, $d_l$ are placed to the extremes of this sequence.

## 2. Systolic Array Implementation for Soft Morphological Filtering

### a. A Systolic Array for a 3 × 3 Structuring Element

A pipelined systolic array capable of computing soft gray-scale dilation/erosion on a 3 × 3-pixel image window using a 3 × 3-pixel structuring element, both of 8-bit resolution, is presented in Figure 10 (Gasteratos et al., 1998b). The central pixel of the structuring element is its core, whereas the other eight pixels constitute its soft boundary. The inputs to this array are the nine pixels of the image window and the nine pixels of the soft morphological structuring element and a control signal MODE. Latches (L1) store the image window,

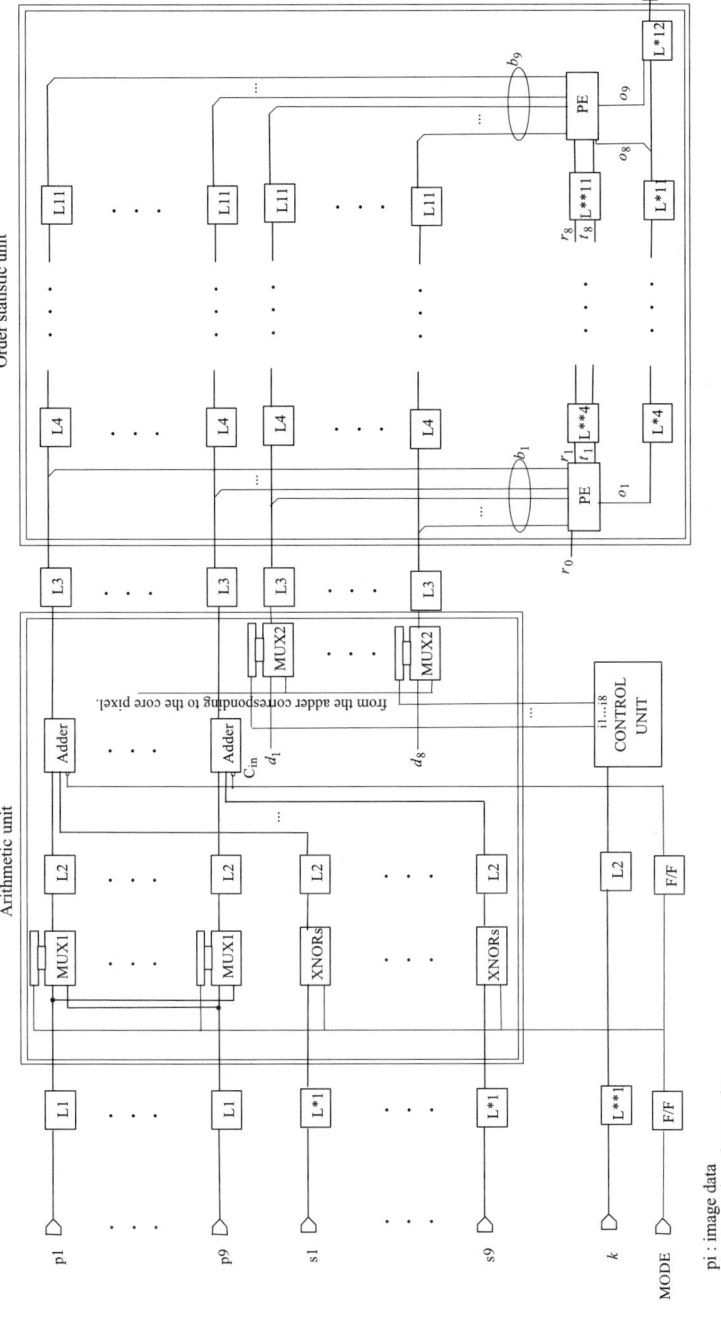

FIGURE 10. Systolic array hardware structure implementing the majority gate technique for soft morphological filtering.

TABLE 1
USE OF DUMMY NUMBERS IN THE COMPUTATION OF WEIGHT ORDER STATISTICS

| k | Sequence of numbers | Dummy numbers |
|---|---|---|
| 1 | 9 | 8 |
| 2 | 10 | 7 |
| 3 | 11 | 6 |
| 4 | 12 | 5 |

latches (L*1) store the structuring element, and latch (L**1) stores the number $k$. Signal MODE is used to select the operation. When this is 1, soft dilation is performed, whereas when it is 0, a soft erosion operation is performed. Image data are collected through multiplexers MUX1, which are controlled by the signal MODE. The pixels of the structuring element remain either unchanged for the operation of dilation or they are complemented (by means of XNOR gates) for the operation of erosion. In the next stage of the pipeline, data are fed into nine adders. In the case of soft erosion, the 2's complements of the pixel values of the structuring element are added to the image pixel values. This is equivalent to the subtraction operation.

According to the constraint $k \leq \min\{Card(B)/2, Card(B_2)\}$, in this case $k$ is in the range $1 \leq k \leq 4$. Table 1 shows the number of the elements of the image data window contained in the list, as well as the number of the dummy elements. For soft dilation all the dummy inputs are pushed to the top, whereas for soft erosion they are pushed to the bottom. Thus, the appropriate result is obtained from the order statistic unit. A control unit controls an array of multiplexers MUX2 (its input is number $k$). This is a decoder, and its truth table is shown in Table 2. It provides the input to the order statistic unit, either a dummy number or a copy of the addition/subtraction result of the core.

TABLE 2
TRUTH TABLE OF THE CONTROL UNIT

| Input k | Outputs | | | | | | | |
|---|---|---|---|---|---|---|---|---|
| | $i1$ | $i2$ | $i3$ | $i4$ | $i5$ | $i6$ | $i7$ | $i8$ |
| 0001 | 0 | 0 | 0 | 0 | 0 | 0 | 0 | 0 |
| 0010 | 1 | 0 | 0 | 0 | 0 | 0 | 0 | 0 |
| 0011 | 1 | 1 | 0 | 0 | 0 | 0 | 0 | 0 |
| 0100 | 1 | 1 | 1 | 0 | 0 | 0 | 0 | 0 |

# EXTENSIONS, ALGORITHMS, AND IMPLEMENTATIONS 43

The order statistic unit consists of identical processing elements (PEs) separated by latches (L**4 to L**11). The resolution of the latches, which hold the addition/subtraction results or the dummy numbers (L3 to L11), decreases by 1 bit at each successive stage, because there is no need to carry the bits, which have been already processed. On the other hand, the resolution of the latches that hold the result (L4* to L*11), increases by 1 bit at each successive stage. The circuit diagram of this PE is shown in Figure 11. In this figure $W' = 4N + 1$; the $2N + 1$ inputs are the numbers $x_i$, whereas the rest

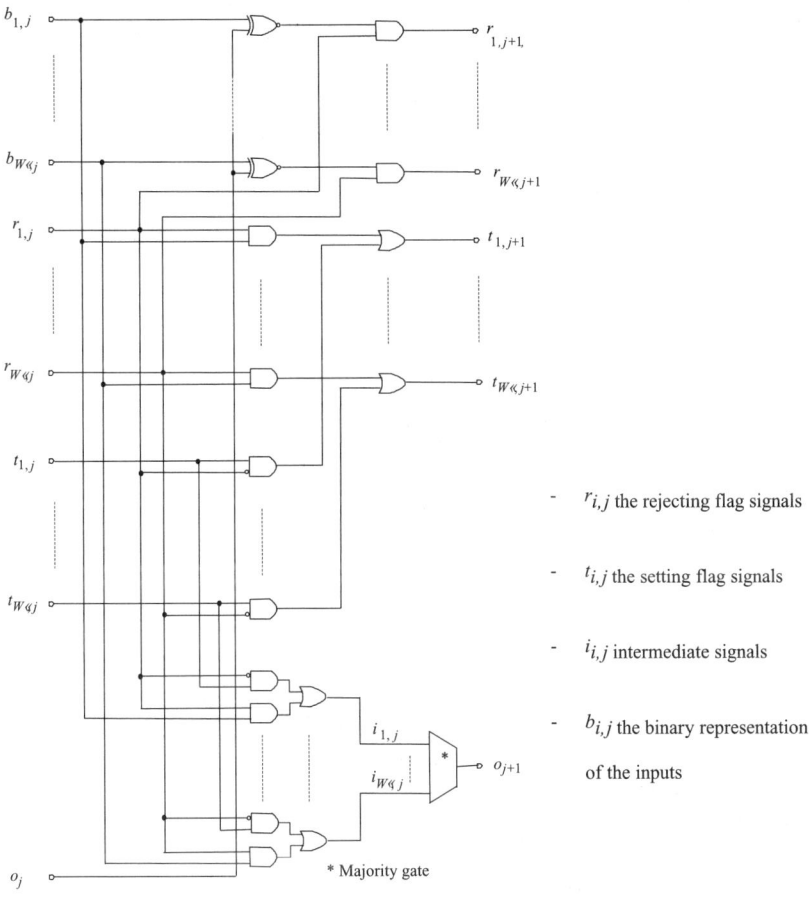

- $r_{i,j}$ the rejecting flag signals
- $t_{i,j}$ the setting flag signals
- $i_{i,j}$ intermediate signals
- $b_{i,j}$ the binary representation of the inputs

\* Majority gate

FIGURE 11. The basic processing element (PE).

are the dummy inputs. Due to its simplicity, it can attain very short processing times, independent of the data window size. Also, it becomes clear that the hardware complexity of the PE is linearly related to the number of its inputs.

b. *Order Statistic Module Hardware Requirements for Other Structuring Elements*

Next, we describe a case study of the hardware requirements for the order statistic unit of a more complex structuring element. The arithmetic unit consists of a number of adders/subtractors equal to the number of pixels of the structuring element. Figure 12a illustrates the structuring element. In this case: Card($B$) = 16, Card($B_1$) = 12, Card($B_2$) = 4, and $k \leq \min\{8, 4\}$, i.e., $1 \leq k \leq 4$. When $k = 4$ the maximum number of the elements of the multiset is Card($B_2$) + $k$ Card($B_1$) = 52. The 49th- (4th-) order statistic of the multiset is sought. Thus, the total number of the inputs to the order statistic unit is 97. The dummy numbers, which are pushed to the top (bottom) in the operation of soft dilation (erosion), are 45. When $k = 3$, the elements of the multiset are 40 and the 38th- (3rd-) order statistic is searched. Now the dummy numbers, which are pushed to the top (bottom), are 46 and to the bottom (top) are 11. In the same way, when $k = 2$ the elements of the multiset are 28 and the 27th- (2nd-) order statistic is searched and the dummy numbers that are pushed to the top (bottom) are 47 and to the bottom (top) are 22. Finally, when $k = 1$ the elements of the multiset are 16 and the 16th- (1st-) order statistic is searched. In this case the dummy numbers that are pushed to the top (bottom) are 48 and to the bottom (top) are 33. For any structuring element, an order statistic unit can be synthesized following this procedure. In this case hardware complexity is linearly related both to the structuring element size and the resolution of the pixels, i.e., the hardware complexity is $O(Nb)$.

3. *Architecture for Decomposition of Soft Morphological Structuring Elements*

An architecture suitable for the decomposition of soft morphological structuring elements is depicted in Figure 13. The structuring element is loaded into the *structuring element management* module. This divides the structuring element into $n$ smaller structuring elements and provides the appropriate one to the next stage. The pixels of the image are imported into the *image window–management* module. This provides an image window, which interacts with the appropriate structuring element, provided by the *structuring element–management* module. Both the previous modules consist of registers

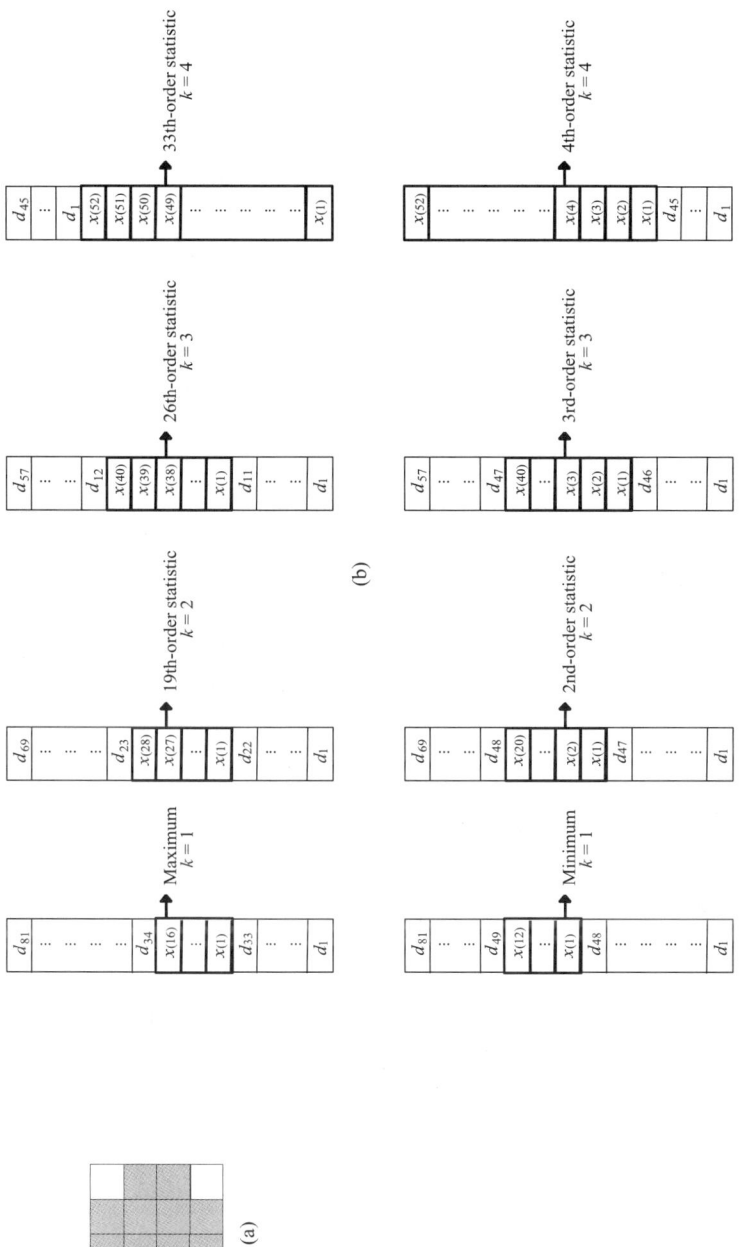

FIGURE 12. (a) Structuring element, (b) arrangement of the dummy numbers in soft morphological dilation using the structuring element of (a), and (c) arrangement of the dummy numbers in soft morphological erosion using the same structuring element.

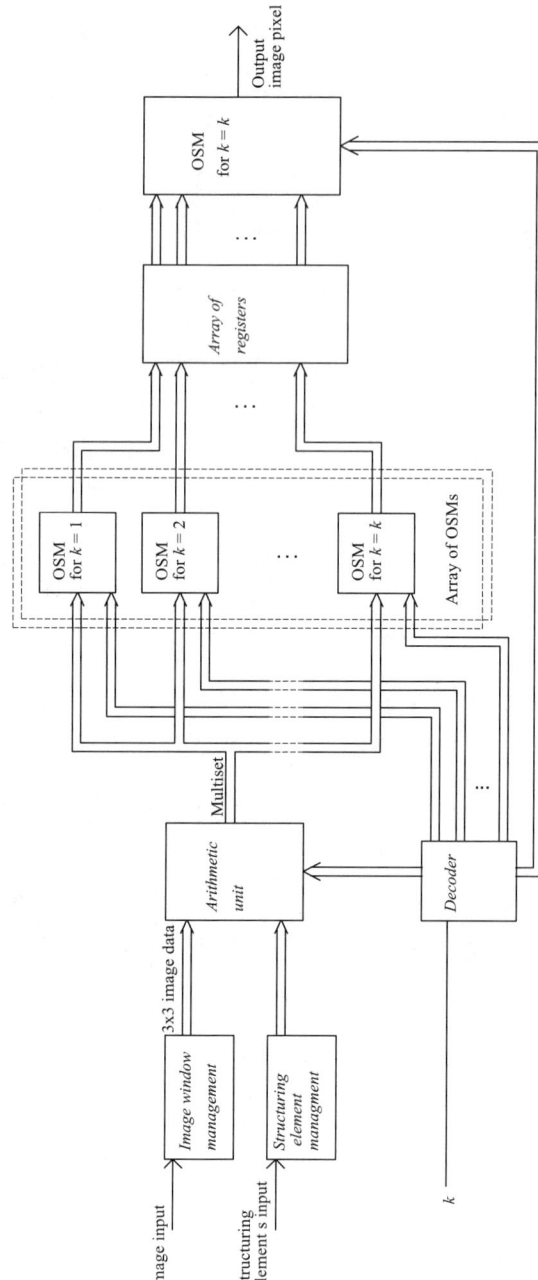

FIGURE 13. Architecture for the implementation of the soft morphological structuring element decomposition technique.

FIGURE 14. Data window management for soft morphological structuring element decomposition.

and multiplexers (MUXs), controlled by a counter mod $n$ (Fig. 14). The second stage, i.e., the *arithmetic unit,* consists of adders/subtractors (dilation/erosion) and an array of MUXs that are controlled by the order index $k$, as the one shown in Figure 11. The MUXs provide the multiple copies of the addition/subtraction results to the next stage, i.e., an array of *order statistic modules* (OSMs). The $\max^{(l)}/\min^{(l)}$ results ($l = 1, \ldots, k$) of every multiset are collected through an *array of registers.* These registers provide the $n \times k$ $\max^{(l)}/\min^{(l)}$ of the $n$ multisets concurrently to the last stage OSM, which computes the final result according to Eqs. (55) and (56).

## C. Histogram Technique

A method for computing an order statistic is to sum the values in the local histogram until the desired order statistic is reached (Dougherty and Astola, 1994). However, instead of adding the local histogram values serially, a successive approximation technique can be adopted (Gasteratos and Andreadis, 1999). This ensures that the result is traced in a fixed number of steps. The number of steps is equal to the number $b$ of the bits per pixel. In the successive approximation technique the result is computed recursively; in each step of the process the $N$ pixel values are compared to a temporal result. Pixel values, which are greater than, less than, or equal to that temporal result, are marked with labels GT, LT, and EQ, respectively. GT, LT and EQ are Boolean variables. Pixel labels are then multiplied by the corresponding pixel weight ($w_j$). The sum of LTs and EQs determines whether the $k$th-order statistic is greater than, less than, or equal to the temporal values.

The pseudocode of the algorithm follows:

*Notation*: $N$: number of pixels; $b$: pixel value resolution (bits); $im_1, im_2, \ldots, im_N$: image pixels; $w_1, w_2, \ldots, w_N$: corresponding weights; $k$: the sought order statistic; temp: temporal result; $o$: output pixel.
**initial**
$o = 0$
$\text{temp} = 2^{b-1}$
**begin**
    **for** $i = 1$ **to** b **do**
        **begin**
        **compare**$(im_1, im_2, \ldots im_N: \text{temp})$
        $\{$**if** $im_j = \text{temp}$ **then** $EQ_j = 1$ **else** $EQ_j = 0$
        **if** $im_j < \text{temp}$ **then** $LT_j = 1$ **else** $LT_j = 0\}$
        **if** $\left(\sum_{j=1}^{N} w_j (EQ_j + LT_j) \geq k\right)$ **AND** $\left(\sum_{j=1}^{N} w_j LT_j < k\right)$
        **then** $o \leftarrow \text{temp}$
        **elsif** $\sum_{j=1}^{N} w_j LT_j \geq k$
        **then** $\text{temp} \leftarrow \text{temp} - 2^{b-1-i}$
        **else** $\text{temp} \leftarrow \text{temp} + 2^{b-1-i}$
        **end**
**end**

A module utilizing standard comparators, adders/subtractors, multipliers, and multiplexers (for the "if" operations) can be used to implement this technique in hardware. Also, there are two ways to realize the algorithm. The first is through a loop, which feeds the temp signal back to the input $b$ times. Such a module is demonstrated in Figure 15. Its inputs are the addition or subtraction results of the image pixel value data with the structuring element pixel values, depending on whether the operation is soft dilation or soft erosion, respectively. Alternatively, $b$ successive modules can be used to process the data in a pipeline fashion. The latter implementation is more hardware demanding but results in a faster hardware structure.

The preceding algorithm requires a fixed number of steps equal to $b$. Furthermore, the number of steps grows linearly according to the pixel value resolution ($O(b)$). Its main advantage is that it can directly compute weighted rank order operations. This means that there is no need to reconstruct the local histogram according to the weights of the image pixels. Comparative experimental results using typical images showed that for $5 \times 5$ and larger image data windows the combined local histogram and successive approximation technique outperforms the existing quicksort algorithm for weighted order statistics filtering (Gasteratos and Andreadis, 1999).

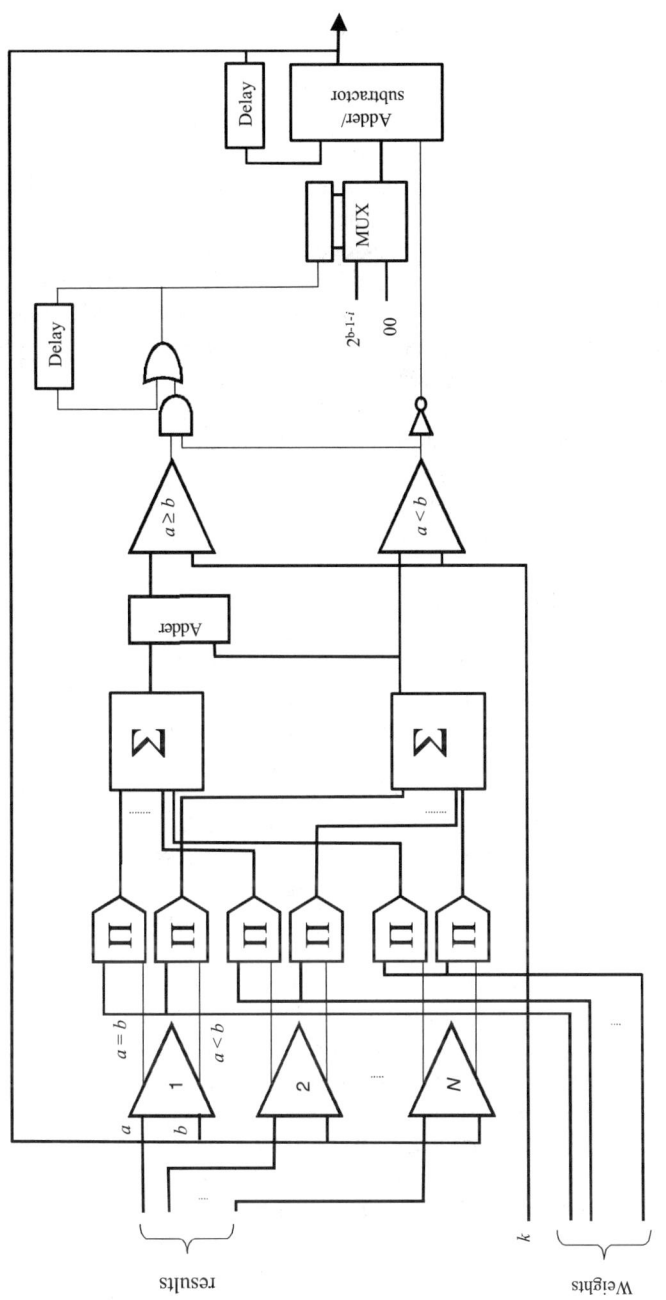

FIGURE 15. Block diagram of a hardware module for the computation of weighted order statistics, based on the local histogram–successive approximation technique.

## D. Vector Standard Morphological Operation Implementation

The block diagram of a new hardware structure that performs vector erosion or dilation of a color image $f$ by a color $3 \times 3$-pixel structuring element $g$, both of 24-bit resolution, is presented in Figure 16. The input image $f$ may be of any dimension. Consider that the $3 \times 3$-pixel window defined by the domain $G$ of $g$ is located at spatial coordinates $x$, including the nine vectors of the input image, $vecim_j$ ($1 \leq j \leq 9$). In each clock cycle the $j$th vector of the input image ($vecim_j$) and the corresponding $j$th vector of the structuring element ($vecse_j$) are imported into the input unit. This unit consists of an array of D-type flip-flops and a MUX, which ensures that after the ninth clock cycle the nine vectors of the structuring element are fedback to the input unit. Thus, the structuring element is introduced only once.

In Figure 16 we assume that $vecim_j = c(him_j, sim_j, vim_j)$ and also $vecim_j = c(him_j, sim_j, vim_j)$. As can be seen, in the next clock cycle the $h$, $s$, and $v$ components of the $j$th vectors under consideration are pairwise subtracted or added in the summation/subtraction units, according to the value of the select operation input signal, which determines the vector morphological operation (erosion or dilation) that is carried out.

In the next stage, the $h$, $s$, and $v$ differences or sums are normalized to the upper or the lower bound of each vector component (see constraints in Eqs. (29) and (30)) in the three normalization units. These three differences or sums are the components of the $j$th vector ($vector_j$) in $\{f(x+y) - g(y)\}$ or in $\{f(x-y) + g(y)\}$, respectively (see Eqs. (27) and (28)).

Consequently, $vector_j$ is loaded into the supremum or infimum finding module. The heart of this module is a mod-9 counter, which ensures that the nine vectors in $\{f(x+y) - g(y)\}$ or in $\{f(x-y) + g(y)\}$ will be compared during nine clock cycles and the infimum or the supremum of these nine vectors will be the output of the module at the tenth clock cycle. The module also includes an array of D-type flip-flops, which transmit the $j$th and the $(j+1)$st vectors ($vector_j$ and $vector_{j+1}$) simultaneously to an array of comparators (each comparator of the array compares only one component of the vectors under consideration). The outputs of the comparators are then fed to the control unit (a combinational circuit), which, according to the select operation signal, determines whether the $j$th or the $(j+1)$st vector will be compared to the next input of the module, i.e., the $(j+2)$nd vector ($vector_{j+2}$). The result of the ninth comparison is forwarded to the output of the module, through an array of MUXs.

The whole procedure is fully pipelined. After initialization, the hardware structure of Figure 16 produces an output result every nine clock cycles.

The proposed circuit was designed and successfully simulated by means of the MAX+plus II software of Altera Corporation. The FPGA used is the

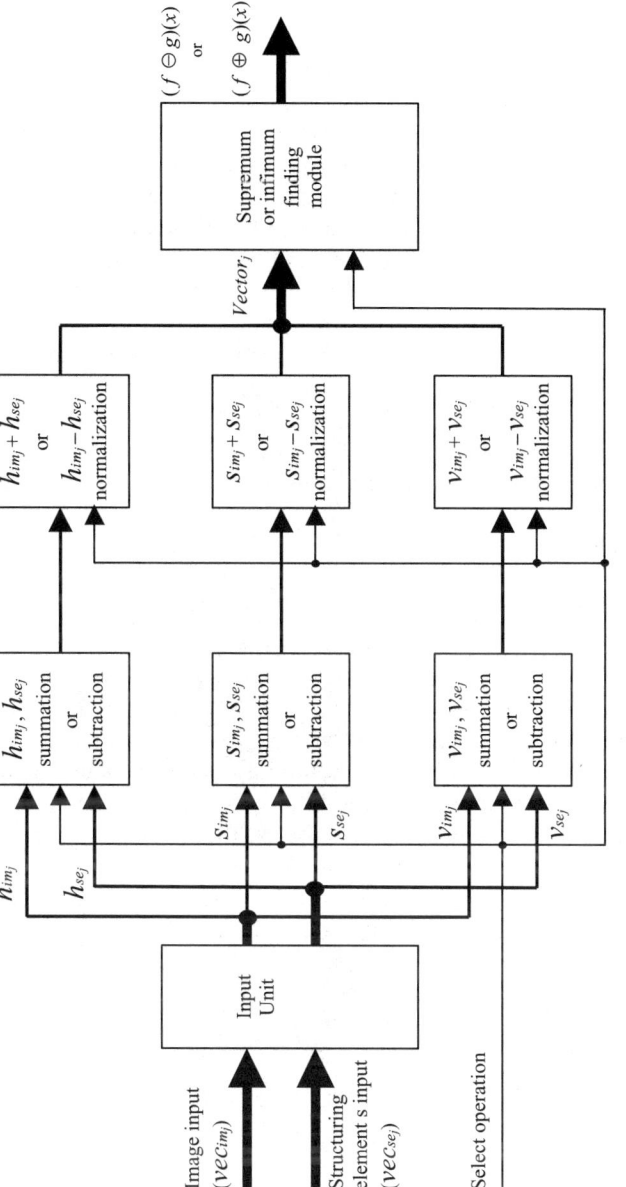

FIGURE 16. Block diagram of a hardware structure that performs vector erosion or dilation of a color input image $f$ by a $3 \times 3$-pixel color structuring element $g$.

EPF10K30EQC208-1 device of the FLEX10KE Altera device family. The typical system clock frequency is 40 MHz.

## VII. Conclusions

Soft morphological filters are a relatively new subclass of nonlinear filters. They were introduced to improve the behavior of standard morphological filters in noisy environments. In this paper the recent descriptions of soft morphological image processing have been presented. Vector soft mathematical morphology extends the concepts of gray-scale soft morphology to color image processing. The definitions of vector soft morphological operations and their properties have been provided. The use of vector soft morphological filters in color impulse noise attenuation has been also demonstrated. Fuzzy soft mathematical morphology applies the concepts of soft morphology to fuzzy sets. The definitions and the algebraic properties have been illustrated through examples and experimental results. Techniques for soft morphological structuring element decomposition and its hardware implementation have been also described.

Soft morphological operations are based on weighted order statistics. Algorithms for implementation of soft morphological operations include the well-known mergesort and quicksort algorithms for weighted order statistics computation. An approach based on local histogram and a successive approximations technique has been also described. This algorithm is a great improvement in speed for a $5 \times 5$ image data window or larger. Soft morphological filters can be implemented in hardware using the threshold decomposition and the majority gate techniques. The threshold decomposition technique is fast, but its hardware complexity is exponentially related both to the structuring element size and the resolution of the pixels. In the majority gate algorithm the hardware complexity is linearly related both to the structuring element size and the resolution of the pixels. A hardware structure that performs vector morphological operations has been also described.

### References

Barnett, V. (1976). *J. R. Statist. Soc. A* **139**, 318–355.
Bloch, I., and Maitre, H. (1995). *Pattern Recognition* **28**, 1341–1387.
Comer, M. L., and Delp, E. J. (1998). *The Colour Image Processing Handbook,* Sangwine and Horne, eds., pp. 210–224. London, Chapman & Hall.
David, H. A. (1981). *Order Statistics,* New York, Wiley.
Dougherty, E. R., and Astola, J. (1994). *Introduction to Nonlinear Image Processing,* Bellingham, Wash., SPIE.

Gasteratos, A., and Andreadis, I. (1999). *IEEE Signal Proc. Letters* **6**, 84–86.
Gasteratos, A., Andreadis, I., and Tsalides, Ph. (1997a). *Pattern Recognition* **30**, 1571–1576.
Gasteratos, A., Andreadis, I., and Tsalides, Ph. (1998a). *IEE Proceedings—Vision Image and Signal Processing* **145**, 40–49.
Gasteratos, A., Andreadis, I., and Tsalides, Ph. (1998b). *IEE Proceedings—Circuits Devices and Systems* **145**, 201–206.
Gasteratos, A., Andreadis, I., and Tsalides, Ph. (1998c). In *Mathematical Morphology and its Applications to Image and Signal Processing,* H. J. A. M. Heijmans and J. B. T. M. Roerdink, eds., pp. 407–414. Dordrecht, The Netherlands, Kluwer Academic Publishers.
Giardina, C. R., and Dougherty, E. R. (1988). *Morphological Methods in Image and Signal Processing,* Upper Saddle River, N. J. Prentice Hall.
Goetcharian, V. (1980). *Pattern Recognition* **12**, 7–15.
Goutsias, J., Heijmans, H. J. A. M., and Sivakumar, K. (1995). *Computer Vision and Image Understanding* **62**, 326–346.
Haralick, R. M., Sternberg, R., and Zhuang, X. (1986). *IEEE Trans. Pattern Analysis and Machine Intelligence* **PAMI-9**, 532–550.
Harvey, N. R. (1998). *http://www.spd.eee.strath.ac.uk/~harve/bbc_epsrc.html.*
Koskinen, L., Astola, J., and Neuvo, Y. (1991). *Proc. SPIE Symp. Image Algebra and Morphological Image Proc.* **1568**, 262–270.
Koskinen, L., and Astola, J. (1994). *J. Electronic Imag.* **3**, 60–70.
Kuosmanen, P., and Astola, J. (1995). *J. Mathematical Imag. Vision* **5**, 231–262.
Lee, C. L., and Jen, C. W. (1992). *IEE Proc.-G* **139**, 63–71.
Matheron, G. (1975). *Random Sets and Integral Geometry,* New York, Wiley.
Pitas, I., and Venetsanopoulos, A. N. (1990). *Proceedings of the IEEE* **80**, 1893–1921.
Plataniotis, K. N., Androutsos, D., and Venetsanopoulos, A. N. (1999). *Proceedings of the IEEE* **87**, 1601–1622.
Pu, C. C., and Shih, F. Y. (1995). *Graphical Models and Image Processing* **57**, 522–526.
Serra, J. (1982). *Image Analysis and Mathematical Morphology: Vol. I.* London, Academic Press.
Shih, F. Y., and Pu, C. C. (1995). *IEEE Trans. Signal Proc.* **43**, 539–544.
Shinha, D., and Dougherty, E. R. (1992). *J. Visual Commun. Imag. Repres.* **3**, 286–302.
Talbot, H., Evans, C., and Jones, R. (1998). In *Mathematical Morphology and Its Applications to Image and Signal Processing,* H. J. A. M. Heijmans and J. B. T. M. Roerdink, eds., pp. 27–34. Dordrecht, The Netherlands, Kluwer Academic Publishers.
Tang, K., Astola, J., and Neuvo, Y. (1992). In *Proceedings of 6th European Signal Processing Conference,* pp. 1481–1484. Brussels, Belgium.
Wendt, P. D., Coyle, E. J., and Gallagher, N. C., Jr. (1985). *IEEE Trans. Acoustics Speech and Signal Proc.* **ASSP-34**, 898–911.

# Still Image Compression with Lattice Quantization in Wavelet Domain

## MIKHAIL SHNAIDER

*Telstra Research Laboratories, Ormond, Victoria 3204, Australia*

## ANDREW P. PAPLIŃSKI

*School of Computer Science and Software Engineering*
*Monash University, Clayton, Victoria 3168, Australia*

|     |     |
| --- | --- |
| I. Introduction | 56 |
| II. Quantization of Wavelet Coefficients | 57 |
|    A. Fundamentals of the Quantization Process | 59 |
|    B. Information Distribution across the Coefficient Matrix | 60 |
|    C. Optimal Bit Allocation | 65 |
|       1. Distortion Function | 65 |
|       2. A Statistical Model of Wavelet Coefficients | 66 |
|       3. Minimization of the Distortion Function | 69 |
| III. Lattice Quantization Fundamentals | 70 |
|    A. Codebook for Quantization | 71 |
|    B. Distortion Measure and Quantization Regions | 71 |
|    C. Optimal Quantizer for Wavelet Coefficients | 73 |
| IV. Lattices | 75 |
|    A. Root Lattices | 76 |
|       1. Root Systems | 76 |
|       2. Cartan Matrices | 80 |
|    B. Construction of Root Lattices | 81 |
|    C. Laminated Lattices | 82 |
| V. Quantization Algorithms for Selected Lattices | 85 |
|    A. The Closest Point of a Dual Lattice | 87 |
|    B. $Z_n$ Lattice | 88 |
|    C. $D_n$ Lattice and Its Dual | 88 |
|    D. The Laminated $\Lambda_{16}$ Lattice | 89 |
| VI. Counting the Lattice Points | 89 |
|    A. Estimation of the Number of the Lattice Points within a Sphere | 89 |
|    B. The Number of Lattice Points on a Sphere | 91 |
|    C. Relationship Between Lattices and Codes | 93 |
|       1. Construction A | 95 |
|       2. Construction B | 96 |
|       3. Construction C | 97 |
| VII. Scaling Algorithm | 97 |
| VIII. Selecting a Lattice for Quantization | 100 |
| IX. Entropy Coding of Lattice Vectors | 105 |
| X. Experimental Results | 111 |

XI. Conclusions . . . . . . . . . . . . . . . . . . . . . . . . . . . . . . . . 116
XII. Cartan Matrices of Some Root Systems . . . . . . . . . . . . . . . . . 117
    References . . . . . . . . . . . . . . . . . . . . . . . . . . . . . . . . 119

## I. Introduction

A vast variety of approaches to deal with the task of image coding and compression have been developed. Each of the methods is characterized by parameters such as compression ratio, speed of coding, quality of decoded images, and simplicity of implementation. In practice it is extremely difficult to satisfy all these criteria, so usually a trade-off between them is necessary. For instance, with an increase in the compression ratio the quality of a decoded image is in general degraded. Also, an increase in the speed of coding is typically achieved by simplifying the compression algorithm, which, in turn, results in a reduction in the quality of the decoded image.

A practical image compression system is built from a number of components, the first being a transform unit followed by a quantization unit and a bit allocation unit. We emphasize these three generic components of a compression system because our considerations will be primarily related to them.

Although quantization is a very effective technique for data compression, its application directly to raw image data has some drawbacks, which are, in general, independent of the specific method of quantization employed. For example, a significant disadvantage of vector quantization (VQ) is the well-known *blocking* effect, which is especially visible at low and very low bit rates. This can be overcome by applying VQ in the frequency domain instead of the space domain. The domain change is achieved by an initial mapping of the raw image data onto a frequency plane using one of the transforms, such as Fourier or DCT, and then quantizing the resulting transform coefficients.

Among the techniques for space-frequency transformation developed in recent years, the wavelet transform has attracted remarkable attention. It has already found applications in a wide range of areas spanning new methods of solving partial-differential equations to the analysis of geological data.

By projecting image data onto a set of independent subspaces, the wavelet transform captures essential space-frequency characteristics of an image. An appropriate quantization of the projections obtained as a result of this process may lead to significant compression of the original image data, while introducing only minimal distortion.

The combination of the wavelet transform and scalar or vector quantization forms a coding method which has become extremely attractive for image compression. The seminal work in this area of research was published by Mallat (1989). The idea was further developed by Zettler *et al.* (1990), Antonini *et al.*

(1992), Lewis and Knowles (1992), DeVore *et al.* (1992), and others (Shapiro, 1993; Wang *et al.*, 1996). However, the main emphasis of this study, *lattices and their application for quantization* of wavelet coefficients, appears to have been beyond the scope of most studies. Only recently some results on the application of lattices have been published (Barlaud *et al.*, 1993, 1994; Sampson *et al.*, 1994; Chen *et al.*, 1997; Shnaider, 1999).

The focal point in this study* is lattice vector quantization based on the theory of multidimensional lattices. In this work we present a methodical investigation of the main concepts in lattice theory, which emerge across a number of areas in mathematical and information sciences. In particular, we formally show that uniform lattices are highly suitable for quantization of the wavelet coefficients of real images. Moreover, based on existing studies, it is shown that these lattices constitute optimal quantizers for wavelet coefficients in vector spaces of corresponding dimensions. We discuss the basic properties of a variety of lattices, in particular, the root and laminated lattices. The problem of counting the number of available lattice points essential from the point of view of quantization is also addressed.

The vast variety of lattices opens a possibility to select one lattice over another for a particular application. In this papers we present a solution for lattice selection for quantization of wavelet coefficients of real images.

A number of compression test results pertaining to the application of the theoretical aspects of image coding using multidimensional lattices are presented in the experimental part of this paper. In the context of a practical image-coding system, we also present the details of the method for the transmission of Huffman codebook, which was originally proposed in Shnaider and Paplinski (1997). The presentation is summarized with a conclusion.

## II. Quantization of Wavelet Coefficients

We assume reader's familiarity with the wavelet transform. From the literature devoted to this subject, Daubechies (1992) and Meyer (1993) can be recommended as a good starting point. A single step of the forward wavelet transform of a one-dimensional signal consisting of $2N$ samples generates two vectors of low- and high-frequency wavelet coefficients, each of size $N$. For two-dimensional signals, such as images, the structure of the coefficient matrix can be complex even for a small number of expansion levels.

In this section we briefly introduce some technicalities related to the two-dimensional (2D) wavelet transform. We observe first that in practice, a one-dimensional wavelet transform is performed by successive application

---

*The current text is an updated version of that one appeared in Shnaider and Paplinski (1999). In particular, Section IX has a number of additions from Shnaider and Paplinski (2000).

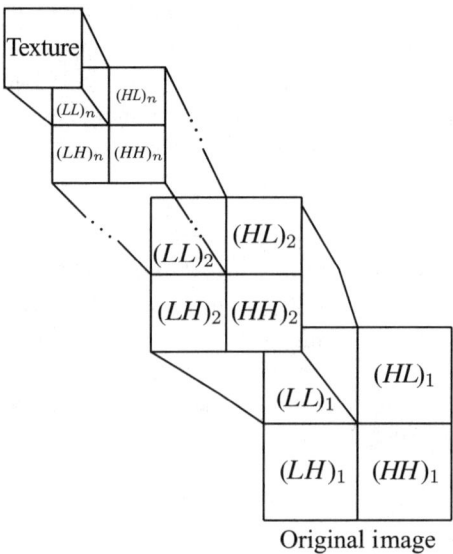

FIGURE 1. Typical expansions steps for the wavelet transform of an image.

of suitably selected high-pass and low-pass filters and an operation of subsampling by a factor 2. In order to obtain coefficients of the wavelet transform of an image, a one-dimensional expansion is performed first on the image rows, followed by an expansion by columns. The set of wavelet coefficients reflects the space-frequency content of an input image.

Typical expansion steps for the discrete wavelet transform (DWT) of images are depicted in Figure 1, and resulting structure of the wavelet coefficients is shown in Figure 2. The wavelet coefficient matrix is of the same size as the original image. The upper-diagonal quarter of the coefficient matrix

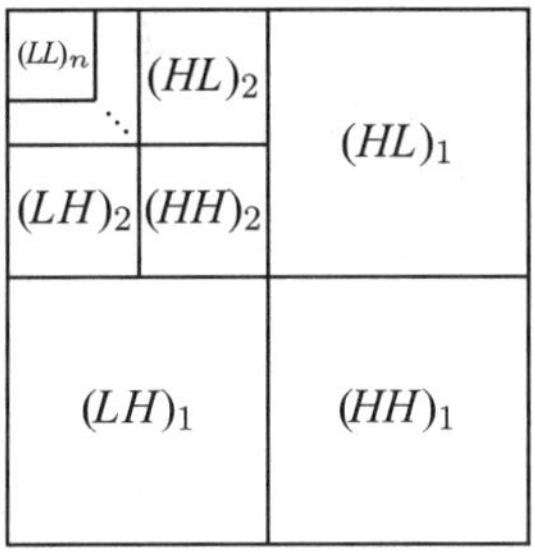

FIGURE 2. A structure of coefficients of a wavelet transform of an image.

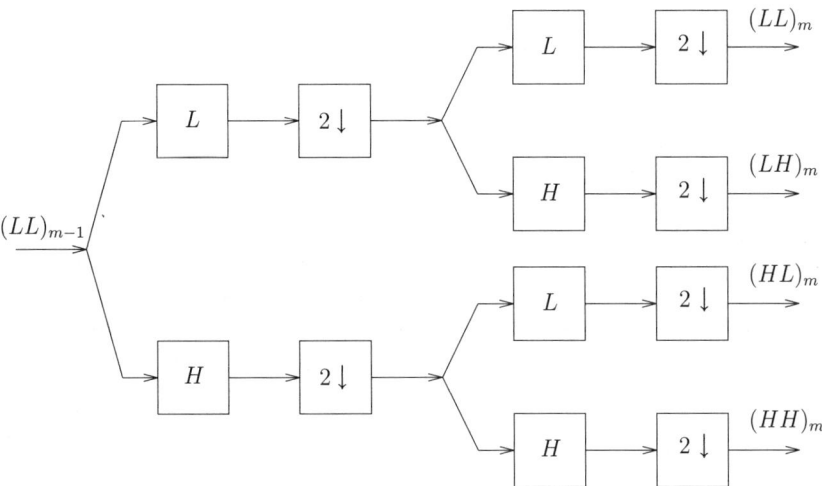

FIGURE 3. A single expansion step of a 2-D DWT.

is expanded further in successive steps. In detail, operations performed during a single expansion step are illustrated in Figure 3. In Figure 3 the 2D input signal $(LL)_{m-1}$ represents a matrix of low-frequency coefficients calculated in the previous step, for example, the matrix $(LL)_2$ in Figure 1. Each row of the matrix $(LL)_{m-1}$ is decomposed using the low-pass ($L$) and high-pass ($H$) filters and is subsampled by a factor of 2. As a result, we obtain two submatrices, each having half the number of columns of the matrix $(LL)_{m-1}$. Next, each column of these two submatrices is decomposed using again the low-pass and high-pass filters and the subsampling operation. Each of the four resulting matrices, $(LL)_m$, $(LH)_m$, $(HL)_m$, and $(HH)_m$, has the quarter of the matrix $(LL)_{m-1}$ size, as shown in Figures 1 and 2, because dilution factor is equal to 2. The low-frequency matrix, $(LL)_m$, is then further expanded as described earlier until the required number of expansion steps, $n$, is reached.

## A. Fundamentals of the Quantization Process

Quantization of the wavelet coefficients is the next step in image coding after the wavelet expansions is performed on an image. Because quantization allows a certain amount of flexibility in bit allocation to the matrix of coefficients, one needs to select optimal parameters for a quantizer, such as the block size and the length of the codebook, in order to achieve a satisfactory image quality for a given bit rate.

The problem of optimal quantization of wavelet coefficients of different frequencies has been extensively studied (Vetterli, 1984; Woods and O'Neil, 1986; Charavi and Tabatabai, 1988; Senoo and Girod, 1992; Li and Zhang). The latest overview of some of the existing quantization algorithms was conducted by Cosman *et al.* (1996). In this survey, the authors point out that the *equal-slope* algorithm is one of the most popular. There exist a number of variations of the basic equal-slope algorithm. The goal of this algorithm is to determine the optimal quantizer for a given source. The idea is to initially collect the distortion information by encoding every subband with all candidate quantizers. Then, for each subband, graphs are plotted to show the quantizer versus the level of distortion it introduces. The optimal quantizers are found on the parts of the curves with equal slopes corresponding to the required overall bit rate. However, in most cases in order to generate an optimal solution, this method requires an extensive amount of calculations due to the necessity of forward and inverse quantization for all candidate quantizers across the solution space. The term candidate/available quantizers does not necessarily imply that the referred quantizers rely on different methods or algorithms. A simple modification of the parameter of the quantizer, such as the size of a quantization block and the length of the codebook, distinguishes that quantizer from similar quantizers from the same class.

In addition to the studies mentioned in Cosman *et al.* (1996), some research has been conducted on the sensitivity of the human eye to different spatial frequencies. This research shows that our eyes are not equally perceptive across the frequency range (Campbell and Robson, 1968). This feature of human eyes is particularly useful in terms of wavelet expansion of images because each level of wavelet coefficients represents a certain spatial frequency of the image. We can take advantage of this information and encode the wavelet coefficients with a bit rate that produces minimal subjective distortions.

The preceding two methods, namely, the 'equal-slope' method and the one based on human visual perception, can be used for bit allocation across the matrix of wavelet coefficients. Although both methods give acceptable results, it seems possible that incorporating them into a single algorithm could exploit the advantages of both. Such an algorithm will be presented later in this section. However, before being able to do so, we need to estimate the distribution of the information among the blocks within a matrix of wavelet coefficients. This task is carried out in the next section.

## B. Information Distribution across the Coefficient Matrix

As mentioned earlier, the result of the wavelet transform of an image is a matrix that consists of blocks of wavelet coefficients corresponding to different

FIGURE 4. Image "Lena"—the original image.

frequency components of the image. From the point of view of subsequent quantization, it is desirable to know the influence of each block of coefficients on the overall reconstruction quality of images. In this section we address and illustrate this problem by encoding and decoding a number of test images (high-resolution scans of two of them, "Lena" and "Coast," are given in Figs. 4 and 5).

For simulation purposes we used MATLAB (1992) with the wavelet toolbox described in Shnaider and Paplinski (1995). The test images were expanded by means of the (five-level DWT using a set of biorthogonal low-pass and high-pass wavelet filters (Daubechies, 1992). Applying the steps as previously described, we can obtain the structure of wavelet coefficients as depicted in Figure 6. Such a structure of the wavelet coefficient matrix makes it possible to cover a wide range of spatial frequencies of the original images.

In order to study the influence of the wavelet coefficients at each expansion level, we have consecutively canceled one level of coefficients at each step; i.e., at the $i$th step we have assigned to zero coefficients $((LH)_i, (HL)_i, (HH)_i)$. Then, after performing the inverse wavelet transform, we measure the quality of the reconstructed image. We use the signal-to-noise ratio (SNR) as the measure of the quality of the reconstructed images. The obtained results are presented in Figure 7, where a contribution of each level of wavelet coefficients, from the lowest frequency (level 1) to the highest frequency (level 5), is plotted.

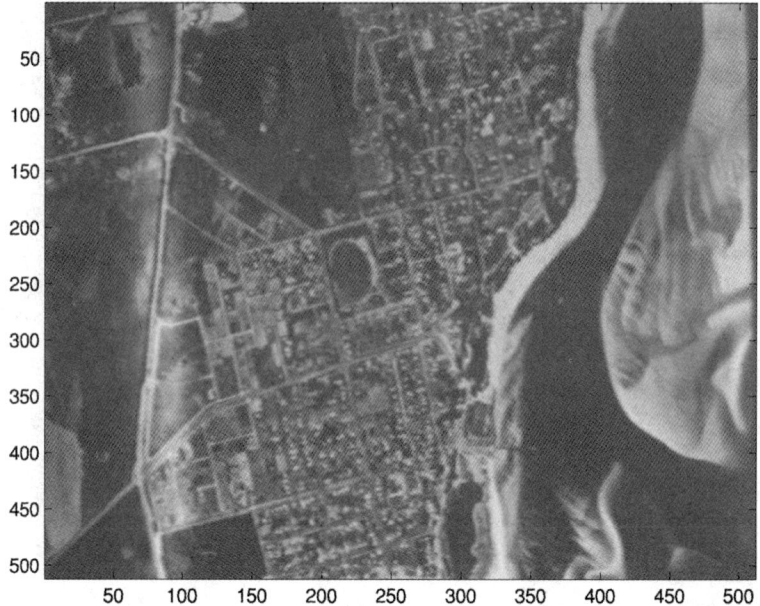

FIGURE 5. Image "Coast"—the original image.

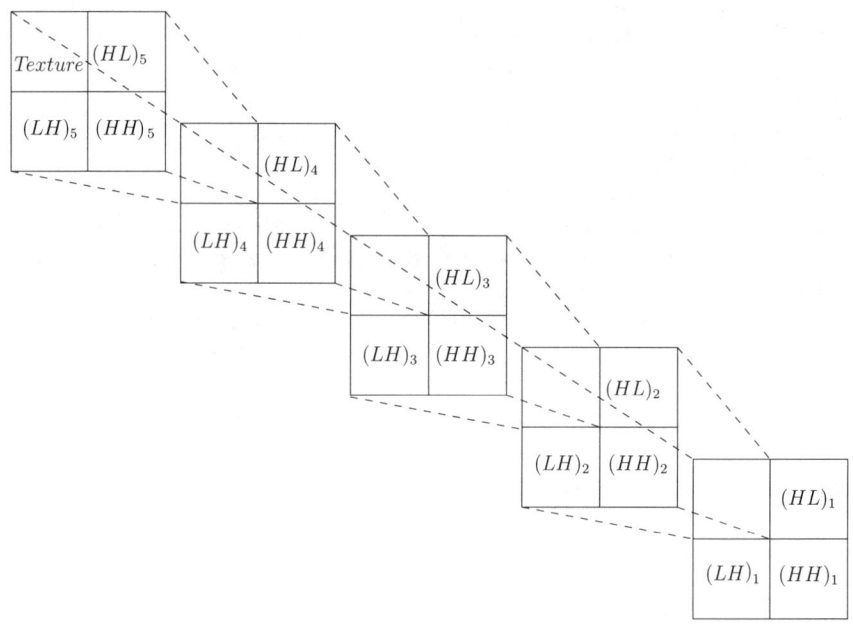

FIGURE 6. Test expansion.

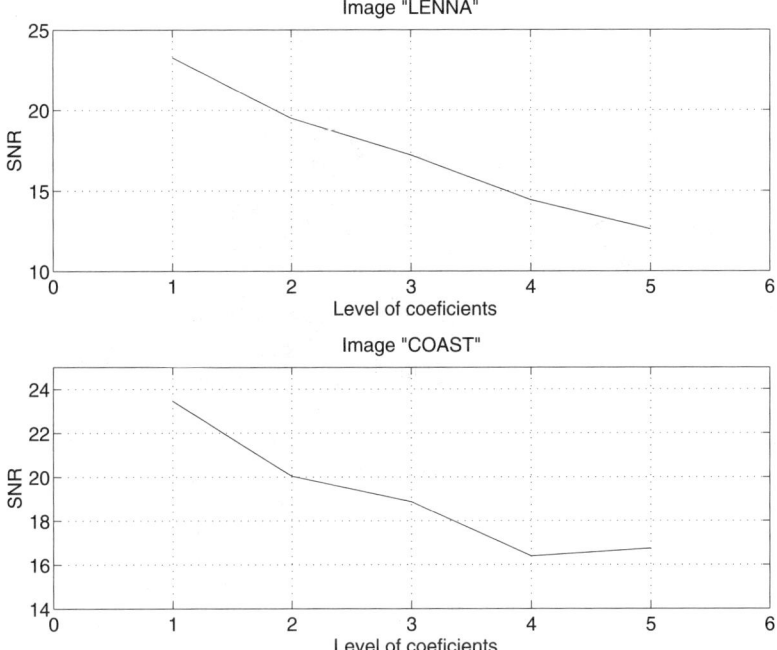

FIGURE 7. Test expansions: level of coefficients vs. SNR.

From Figure 7 one can see that for both test images, increasing the frequency reduces the effect on overall quality. In other words, most of the real images is the texture with low-frequency edges. These statements are general and may not be applicable to a specific image. However, for the purpose of this study and because we are mainly interested in building an image compression system acceptable for an "average" image, this conclusion is important.

Another set of experiments was performed on different submatrices of the same level of expansion. It is known that each submatrix of the wavelet coefficients reflects the information on horizontal, vertical, or diagonal edges of the input image. The submatrix $(HL)_i$ captures vertical edges of the input image, the submatrix $(LH)_i$ provides horizontal edges, and diagonal edges are located in $(HH)_i$.

We have consecutively set each submatrix of the wavelet coefficients to zeros and measured the quality of reconstruction. The results of the experiments for the test images "Lena" and "Coast" are given in Figures 8 and 9. In these figures the relative SNR, which is defined as maximum SNR for the image minus the SNR with a submatrix set to 0, is given in the vertical axis and is plotted against the matrix structures, as in Figure 6.

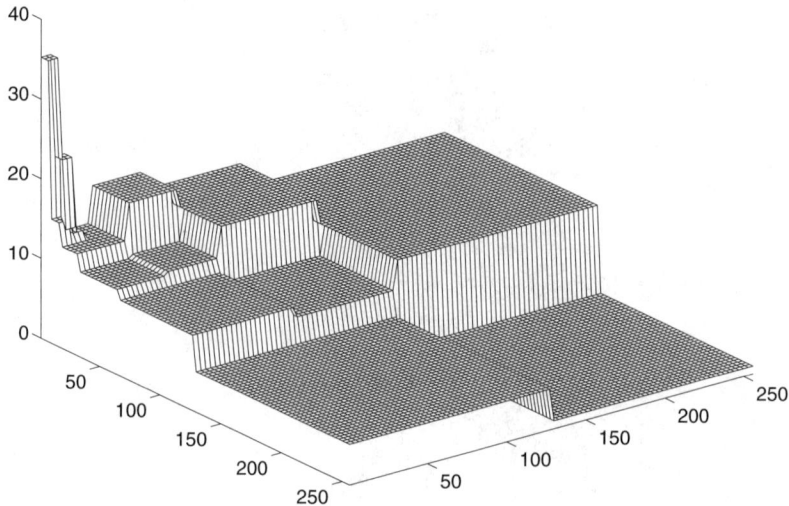

FIGURE 8. Test expansions of the "Lena" image: blocks of wavelet coefficients vs. relative SNR.

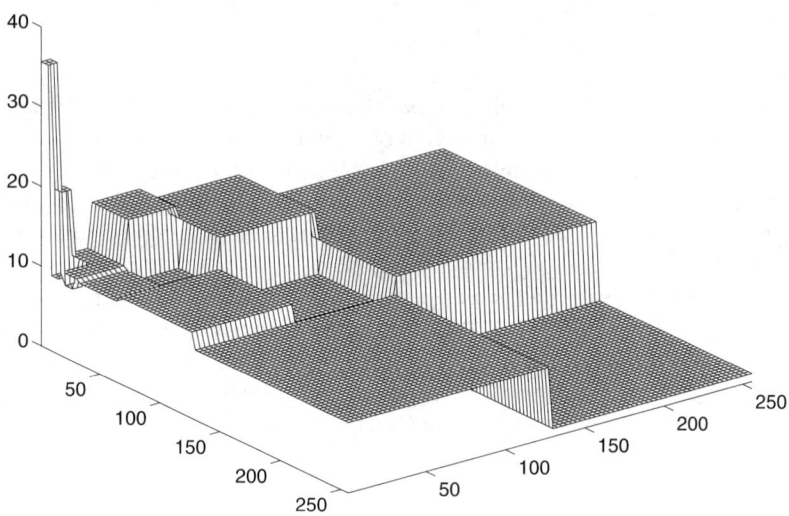

FIGURE 9. Test expansions of the "Coast" image: blocks of wavelet coefficients vs. relative SNR.

Examination of Figures 8 and 9 shows that in general the overall performance depends less on diagonal orientation than on vertical or horizontal orientations within the same level. Furthermore, the effect of cancellation of the vertical or horizontal edges of a level is close to the effect of cancellation of the diagonal edges of the next level up.

We can observe that most of the energy of the real images is typically concentrated in the low- and medium-frequency components (higher block indexes), whereas the high-frequency components formed by very sharp edges are not as significant. In terms of wavelet coefficients, this means that, following the general convention, the coefficients containing most of the image data are concentrated in the top left corner of the coefficient matrix. Therefore, it is desirable that the parameters of vector quantization for encoding the wavelet coefficients reflect this property.

An uneven distribution of energy across a range of frequencies in images is taken into account in the algorithm of an optimal bit allocation for quantization of wavelet coefficients presented next. This algorithm takes advantage of human perception of images and keeps the amount of calculations low due to fact that it relies on an appropriate model of the probability distribution function of wavelet coefficients instead of on the coefficients directly.

## C. Optimal Bit Allocation

The process of quantization unavoidably introduces some distortion to the image being compressed. The appropriate shaping of the quantization noise across the matrix of wavelet coefficients can reduce undesirable distortion in the reconstructed images. In this section we discuss an algorithm for an optimal bit allocation across the matrix of wavelet coefficients. This algorithm is based on modeling the probability distribution function (PDF) of the wavelet coefficients.

### 1. Distortion Function

Following the test results presented in Section II.B we can conjecture that quality of image coding depends highly on the appropriate shaping of the bit rate across the coefficient matrix (Shnaider and Paplinski, 1994). Therefore, an optimal bit allocation routine that minimizes the overall distortion $D$ for a given bit rate $R$ is required.

Let us specify the total distortion $D$ for a given bit rate $R$ as follows:

$$D(R) = D_N(r_N) + \sum_{i=1}^{N-1} a_i b_i 2^{-cr_i} \qquad (1)$$

where $D_N(r_N)$ is the distortion produced by quantization of the lowest-frequency block (texture), $r_i$ is a bit rate for a block $i$, $(1 \le i \le N)$, and $c$ is a parameter that is set to 2 for the mean squared error distortion measure. The remaining variables are specified in the following way:

$$a_i(c) = G(n_i, 2) B_i \left( \int_{-\infty}^{\infty} p_i^{n_i/(n_i+2)} \bigg/ dx \right)^{(n_i+2)/n_i} \quad (2)$$

where $G(n_i, 2)$ is a precalculated value of the average mean squared error of the optimal $n_i$-dimensional quantizer (Gersho, 1979), $p_i$ is the PDF of the $i$th block of the wavelet coefficient matrix, and $k_i$ is the size of the $i$th block. $B_i$, a human perception factor (Campbell and Robson, 1968), is defined as

$$B_i = \gamma^{q_i} \beta \log \sigma_i \quad (3)$$

where $q_i$ is a relative position of the $i$th block in the frequency plane and $\sigma_i$ is the variance of the wavelet coefficients. Finally, the parameter $b_i$ from Eq. (1) is specified as follows:

$$b_i = 2^{-(l_i^h + l_i^v)} \quad (4)$$

where $l_i^h$ and $l_i^v$ are the levels of the transform in the horizontal and vertical directions, respectively. Equation (1) is a modified version of that given in (Antonini et al., 1992). The modification takes into account wavelet packets as well as the traditional wavelets.

One can see that the total distortion $D(R)$ given by Eq. (1) depends on the probability distribution of the wavelet coefficients. The straightforward approach is to compute the PDF for each block within the wavelet coefficient matrix. Alternatively, one can use a prior known function to model the required PDF. Although the model gives only an approximation of the real PDF, this approach is significantly faster, which is important for a number of applications, such as real-time image compression. In the next section we will develop a PDF model of wavelet coefficients using their statistical properties.

## 2. A Statistical Model of Wavelet Coefficients

Before we proceed to the modeling the PDF of wavelet coefficients, we need to reexamine the structure of the single-stage two-dimensional wavelet expansion given in Figure 3. As we have described it already, at each level of the 2-D wavelet expansion, the input matrix is split into four matrices, each of which represents certain frequencies of the input matrix. For an input matrix $(LL)_{m-1}$, the output of the 2-D wavelet expansion is $(LL)_m$, $(LH)_m$, $(HL)_m$, and $(HH)_m$, where $L$ and $H$ refer to application of the low- and high-pass filters, respectively. In such a way, while matrices $(LH)_m$, $(HL)_m$, and $(HH)_m$ represent high-frequency content, such as edges, the matrix $(LL)_m$ gives the low-frequency content of the input image, such as textures and shapes. Accordingly, we can classify the complete set of the wavelet coefficients into high-frequency and low-frequency coefficients.

a. *High-Frequency Coefficients*

It was shown in Antonini *et al.* (1992) that for real images, the high-frequency wavelet coefficients fluctuate around zero with a variance that depends on the density of discontinuities or edges in the input image. In order to approximate the probability distribution of the high-frequency wavelet coefficients with an appropriately selected model, we compare the graphs of distribution of wavelet coefficients presented in Figure 10.

The distribution of wavelet coefficients shown in Figure 10 is based on a randomly selected block from the matrix of coefficients of the image "Lena." A direct comparison shows that the Laplacian and generalized Gaussian (GGF with $\xi = 0.7$) functions give a good approximation of the PDF of the

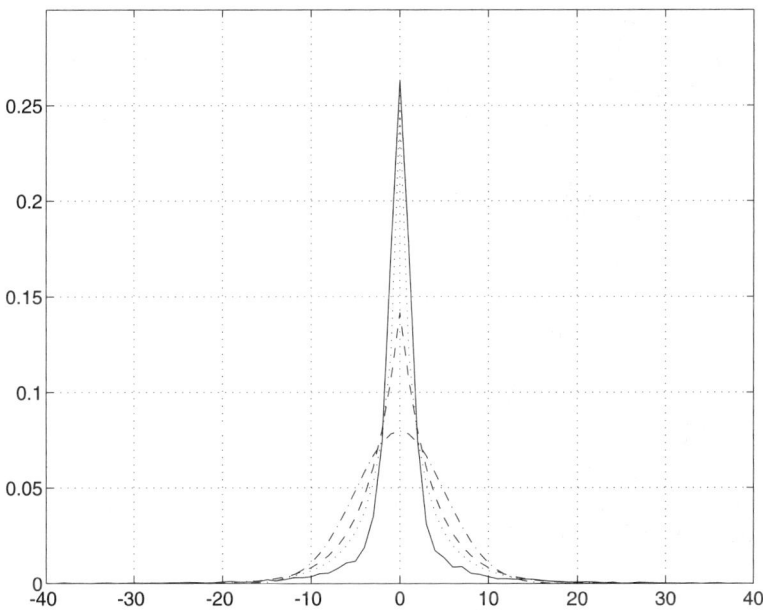

FIGURE 10. Probability distribution functions of wavelet coefficients and generalized Gaussian functions for the high-frequency wavelet coefficients (——), the generalized Gaussian with $\xi = 0.7$ ($\cdots$) the Laplacian (- -), and the Gaussian (-·-).

higher-frequency coefficients (Antonini *et al.*, 1992; Barlaud *et al.*, 1994). The generalized Gaussian function (GGF) is given by

$$p(x) = a \, \exp(-|bx|^\xi) \tag{5}$$

where

$$a = \frac{b\xi}{2\Gamma(1/\xi)} \quad \text{and} \quad b = \frac{\Gamma(3/\xi)^{1/2}}{\sigma \Gamma(1/\xi)^{1/2}} \tag{6}$$

Here, $\Gamma(\cdot)$ denotes the Gamma function, $\sigma$ is the standard deviation, and $\xi$ is a parameter. If $\xi$ is set to 1, Eq. (6) becomes the Laplacian distribution. For $\xi$ equals 2, it leads to the standard Gaussian distribution.

b. *Low-Frequency Coefficients*

The statistics of low-frequency coefficients are similar to the statistics of the image data. Because we do not restrict ourselves to any particular type of images, it is not possible to give a general model of the PDF valid for any image. However, most real images exhibit significant correlation in the space domain, which can be exploited by the familiar predictive coding (Gersho and Gray, 1992). Typically, the probability distribution of the errors of prediction is a bell-shaped function that can be approximated by the generalized Gaussian function with $\xi = 2$, i.e., the standard Gaussian distribution with zero mean.

Therefore, by using predictive coding it is possible to utilize a single function (GGF) for modeling the PDF of all wavelet coefficients. For low-frequency coefficients, the predictive coding error vectors are approximated by GGF with $\xi = 2$, whereas for high-frequency wavelet coefficients, the GGF approximation with $\xi = 0.7$ is used (Shnaider and Paplinski, 1995). Formally, we can write

$$p_i = \begin{cases} \text{GGF}(\xi = 0.7), & \text{if } 1 \le i \le N-1 \\ \text{GGF}(\xi = 2), & \text{if } i = N \end{cases} \tag{7}$$

where $i$ is the index of the block, i.e., the level of the transform within the matrix of wavelet coefficients.

Given this probability distribution model, function (1), describing the total distortion after quantization of the wavelet coefficients, can be expressed in a homogenous form as

$$D(R) = \sum_{i=1}^{N} a_i b_i 2^{-cr_i} \tag{8}$$

Now, this function needs to be minimized for a given total bit rate.

## 3. Minimization of the Distortion Function

As shown in the previous section, the initial problem of optimal distribution of the quantization bits across the matrix of wavelet coefficients narrows down to minimization of the total distortion for a given overall bit rate or, in other words, compression ratio. That is, we need to minimize the distortion function (8) subject to

$$R = \sum_{i=1}^{N} b_i r_i \qquad (9)$$

The first approach to the minimization is to use the Lagrange multiplier technique. In this case the following derivatives with the respect to $r_i$ must attain zero:

$$\frac{\partial}{\partial r_i}\left[\sum_{i=1}^{N} a_i b_i 2^{-cr_i} - \lambda\left(R - \sum_{i=1}^{N} b_i r_i\right)\right] = 0 \qquad (10)$$

From this we obtain the following equation for $r_i$:

$$-c \ln(2) a_i b_i 2^{-cr_i} + \lambda b_i = 0 \quad \text{or} \quad 2^{-cr_i} = \frac{\lambda}{c \ln(2) a_i}$$

Solving for $r_i$ yields

$$r_i = -\frac{1}{c} \log_2 \frac{\lambda}{c \ln(2) a_i} \qquad (11)$$

or, equivalently,

$$r_i = \frac{1}{c}\left(\log_2(\ln(2) a_i) - \log_2 \frac{c}{\lambda}\right) \qquad (12)$$

By substituting Eq. (12) into the constraint equation, Eq. (9), we have

$$cR = \sum_{i=1}^{N} b_i \left(\log_2(\ln(2) a_i) - \log_2 \frac{c}{\lambda}\right) \qquad (13)$$

or

$$\log_2 \frac{c}{\lambda} = B\left(\sum_{i=0}^{N} \log_2(\ln(2) a_i)^{b_i} - cR\right) \quad \text{where } B = \left(\sum_{i=1}^{N} b_i\right)^{-1} \qquad (14)$$

Now we can substitute Eq. (14) into Eq. (12) to obtain the following expression for the bit rate:

$$r_i = BR + \frac{1}{c}\left(\log_2(\ln(2) a_i) - \sum_{i=1}^{N} \log_2(\ln(2) a_i)^{Bb_i}\right) \qquad (15)$$

Finally, the bit rate for the $i$th block can be calculated as

$$r_i = BR + \frac{1}{c} \log_2 \frac{\ln(2)a_i}{\prod_{i=1}^{N}(\ln(2)a_i)^{Bb_i}} \qquad (16)$$

However, the direct calculation of the derivative using Eq. (10) does not guarantee that the values of $r_i$ are within acceptable range. The alternative solution to the minimization problem specified in Eq. (8) is the utilization of one of the numerical optimization methods. We use the algorithm developed in Riskin (1991), which gives the optimal bit allocation and allows the control of the values of bit rates.

Initially, $r_i = 0$ ($1 \leq i \leq N$). Then, some additional bits, $\Delta r$, are allocated at each iteration to the $r_i$, which attain the minimal overall distortion for all $i$ ($1 \leq i \leq N$). The procedure continues until the desired bit rate in Eq. (9) is reached. In our calculations $\Delta r$ has been set to 0.01 bpp, and the parameters in Eq. (3), $\beta$ and $\gamma$ have been set to 1.05 and 1, respectively.

In conclusion, the algorithm for an optimal bit allocation for quantization of wavelet coefficients presented in this section enables to keep the number of calculations low by taking advantage of the GGF model. Numerical minimization algorithm for the distortion function has been employed.

## III. Lattice Quantization Fundamentals

Examination of various vector quantization (VQ) techniques shows that a majority of them suffer from a high computational cost associated with the generation of an appropriate codebook. This implies that they require either a long time to encode images or complex hardware to facilitate the computation in an acceptable time. It is possible, however, based on the theory of $n$-dimensional lattices to develop a family of quantizers that are free from this drawback. The resulting *lattice quantizers* offer computational simplicity for image coding and will be examined in detail in this section.

We begin our discussion by considering the motivation for lattice quantization. Then, we give a formal definition of lattices, followed by the description of two classes of lattices especially important in the context of quantization: root and laminated lattices. Successful application of a lattice is possible only if there exist enough lattice points available for quantization. Therefore, we also address the problem of counting the lattice points. First, an approximate solution to this problem is presented. Then, by using theta functions, we show how the exact number of lattice points available can be computed.

## A. Codebook for Quantization

Let us examine the process of quantization. The quantization scheme requires a knowledge of the codebook by both the encoder and the decoder prior to the process of quantization. In most cases, particularly in VQ, to satisfy this requirement the codebook is also sent through the communication channel or included in the compressed file. As a result, the overall compression ratio may be significantly reduced, especially for large codebooks or high-dimension vectors, in the case of VQ.

Lattice quantization (LQ), discussed shortly, is free from this drawback because it does not require transmission of the complete codebook due to its regular structure; as a result, it is possible to generate the codebook independently at both the encoder and decoder ends. In addition, LQ delivers a superior speed compared with other types of quantization, such as learning vector quantization, due to the simplicity of generating the codebook.

## B. Distortion Measure and Quantization Regions

Let us consider an input stream of continuous random vectors $x = (x_1, x_2, \ldots, x_N)$ with the probability density function $p(x)$. Let us also consider an output stream $y = (y_1, y_2, \ldots, y_N)$ related to an input stream $x$ by the conditional probability density function $q(y|x)$. Assuming that the stream $x$ is the input to a quantizer, then the corresponding values from the output stream $y$ are obtained by a reference to the quantizer codebook.

According to Shannon (1948) the amount of uncertainty $R(x, y)$ of a value of $x$ when we receive its quantized counterpart $y$ from a transmission channel is given by

$$R(x, y) = h(x) - h(x|y) \quad (17)$$

where $h(x) = -\int p(x) \log p(x) dx$ is the differential entropy of a variable $x$. The value of $R(x, y)$ is, effectively, the actual transmission rate. With this in mind, we can pose the problem of optimal quantization in terms of minimization of the amount of uncertainty, i.e.,

$$R(D) = \min_{q \in Q_D}(R(x, y)) \quad (18)$$

with $Q_D$ being a set specified by the conditional probability $q(y|x)$ so that

$$Q_D = \{q(y|x) : E[d(x, y)] \leq D\} \quad (19)$$

where $d(\cdot, \cdot)$ is a distance function, defined next. The substitution of Eq. (17) in Eq. (18) leads to

$$R(D) = \min_{q \in Q_D}(h(x) - h(x|y)) \tag{20}$$

Because the entropy $h(x)$ is independent of the conditional probability $q(y, x)$ and the entropy $h(x|y) = h(x - y|y) \leq h(x - y)$, we can further modify Eq. (20) in the following way:

$$R(D) = h(x) - \max_{q \in Q_D}(h(x|y))$$
$$= h(x) - \max_{q \in Q_D}(h(x - y|y))$$
$$\geq h(x) - \max_{q \in Q_D}(h(x - y)) \tag{21}$$

In information theory the last equation is known as the *Shannon lower bound* (Berger, 1971).

Now let us return to the process of quantization. We collect the input vectors $x_j = (x_1, x_2, \ldots, x_N)$ in blocks $X = (x_1, x_2, \ldots, x_L)$ and, for each input block, the quantizer finds a corresponding output block $Y = (y_1, y_2, \ldots, y_L)$ consisting of vectors $y_j = (y_1, y_2, \ldots, y_N)$. The average bit rate per vector for a vector quantizer can be calculated as

$$h_Q(Y) = \frac{1}{L}h(Y) = -\frac{1}{L}\sum_j q(Y_j)\log q(Y_j) \tag{22}$$

Assuming that the distribution of $x$ is uniform within each quantization region $V$ and that all regions are simple translations of each other, we have the following approximation for the average bit rate per vector

$$h_Q(Y) \approx \frac{1}{L}h(X) - \frac{1}{L}\log V \tag{23}$$

If the input vectors $x$ are independent, we have

$$h(x) = \frac{1}{L}h(X) \quad \text{and} \quad h(x - y) = \frac{1}{L}h(X - Y).$$

Furthermore, it can be shown (Berger 1971) that for sufficiently high dimensionality of vectors, the following equality is satisfied:

$$h(X - Y) = \log V \tag{24}$$

Thus, we can conclude that the performance of a *uniform vector quantizer* asymptotically achieves the Shannon lower bound for large dimensions.

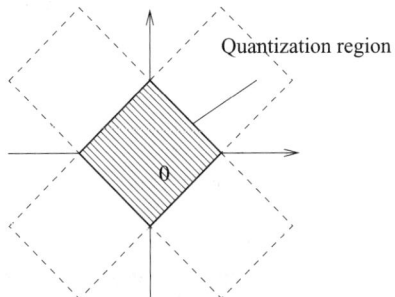

FIGURE 11. A pyramidal quantization region for $m = 1$.

An average distortion of a quantizer can be defined in the following way:

$$D_Q^{(m)} = \int_{-\infty}^{\infty} d^{(m)}(x)p(x)\,dx \qquad (25)$$

where for every input $x$ with the **PDF** $p(x)$ the quantizer produces an output $y$. The $m$th-power distortion $d^{(m)}$ introduced by this operation is defined by

$$d_k^{(m)}(x, y) = |y - x|^m \qquad (26)$$

By varying $m$, different distortion measures can be obtained. Closely related to the distortion measure is an optimal shape of the quantization region. It is easy to see that for a uniform source, the optimal shape of quantization region is a pyramid if $m$ is selected to be 1, as depicted in Figure 11. For $m$ equal to 2, Eq. (26) gives rise to the well-known mean-squared error (MSE) distortion measure, the optimal quantization region for an uniform source being a *sphere*. Furthermore, the radius of the sphere is equal to $\sqrt{LD}$, where $D$ and $L$ are the target distortion and dimensionality of quantization space, respectively.

From the preceding considerations we can conjecture that, assuming the MSE distortion measure, an optimal quantizer for an $L$-dimensional uniform source constitutes a set of uniformly distributed spheres with code vectors as centroids of the spheres.

### C. Optimal Quantizer for Wavelet Coefficients

In order to generalize the statistics of the wavelet coefficients, it is necessary to approximate their probability distribution functions (PDF) by an appropriately selected model.

As discussed in Section II, for real images the PDF of wavelet coefficients can be well approximated by the generalized Gaussian functions (GGF) given by Eq. (5) with appropriately chosen exponents, as follows:

$$p_i = \begin{cases} \text{GGF}(\xi = 0.7), & \text{if } 1 \leq i \leq N-1 \\ \text{GGF}(\xi = 2), & \text{if } i = N \end{cases} \quad (27)$$

Here $i$ is the index of the block of wavelet coefficients. If $i$ is set to 0, this corresponds to the low-frequency block, located in the upper-left corner of the matrix of coefficients.

Assuming that the PDF of wavelet coefficients of an image can be well approximated by Gaussian-type functions and using the mean-squared error as a measure of the quality of this approximation, it can be shown that in order to achieve an optimal quantization of the wavelet coefficients, the codevectors should be uniformly distributed on a hypersphere with the center at the origin.

Consider a continuous random variable $x$ with the Gaussian PDF $p(x)$. Without loss of generality, let us assume that $p(x)$ can be well approximated by a staircase function, $p'(x)$. For some small $\epsilon_0$ and $\epsilon_1$, we can write

$$p'(x) = \text{const} \quad \text{and} \quad |p(x) - p'(x)| \leq \epsilon_0 \, \forall \, |x - x_0| \leq \epsilon_1 \quad (28)$$

Following the preceding assumption, within each of the constant probability segments of our approximation function $p'(x)$, we have uniformly distributed values of $x$. For example, in a 2D space the regions of constant probabilities form rings around the origin (Fig. 12).

As shown in Section III.B, for optimal quantization of an uniformly distributed source we need a uniform vector quantizer with spherical quantization regions. This is true provided that the chosen distortion measure is the mean-squared error. If the source has the Gaussian PDF, e.g., wavelet coefficients,

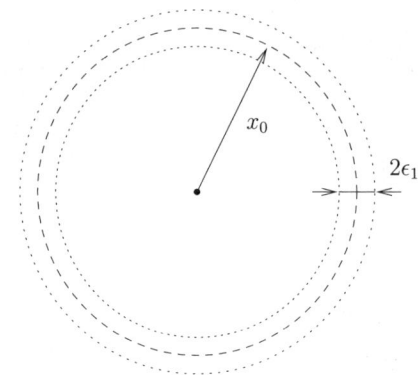

FIGURE 12. Regions of constant probabilities in a 2D space.

which can be approximated as in Eq. (28), then the spherical quantization regions must be placed so that their centroids located on a hypersphere centered at the origin of the source distribution.

As will be shown next, this result directly leads to the utilization of lattices for quantization of the wavelet coefficients.

## IV. LATTICES

A lattice $L$ can be defined by an integral combination of linearly independent basis vectors of dimension $n$ collected in a *generator* matrix $M$ (Birkhoff, 1967; Conway and Sloane, 1993; Gibson and Sayood, 1988). Assuming a vector $\xi \in Z_n$, the lattice $L$ is generated by the product $M\xi$. The simplest lattice is the $Z_n$ lattice, which consists of all the integral points in an $n$-dimensional space.

Now, let us consider an $n$-dimensional lattice $L$ specified by the generator matrix $M$, consisting of $n$ basis vectors of the form

$$M = [\boldsymbol{b}_1 \cdots \boldsymbol{b}_n] = \begin{bmatrix} b_{1,1} & \cdots & b_{n,1} \\ \vdots & \ddots & \vdots \\ b_{1,n} & \cdots & b_{n,n} \end{bmatrix} \quad (29)$$

Then a vector $x = (x_1, \ldots, x_n)^T$ belonging to this lattice can be expressed as a linear combination of basis vectors $\boldsymbol{b}_i$:

$$x = \xi_1 \boldsymbol{b}_1 + \xi_2 \boldsymbol{b}_2 + \cdots + \xi_n \boldsymbol{b}_n = M\xi \quad (30)$$

The squared norm of a vector $x$ can be defined as a function of $\xi$:

$$\|x\|^2 = \langle x, x \rangle = \sum_i \sum_j \xi_i \xi_j \langle \boldsymbol{b}_i, \boldsymbol{b}_j \rangle$$

$$= \xi^T M^T M \xi = \xi^T A \xi = f_L(\xi) \quad (31)$$

The matrix $A = M^T M$ is called a *Gram matrix* of the lattice $L$. The $(i, j)$ entry of this matrix is the inner product of the corresponding basis vectors $\langle \boldsymbol{b}_i, \boldsymbol{b}_j \rangle$. In Eq. (31) the function $f_L(\cdot)$ is known as a *quadratic form* associated with lattice $L$.

For each lattice $L$ there exists a *dual lattice* $L^*$ defined as follows. If a vector $j$ of dimension $n$ belongs to the dual lattice $L^*$, then for any vector $i$ from the subspace spanned by the lattice $L$ the inner product $\langle i, j \rangle$ is an integer, i.e.,

$$L^* = \{j \in \mathbb{R}^n : \langle i, j \rangle \in \mathbb{Z} \, \forall i \in L\}. \quad (32)$$

A dual lattice $L^*$ may be also defined as

$$L^* = \bigcup_{i=0}^{k-1} (r_i + L) \qquad (33)$$

where $r_i$ are the representatives of $L$ in $L^*$ known as *glue vectors*. If $r_i = 0$, then the lattice $L$ is integral because it is contained in its dual lattice $L^*$.

## A. Root Lattices

In this section we examine some concepts from Lie algebras—in particular, root systems—and their utilization for generation of lattices. The material presented in this section is mainly due to (Gibson and Sayood, 1988; Conway and Sloane, 1993; Humphreys, 1972; Grove and Benson, 1972; Gilmore, 1974; Coxeter, 1973; Shnaider, 1997).

As shown before, with each lattice we can associate a quadratic form (31). The elements in the Gram matrix $A$ are the inner products of the basis vectors of a lattice $a_{ij} = \langle b_i, b_j \rangle$. The Gram matrix of a root lattice is closely related to the so-called Cartan matrix of the root system corresponding to this lattice.

### 1. Root Systems

Let us begin with the definition of an orthogonal transformation $S \in \Gamma(\mathbf{R}^n)$, where $\Gamma(\mathbf{R}^n)$ is the set of all orthogonal transformations over an $n$-dimensional real Euclidean space $\Xi$. Consider a reflection transformation in a two-dimensional space $\Xi(\mathbf{R}^2)$ shown in Figure 13. The reflection transformation $S_r$ converts every vector $x$ from the space $\mathbf{R}^2$ to its mirror image $S_r x$ with respect to the line $l$. Thus, if $r \perp l$ and $\hat{r} \in l$, we can write $x = a_0 \hat{r} + a_1 r$ and, consequently, $S_r x = a_0 \hat{r} - a_1 r$. In general, for an $n$-dimensional space the transformation $S_r$ is called the reflection through a hyperplane $l$, or the reflection along $r$. Mathematically, such a transformation $S_r$ can be formulated as

$$S_r x = a_0 \hat{r} - a_1 r = x - 2 \frac{\langle x, r \rangle}{\langle r, r \rangle} r \qquad (34)$$

Assuming that there exists a subgroup of transformations $G \leq \Gamma(\mathbf{R}^2)$ and $S_r \in G$, we can define the roots of $G$ as $\pm r$. Formally, a subset $R$ of the Euclidean space $\Xi(\mathbf{R}^n)$ is called a *root system* of this space if

1. $R$ is finite, spans $\mathbf{R}^n$ and does not contain 0;
2. For all $r \in R$, $-r \in R$;
3. $R$ is invariant over all reflections $S_r$ for all $r \in R$;

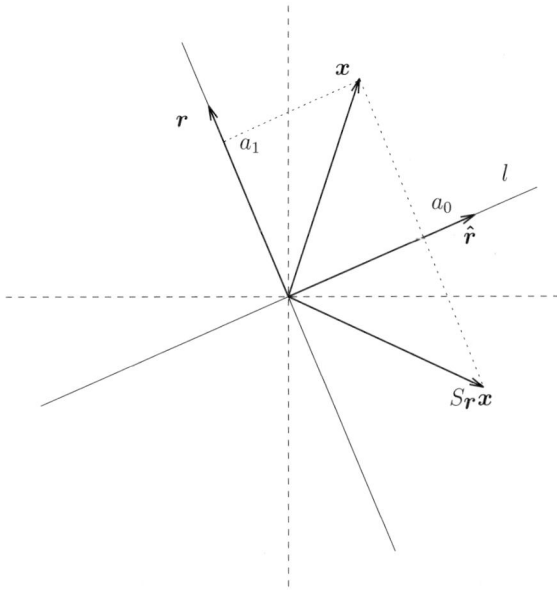

FIGURE 13. The reflection transformation $S_r$.

4. For all $r_i, r_j (i \neq j) \in R$,

$$\frac{2\langle r_i, r_j \rangle}{\langle r_i, r_i \rangle} \quad \text{is integer} \tag{35}$$

In Lie algebras the number

$$\frac{2\langle r_i, r_j \rangle}{\langle r_i, r_i \rangle}$$

often has a separate notation, namely,

$$(r_i, r_j) \stackrel{\text{def}}{=} \frac{2\langle r_i, r_j \rangle}{\langle r_i, r_i \rangle}. \tag{36}$$

There exist a number of root systems that satisfy the definition given in this section. They are denoted by uppercase letters with a subscript referring to the dimensionality of the space they span, for example, $A_n$, $B_n$, $C_n$.

The set of roots in the root system can be classified into two groups. For any vector $t \in \Xi$ such that $\langle t, r \rangle \neq 0$, a root is referred to as either negative or positive depending on the sign of the inner product of this vector $t$ with the corresponding root:

$$R_t^+ = \{r \in R : \langle t, r \rangle > 0\} \quad \text{(positive)} \tag{37}$$

and

$$R_t^- = \{r \in R : \langle t, r \rangle < 0\} \quad \text{(negative)} \tag{38}$$

This classification can be geometrically interpreted as two subsets of the root system lying on opposite sides of the hyperplane perpendicular to the vector $t$.

A subset $\Delta$ of $R$ is called a base if

1. $\Delta$ forms a basis for $\Xi$;
2. Each root $\alpha \in R$ can be expressed as

$$\alpha = \sum_i k_i r_i \quad \text{with} \quad r_i \in \Delta \tag{39}$$

where $k_i$ are either all negative integers for negative roots or all positive integers for positive roots.

The roots in the base $\Delta$ are called simple.

Consider an example of a root system shown in Figure 14. This system comprises eight roots given by

$$R = \{\pm(1, 0), \pm(0, 1), (\pm 1, \pm 1)\}. \tag{40}$$

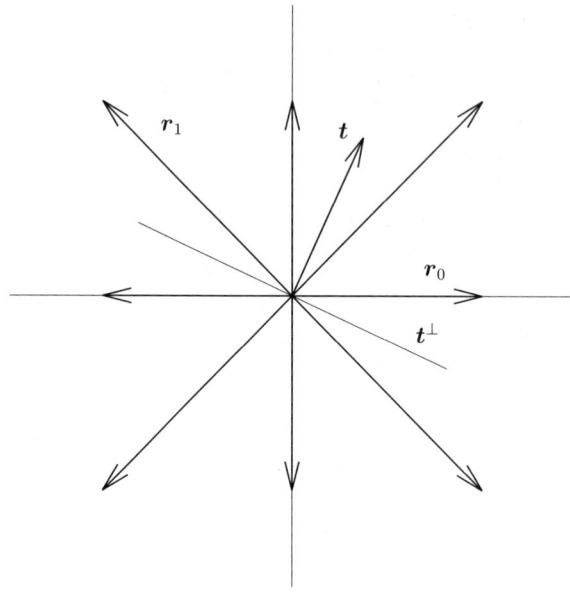

FIGURE 14. Example of a root system.

By choosing $t = (\cos\beta, \sin\beta)$ with $\pi/4 < \beta < \pi/2$, the positive roots are

$$R_t^+ = \{(1, 0), (0, 1), (1, 1), (-1, 1)\} \tag{41}$$

Finally, the simple positive roots forming a base are

$$\Delta_t^+ = \{(1, 0), (-1, 1)\} \tag{42}$$

Let us analyze the possible angles between the roots in a system. Condition (35) restricts these angles to a limited set of values. Because we know that

$$\langle r_i, r_j \rangle = \|r_i\| \|r_j\| \cos\theta_{ij} \tag{43}$$

we have the following expression for the Lie number $(r_i, r_j)$:

$$(r_i, r_j) = \frac{2\langle r_i, r_j \rangle}{\langle r_i, r_i \rangle} = 2\frac{\|r_j\|}{\|r_i\|} \cos\theta_{ij} \tag{44}$$

and, consequently,

$$(r_i, r_j)(r_j, r_i) = 4\cos^2\theta_{ij} \tag{45}$$

The last equation can be also written as

$$\frac{\langle r_i, r_j \rangle}{\langle r_i, r_i \rangle} \frac{\langle r_j, r_i \rangle}{\langle r_j, r_j \rangle} = \cos^2\theta_{ij} \tag{46}$$

Now, with respect to the condition (35), which restricts the values of the Lie number $(r_i, r_j)$ to integers and considering also that $0 \leq \cos^2\theta_{ij} \leq 1$, we can compile all possible angles $\theta_{ij}$ between roots $r_i$ and $r_j$, as well as their relative lengths. The results are given in Table 1.

The two specific choices of roots, $\{(r_i, r_j) = \pm 4, (r_j, r_i) = \pm 1\}$ and $\{(r_i, r_j) = \pm 1, (r_j, r_i) = \pm 4\}$ when $\cos^2\theta_{ij} = 1$, which also satisfy Eq. (45), are invalid due to a contradiction with the definition of roots. This contradiction can be seen as follows. Because $\cos^2\theta_{ij} = 1$, the angles $\theta_{ij}$ are 0° or 180°. As

TABLE 1
ANGLES BETWEEN THE ROOTS AND THEIR RELATIVE LENGTHS

| $\cos^2\theta_{ij}$ | $\theta_{ij}$ | $(r_i, r_j)$ | $(r_j, r_i)$ | $\langle r_i, r_i \rangle / \langle r_j, r_j \rangle$ |
|---|---|---|---|---|
| 1 | 0°, 180° | ±2 | ±2 | 1 |
| $\frac{3}{4}$ | 30°, 150° | ±3 | ±1 | $\frac{1}{3}$ |
|  |  | ±1 | ±3 | 3 |
| $\frac{2}{4}$ | 45°, 135° | ±2 | ±1 | $\frac{1}{2}$ |
|  |  | ±1 | ±2 | 2 |
| $\frac{1}{4}$ | 60°, 120° | ±1 | ±1 | 1 |
| 0 | 90° | 0 | 0 | Undetermined |

a result, we have two roots in the same or opposite directions; by definition, these roots must be of the same length. This does not agree with the preceding choices of roots.

It can be shown (Humphreys, 1972) that for any two distinctive simple roots the following inequality must be satisfied:

$$\langle r_i, r_j \rangle \leq 0 \qquad \forall r_i \in \Delta \quad \text{and} \quad i \neq j \qquad (47)$$

which together with Eq. (43) limits the values of $\theta_{ij}$ between the simple roots $r_i, r_j$ $(i \neq j)$ to $\pi/2 \leq \theta_{ij} \leq \pi$. Furthermore, considering the results in Table 1, the angles between simple roots are

$$\theta_{ij} = \{90°, 120°, 135°, 150°\} \qquad \forall r_i, r_j \in \Delta \quad \text{and} \quad i \neq j \qquad (48)$$

This result, which is valid for any dimension $n \geq 2$, will allow us to identify a subset of simple roots in a root system.

## 2. Cartan Matrices

Let us consider a set of simple roots $\{r_i : r_i \in \Delta\}$ in the root system $R$ defined previously. The matrix of the Lie numbers $[(r_i, r_j)]$ is called the *Cartan matrix* of $R$. The Cartan matrix completely determines the corresponding root system $R$.

Next, we consider some important properties of Cartan matrices. From Eq. (36) we have $(r_i, r_i) = 2$, which is valid for all roots, including simple positive ones. Thus, the elements on the the main diagonal of a Cartan matrix are always equal to 2. Table 1 shows that in the case of any roots, the values of the Lie number, $(r_i, r_j)$, are $0, \pm 1, \pm 2, \pm 3$. Considering conditions (47) and (48), we can narrow the range of values for simple roots belonging to the base $\Delta$ to nonpositive values. Therefore, we have $(r_i^+, r_j^+) = \{0, -1, -2, -3\}$; consequently, the entries of a Cartan matrix located off the main diagonal also take on only these values.

We mentioned in Section IV.A.1 that there are a number of different root systems. As examples, the Cartan matrices for some of those root systems, namely $A_n, B_n, C_n, D_n, E_6, E_7, E_8, F_4$, and $G_2$, are given in Section XII. Next, we consider an example of construction of the root system $D_n$. For the construction of the remaining root systems the reader is referred to (Gibson and Sayood, 1988).

### a. $D_n$ $(n \geq 4)$ Root System

The $D_n$ root system is obtained by letting

$$\Xi = \mathbf{R}^n \quad \text{and} \quad R = \{r \in Z : \langle r, r \rangle = 2\} = \{\pm(e_i \pm e_j), i \neq j\}$$

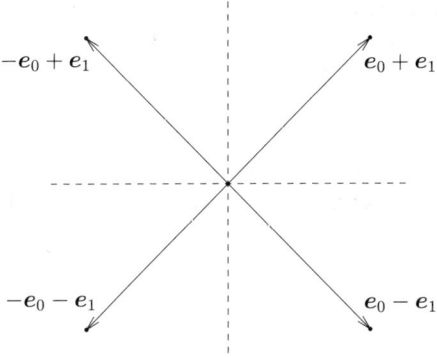

FIGURE 15. $D_2$ root system.

Note that the dimensionality $n$ of root systems of type $D$ is 4 or greater. This is due to the fact that in low-dimensional spaces, some of the root systems are equivalent. The $D_2$ root system, for example, can be represented by the product $A_1 \times A_1$ and is shown in Figure 15.

The 3D version of the $D$ type of root system, $D_3$, is equivalent to $A_3$.

### B. Construction of Root Lattices

As mentioned before, the Gram matrix of a root lattice and the Cartan matrix of a root system are interrelated. In fact, for a root system the elements in the normalized Cartan matrix, i.e., divided by a factor of 2, equal the elements of the Gram matrix of the lattice corresponding to this system.

Therefore, starting from the Cartan matrix of a root system, we can generate a set of basis vectors of the root lattice. However, as shown in Section IV, a lattice can be defined by a number of *distinct* sets of basis vectors. The necessary orientation of the lattice basis can be achieved by setting the direction of one of the basis vectors. As suggested in Gibson and Sayood (1988), this can be done by reorganizing the generator matrix $M$ of a lattice to its upper triangular form, as follows:

$$M = \begin{bmatrix} b_{11} & b_{21} & b_{31} & \cdots & b_{n1} \\ 0 & b_{22} & b_{32} & \cdots & b_{n2} \\ \vdots & \vdots & \vdots & \ddots & \vdots \\ 0 & 0 & 0 & \cdots & b_{nn} \end{bmatrix} = [\boldsymbol{b}_1 \; \boldsymbol{b}_2 \; \boldsymbol{b}_3 \; \cdots \; \boldsymbol{b}_n] \quad (49)$$

In Eq. (31) the Gram matrix $A$ was defined as

$$A = M^T M \quad (50)$$

Let us now consider the normalized Cartan matrix $\Phi = \{\phi_{ij} : 0 \leq i, j \leq n-1\}$. From Eqs. (49) and (50) the first element of the Cartan matrix can be expressed as

$$\phi_{11} = \langle \boldsymbol{b}_1, \boldsymbol{b}_1 \rangle \tag{51}$$

where $\boldsymbol{b}_i$ is the $i$th basis vector in the generator matrix $M$. For the triangular generator matrix $M$, as defined in Eq. (49), we can write that

$$b_{11}^2 = \langle \boldsymbol{b}_1, \boldsymbol{b}_1 \rangle \tag{52}$$

Thus, we have

$$b_{11} = \sqrt{\phi_{11}} \tag{53}$$

For the next basis vector we can write a set of two equations:

$$\begin{cases} \phi_{21} = \langle \boldsymbol{b}_2, \boldsymbol{b}_1 \rangle \\ \phi_{22} = \langle \boldsymbol{b}_2, \boldsymbol{b}_2 \rangle \end{cases} \tag{54}$$

Similarly, using Eqs. (53) and (54) it is possible to deduce the following results for the next basis vector, $\boldsymbol{b}_2$:

$$\boldsymbol{b}_2 = \left( \frac{\phi_{21}}{\sqrt{\phi_{21}}} \left( \frac{\phi_{22} - \phi_{12}^2}{\phi_{21}} \right)^{1/2} 0 \ldots 0 \right)^T \tag{55}$$

Assuming that $\sum_{k=a}^{b} x_k = 0$ for all $x_k$ if $a > b$, the procedure of calculating the basis vectors can be generalized in the form of the following algorithm (Sayood, Gibson, and Rost, 1984):

1. Set $i = 1$ and calculate $b_{11} = \sqrt{\phi_{11}}$.
2. Set $i = i + 1$ and calculate the off-diagonal elements

$$b_{ij} = \frac{1}{b_{jj}} \left( \phi_{ij} - \sum_{k=1}^{j-1} b_{ik} b_{jk} \right) \quad \text{for } j = 1, 2, \ldots, i-1 \tag{56}$$

3. Calculate the diagonal element

$$b_{ii} = \left( \phi_{ii} - \sum_{k=1}^{i-1} b_{ik}^2 \right)^{1/2} \tag{57}$$

4. If $i < n$, go to 2

## C. Laminated Lattices

In this section we examine another class of lattices known as laminated lattices.

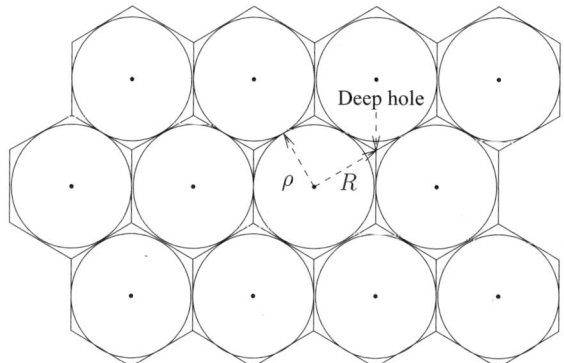

FIGURE 16. The laminated lattice $\Lambda_2$.

One of the classical mathematical problems is that of packing an $n$-dimensional container with identical spheres as densely as possible. This problem is completely solved for one- and two-dimensional spaces. In such spaces, the centers of the spheres must coincide with points belonging to the lattices

$$\Lambda_1 \cong Z_1 \cong A_1 \cong A_1^*$$
$$\Lambda_2 \cong A_2 \cong A_2^*$$

where $\Lambda$ denotes a laminated lattice.

The one-dimensional laminated lattice $\Lambda_1$ consists of all even integral points and is equivalent to the lattices $Z_1, A_1$, and $A_1^*$. In a 2D space the lattices $\Lambda_2, A_2$, and $A_2^*$ are equivalent. If we draw unit radius circles around each point, then by placing exact copies of the resulting row of circles next to each other as closely as possible, we form the 2D laminated lattice $\Lambda_2$. This lattice, equivalent to the so-called hexagonal lattice $A_2$, is shown in Figure 16.

For the dimensions from 2 to 8, the densest sphere packings are known only among lattices. The lattices that form the basis for the densest sphere packings in the spaces of 2,..., 8, dimensions are the following Conway and Sloane (1993):

$$\Lambda_3 \cong A_3 \cong D_3 \qquad \Lambda_6 \cong E_6$$
$$\Lambda_4 \cong D_4 \cong D_4^* \qquad \Lambda_7 \cong E_7$$
$$\Lambda_5 \cong D_5 \qquad \Lambda_8 \cong E_8 \cong E_8^*$$

where $A_n, D_n$, and $E_n$ are the root lattices discussed in the previous sections.

The laminated lattices are known to be the densest sphere packings in dimensions up to 8. This is also true up to 29-dimensional space, with the exception of 10- to 13-dimensional spaces, where the so-called $K$-type lattices give better results.

Half of the minimal distance $\rho$ between two distinct lattice points (see Fig. 16) is called the *packing radius*. It can also be defined as the largest number $\rho$ such that spheres of radius $\rho$ centered at the lattice points do not overlap. The packing radius of laminated lattices equals unity. The points on the plain that are furthest from the lattice points are called *deep holes*. $R$ is called the *covering radius*, which is equal to the distance from a lattice point to a deep hole. For the lattice in Figure 16 the covering radius is $R = 2\rho/\sqrt{3}$. The covering radius is related to the covering problem, which is to find the least dense covering of the space $R^n$ by overlapping spheres. The covering problem is the dual of the packing problem defined at the beginning of this section. The covering radius $R$ is the smallest number $\rho$ such that spheres of radius $\rho$ centered at the lattice points cover the whole $R^n$.

The two-dimensional construction of sphere packing illustrated in Figure 16 can be further extended into 3D space by replacing the circles with 3D spheres of the same radius as circles and stacking the obtained layers of the $\Lambda_2$ laminated lattices as densely as possible in the third dimension. In this way we obtain the $\Lambda_3$ lattice that is equivalent to the $A_3$ and $D_3$ lattices. The $\Lambda_3$ is known as the face-centered cubic lattice.

Similarly, the $\Lambda_n$ lattice can be recursively constructed from the $\Lambda_{n-1}$ lattices. The $\Lambda_n$ lattice is obtained by placing the $\Lambda_{n-1}$ lattices as close as possible to each other. In such a way every $\Lambda_n$ lattice includes a number of the $\Lambda_{n-1}$ lattices. This relationship is depicted in Figure 17. It can be seen in this figure that for

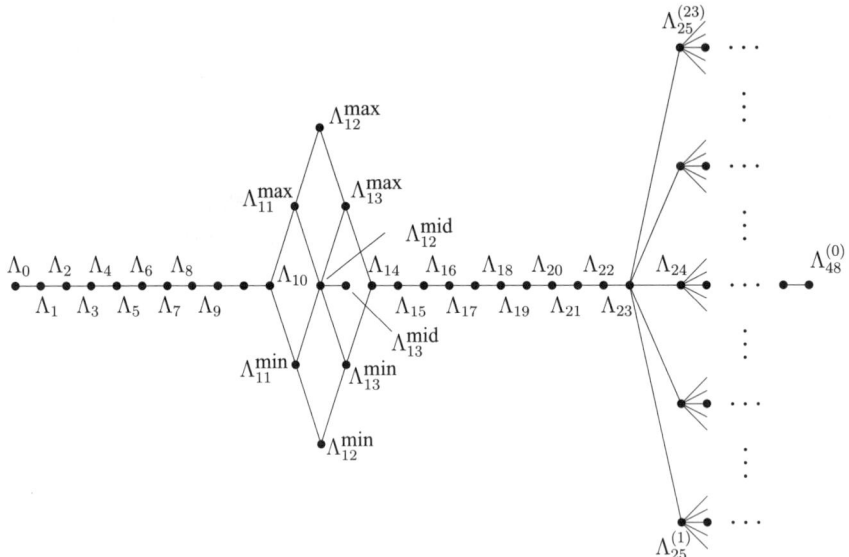

FIGURE 17. Laminated lattices.

spaces of some dimensionalities, the result of the preceding construction is not unique. That is, more than one laminated lattice can be generated in such spaces.

From the point of view of image compression, the most interesting of the high-dimensional laminated lattices is the $\Lambda_{16}$ lattice with the generator matrix

$$M_{\Lambda_{16}} = \frac{1}{\sqrt{2}} \begin{bmatrix} 4 & 2 & 2 & 2 & 2 & 2 & 2 & 2 & 2 & 2 & 2 & 1 & 0 & 0 & 0 & 1 \\ 0 & 2 & 0 & 0 & 0 & 0 & 0 & 0 & 0 & 0 & 0 & 1 & 1 & 0 & 0 & 1 \\ 0 & 0 & 2 & 0 & 0 & 0 & 0 & 0 & 0 & 0 & 0 & 1 & 1 & 1 & 0 & 1 \\ 0 & 0 & 0 & 2 & 0 & 0 & 0 & 0 & 0 & 0 & 0 & 1 & 1 & 1 & 1 & 1 \\ 0 & 0 & 0 & 0 & 2 & 0 & 0 & 0 & 0 & 0 & 0 & 0 & 1 & 1 & 1 & 1 \\ 0 & 0 & 0 & 0 & 0 & 2 & 0 & 0 & 0 & 0 & 0 & 1 & 0 & 1 & 1 & 1 \\ 0 & 0 & 0 & 0 & 0 & 0 & 2 & 0 & 0 & 0 & 0 & 1 & 0 & 1 & 1 & 1 \\ 0 & 0 & 0 & 0 & 0 & 0 & 0 & 2 & 0 & 0 & 1 & 0 & 1 & 0 & 1 \\ 0 & 0 & 0 & 0 & 0 & 0 & 0 & 0 & 2 & 0 & 0 & 1 & 1 & 0 & 1 & 1 \\ 0 & 0 & 0 & 0 & 0 & 0 & 0 & 0 & 0 & 2 & 0 & 0 & 1 & 1 & 0 & 1 \\ 0 & 0 & 0 & 0 & 0 & 0 & 0 & 0 & 0 & 0 & 2 & 0 & 0 & 1 & 1 & 1 \\ 0 & 0 & 0 & 0 & 0 & 0 & 0 & 0 & 0 & 0 & 0 & 1 & 0 & 0 & 1 & 1 \\ 0 & 0 & 0 & 0 & 0 & 0 & 0 & 0 & 0 & 0 & 0 & 0 & 1 & 0 & 0 & 1 \\ 0 & 0 & 0 & 0 & 0 & 0 & 0 & 0 & 0 & 0 & 0 & 0 & 0 & 1 & 0 & 1 \\ 0 & 0 & 0 & 0 & 0 & 0 & 0 & 0 & 0 & 0 & 0 & 0 & 0 & 0 & 1 & 1 \\ 0 & 0 & 0 & 0 & 0 & 0 & 0 & 0 & 0 & 0 & 0 & 0 & 0 & 0 & 0 & 1 \end{bmatrix}$$

(58)

The $\Lambda_{16}$ lattice is known as the Barnes-Wall lattice and can be constructed from the first-order Reed-Muller code of length 16 (Conway and Sloane, 1988). The last five columns of $M_{\Lambda_{16}}$ form a generator matrix for the first-order Reed-Muller code.

## V. QUANTIZATION ALGORITHMS FOR SELECTED LATTICES

Quantization can be viewed as a mapping of an input set onto a precalculated set known as a codebook. For each input sample or batch of samples, the quantizer finds the corresponding samples or batch of samples in the codebook according to the minimal distance criterion and sends its index to the output stream. Let us assume that the codebook consists of points belonging to a certain lattice $L_n$. The dimensionality $n$ of the lattice corresponds to the dimensionality of the input vectors of the quantizer. Assuming that the codebook of a quantizer is formed from the lattice points, we need an algorithm to find the closest lattice point for every input vector. For each lattice such an algorithm is required to quantize the input with the codebook in which the codewords are the lattice points. In this section we examine some fast quantization algorithms (Conway and Sloane, 1982; Gibson and Sayood, 1988) for the lattices discussed in the previous sections.

Before we proceed further with the presentation, let us first introduce a few "utility" functions used in quantization algorithms.

- Let $v(x)$ denotes the closest integer to $x$. In the case of a tie, when $x$ is equidistant from both neighboring integers, $v(x)$ is equal to the integer with the smaller absolute value.
- Let $w(x)$ denotes the second closest integer to $x$ distinct from $v(x)$.
- If the input is an $n$-dimensional vector, that is, $\boldsymbol{x} = (x_1, \ldots, x_n)$, the functions $v(\boldsymbol{x})$ and $w(\boldsymbol{x})$ are applied to each vector component separately, i.e.,

$$v(\boldsymbol{x}) = (v(x_1), \ldots, v(x_n)) \quad \text{and} \quad w(\boldsymbol{x}) = (w(x_1), \ldots, w(x_n))$$

- Let us also define the round-off residue function, $d(\cdot)$, as

$$d(\boldsymbol{x}) = \boldsymbol{x} - v(\boldsymbol{x})$$

- Let us also define a coordinate index $k$ in the following way

$$|d(x_k)| \geq |d(x_i)|, \qquad \forall 1 \leq i \leq n \quad \text{and} \quad 1 \leq k \leq n$$

and

$$|d(x_k)| = |d(x_i)| \quad \text{implies} \quad k \leq i$$

In words, the coordinate $x_k$ has the largest absolute value $d(x_k)$ amongst all coordinates $x_i$ in a vector $\boldsymbol{x}$. If it happens that a number of coordinates in $\boldsymbol{x}$ have the same maximum absolute value $d(x_k)$ then $k$ is assigned to the index of the coordinate, amongst those with the maximum $d(\cdot)$, which is located on the leftmost position in the vector $\boldsymbol{x}$.

- Finally, given that $x_k$ is known, we define a function $g(x)$ as

$$g(\boldsymbol{x}) = (v(x_1), \ldots, w(x_k), \ldots, v(x_n))$$

The function $g(\boldsymbol{x})$ effectively equals $v(\boldsymbol{x})$ with the coordinate $v(x_k)$ substituted for $w(x_k)$. The index $k$ is defined as before.

These functions are illustrated in Figure 18.

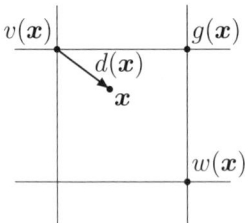

FIGURE 18. Illustration of functions $v(\boldsymbol{x})$, $w(\boldsymbol{x})$, and $d(\boldsymbol{x})$ in a two-dimensional space.

If, for example, $x = (3.6, -2.5, -1.8)$, then the functions $v(\cdot), w(\cdot), d(\cdot)$, and $g(\cdot)$ return the following values:

$$v(x) = (4, -2, -2)$$
$$w(x) = (3, -3, -1)$$
$$d(x) = (-0.4, -0.5, 0.2)$$
$$g(x) = (4, -3, -2)$$

### A. The Closest Point of a Dual Lattice

As mentioned in Section IV, for each lattice $L$ there exists a dual lattice $L^*$, which can be specified in the following way:

$$L^* = \bigcup_{i=0}^{k-1} (r_i + L) \tag{59}$$

where $r_i$ are the glue vectors.

Let us assume that we have a quantization algorithm for the lattice $L$ that assigns a point $y$ from the lattice $L$ to a point $x$ being quantized, i.e.,

$$y = \Phi(x), \quad y \in L$$

Then, the quantization procedure for the corresponding dual lattice, $L^*$, can be defined by the following procedure (Conway and Sloane, 1982):

- Calculate all prospective dual vectors

$$y_i^* = \Phi(x - r_i) + r_i, \quad \forall 0 \le i \le k-1 \tag{60}$$

- Determine the glue vector, $r_j$ for which the distance

$$\text{dist}(x, y_i^*)$$

attains minimum.
- Assign

$$y^* = \Phi(x - r_j) + r_j$$

where $y^* \in L^*$. In other words, $y^*$ is the one closet to the point $x$ being quantized among all other $y_i^*$.

Apart from the dual lattices, this procedure can be used to find the closest points for any lattice that can be represented as a union of cosets of the form given by Eq. (59).

### B. $Z_n$ Lattice

We know that the $Z_n$ lattice consists of all integer points in an $n$-dimensional space. Therefore, for $x \in R^n$, the point $y$ of the lattice $Z_n$ closest to $x$ is given by $v(x)$, i.e.,

$$y = v(x)$$

### C. $D_n$ Lattice and Its Dual

The quantization algorithm for the $D_n$ lattice follows directly from the definition of the corresponding root system. We know that the $D_n$ root system is obtained by letting $\Xi = R^n$ and $R = \{r \in Z : \langle r, r \rangle = 2\} = \{\pm(e_i \pm e_j), i \neq j\}$. Therefore, the lattice $D_n$ is a set of integer points in the $n$-dimensional space with an even sum of coordinates.

For an arbitrary point $x$, the closest point belonging to the lattice $D_n$ is given either by $v(x)$ or $g(x)$, depending on which result has an even sum of coordinates, i.e.,

$$\Phi(x) = \begin{cases} v(x) & \text{if } \sum_i v(x_i) \text{ is even} \\ g(x) & \text{otherwise } (\sum_i g(x_i) \text{ is even}) \end{cases} \quad (61)$$

For example, for a four-dimensional point $x = (2, -3.4, 0.7, 6.1)$, we find that $v(x) = (2, -3, 1, 6)$ and that $g(x) = (2, -4, 1, 6)$. Because the sum of components of $v(x)$ is even and the sum of components of $g(x)$ is odd, then the $D_4$ lattice point closest to $x$ is given by the function $v(x)$.

In general, it can be observed that the function $g(x)$ needs to be calculated only in the case when $\sum_i v(x_i)$ is odd. Otherwise, it is sufficient to calculate only $v(x)$.

The dual lattice $D_n^*$ can be defined as

$$D_n^* = \bigcup_{i=0}^{3} (r_i + D_n) \quad (62)$$

where $r_i$ are the glue vectors specified as follows:

$$r_0 = (\{0\}^n), \quad r_1 = \left(\left\{\begin{matrix} 1 \\ 2 \end{matrix}\right\}^n\right)$$

$$r_2 = (0^{n-1}, 1), \quad \text{and} \quad r_3 = \left(\frac{1}{2}^{n-1}, -\frac{1}{2}\right)$$

Alternatively, the lattice $D_n^*$ can be defined in term of the lattice $Z_n$ in the following way:

$$D_n^* = \bigcup_{i=0}^{1} (r_i + Z_n) \tag{63}$$

with the glue vectors being

$$r_0 = (0^n) \quad \text{and} \quad r_1 = \left(\frac{1^n}{2}\right)$$

Because both definitions (62) and (63) determine the same dual lattice $D_n^*$, one can use any of those definitions in conjunction with Eq. (60), depending on a specific application. However, the latter definition of the lattice $D_n^*$ clearly results in a faster algorithm for obtaining the closest point because of the use of two cosets instead of four, as in Eq. (62). Apart from the smaller number of cosets used in Eq. (63) compared with Eq. (62), quantization of the lattice $Z_n$ in an algorithm based on Eq. (63) is advantageous because it is, on average, faster than quantization of the lattice $D_n$.

### D. The Laminated $\Lambda_{16}$ Lattice

As mentioned before, this lattice can be constructed from the first-order Reed-Muller code of length 16. The quantization procedure relies on this fact. Using the algorithm for the $D_{16}$ lattice, we calculate 32 lattice points, taking the codewords of the Reed-Muller code (MacWilliams and Sloane, 1977) as the coset representatives. The goal point among the 32 lattice points obtained is the one for which the minimal distance from the quantized point is attained.

## VI. COUNTING THE LATTICE POINTS

A problem that seems especially important in the context of lattice quantization is that of counting the lattice points located within a given distance from the origin. In terms of quantization, the number of lattice points corresponds to the size of the codebook used for quantization. Therefore, we need to know how many points of a lattice $L_n$ are located on the surface of an $n$-dimensional sphere centered at the origin. We address first the problem of counting lattice points contained inside a sphere of a given radius.

### A. Estimation of the Number of the Lattice Points within a Sphere

In this section we estimate the number of lattice points contained inside a sphere of a given radius. We will consider first a problem of counting the points of a

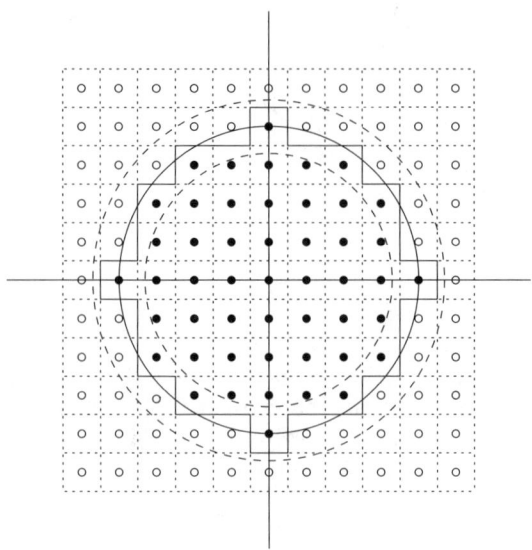

FIGURE 19. Counting the points of the lattice $Z_2$.

two-dimensional lattice $Z_2$ (Krätzel, 1988). The lattice $Z_2$ is the set of all integral points in a two-dimensional space. We would like to estimate the number of lattice points $C_m$ lying within the circle of the radius $\sqrt{m}$ with $m \geq 1$ (Fig. 19):

$$C_m = \#\{(x_0, x_1) : x_0, x_1 \in Z, x_0^2 + x_1^2 \leq m\} \qquad (64)$$

With each lattice point $P_i = (\beta_i, \gamma_i)$, we can associate a square $S_i$, which has the point $P_i$ located at its center. Formally, the square $S_i$ can be defined as

$$S_i = \{(\beta, \gamma) : |\beta - \beta_i| < \tfrac{1}{2}, |\gamma - \gamma_i| < \tfrac{1}{2}\} \qquad (65)$$

Now we can draw a circle of radius $\sqrt{m} + \sqrt{2}/2$ centered at the origin. All squares $S_i$ that lie within this circle satisfy the following inequality:

$$\beta^2 + \gamma^2 \leq \left(\sqrt{m} + \tfrac{1}{2}\sqrt{2}\right)^2 \qquad (66)$$

Therefore, the number of lattice points $C_m$ contained inside the circle of the radius $\sqrt{m}$ can be estimated as

$$C_m \leq \pi\left(\sqrt{m} + \tfrac{1}{2}\sqrt{2}\right)^2 = \pi\left(m + \sqrt{2}\sqrt{m} + \tfrac{1}{2}\right) < \pi m + 2\pi\sqrt{m} \qquad (67)$$

which is true for any $m \geq 1$, as it has been assumed earlier. In order to obtain the lower bound of the $C_m$, we draw a smaller inner circle defined by

$$\beta^2 + \gamma^2 = \left(\sqrt{m} - \tfrac{1}{2}\sqrt{2}\right)^2 \qquad (68)$$

for which the corresponding estimate of the number of lattice points inside this circle is

$$C_m \geq \pi\left(\sqrt{m} - \tfrac{1}{2}\sqrt{2}\right)^2 > \pi m - 2\pi\sqrt{m} \qquad (69)$$

Hence,

$$-2\pi\sqrt{m} < C_m - \pi m < 2\pi\sqrt{m} \qquad (70)$$

or

$$|C_m - \pi m| < 2\pi\sqrt{m} \qquad (71)$$

Thus, we can state that the number of points $C_m$ of the lattice $Z_2$ lying within a circle of radius $\sqrt{m}$ can be closely approximated by the area of this circle. Furthermore, the error of the approximation is of the same order as the length of the circumference of this circle, which is equal to $2\pi\sqrt{m}$.

These considerations can be applied to any close curve rather that just to a circle. Also, they can be applied not only to the two-dimensional space, but also to higher dimensions using corresponding versions of (71). It can also be shown (Krätzel, 1988) that inequality (71) holds for any lattice.

In conclusion, we can say that the number of lattice points within an $n$-dimensional sphere can be closely approximated by the volume of the sphere divided by the volume of a fundamental region of the lattice.

### B. The Number of Lattice Points on a Sphere

In terms of quantization, the points on the surface of a sphere of the radius $\sqrt{m}$, their number being given by $A_m$, will be used as codewords in quantizing Gaussian sources in general or wavelet coefficients in particular (see Sections III.B and III.C). In order to obtain the number of lattice points lying on the surface of a sphere, let us associated with each lattice $L$ a theta series (Ebeling, 1994; Grosswald, 1985; Igusa, 1972; Rankin, 1977; Lang, 1976), as follows:

$$\Theta_L(z) = \sum_{x \in L} e^{\pi i z \|x\|^2} \qquad (72)$$

where $x \in L$, $\|x\|^2 = x_1^2 + x_2^2 + \cdots + x_n^2$ is a squared norm of $x$, and $z$ is a complex variable with $\text{Im}(z) > 0$. Letting $q = e^{\pi i z}$ we can rewrite Eq. (72) as

$$\Theta_L(z) = \sum_{x \in L} q^{\|x\|^2} \qquad (73)$$

After the introduction of a new summation variable $m = \|x\|^2$, we can express the theta series as

$$\Theta_L(z) = \sum_{m=0}^{\infty} A_m q^m \tag{74}$$

where $A_m$ is the number of lattice points with the squared norm of $m$. In other words, assuming that the vector $x$ is of dimension $n$, the summation coefficient $A_m$ gives the number of vectors in the lattice $L$ located on the surface of the $n$-dimensional sphere of radius $\sqrt{m}$ centered at the origin.

The theta functions examined in this study originate from the Jacobi theta function given by

$$\Theta_3(\xi|z) = \sum_{m=-\infty}^{\infty} e^{2mi\xi + \pi i z m^2}, \quad \text{with } \operatorname{Im}(z) > 0$$

where $\xi$ is a complex variable that is permitted to assume any value. From this equation a number of simpler theta functions can be generated. For instance, the functions $\Theta_2(z)$, $\Theta_3(z)$, and $\Theta_4(z)$ are given by the following expressions:

$$\Theta_2(z) = e^{\pi i z/4} \Theta_3\left(\frac{\pi z}{2}\bigg|z\right) = \sum_{m=-\infty}^{\infty} q^{(m+1/2)^2}$$
$$= 2q^{1/4}(1 + q^2 + q^6 + q^{12} + q^{20} + \cdots) \tag{75}$$

$$\Theta_3(z) = \Theta_3(0|z) = \sum_{m=-\infty}^{\infty} q^{m^2}$$
$$= 1 + 2q + 2q^4 + 2q^9 + 2q^{16} + \cdots \tag{76}$$

and

$$\Theta_4(z) = \Theta_3\left(\frac{\pi}{2}\bigg|z\right) = \sum_{m=-\infty}^{\infty} (-q)^{m^2}$$
$$= 1 - 2q + 2q^4 - 2q^9 + 2q^{16} + \cdots \tag{77}$$

The theta series of the lattice of integer points $Z_n$ is

$$\Theta_{Z_n}(z) = \Theta_3^n(z) \tag{78}$$

By substitution of Eq. (76) in Eq. (78), we can calculate the number of points of the lattice $Z_n$ for a given norm $m$. This number is by construction equal to a coefficient $A_m$ of the variable $q$ with the power $m$. For this particular lattice, $A_m$ corresponds to the number of ways of writing $m$ as a sum of $n$ squares.

TABLE 2
NUMBERS OF POINTS ON THE SURFACE OF A SPHERE OF A GIVEN RADIUS
FOR THE LATTICE $Z_3$

| $m$ | 0 | 1 | 2 | 3 | 4 | 5 | 6 | 7 | 8 | 9 | 10 | 11 | 12 | 13 | 14 | 15 |
|---|---|---|---|---|---|---|---|---|---|---|---|---|---|---|---|---|
| $A_m$ | 1 | 6 | 12 | 8 | 6 | 24 | 24 | 0 | 12 | 30 | 24 | 24 | 8 | 24 | 48 | 0 |

An example of the numbers of lattice points for the first several spheres of the lattice $Z_3$ calculated from the theta series $\Theta_3^3(z)$ is given in Table 2.

The next lattice of interest to us is $D_n$. This lattice is formed by points having an even norm. Therefore, its theta series is

$$\Theta_{D_n}(z) = \frac{1}{2}\left(\Theta_3^n(z) + \Theta_3^n\left(\frac{\pi}{2}\bigg|z\right)\right) = \frac{1}{2}\left(\Theta_3^n(z) + \Theta_4^n(z)\right) \quad (79)$$

Among the $D$-type lattices, one of the most interesting is the $D_4$ lattice. The theta series of the $D_4$ lattice is specified in Table 3. The dual lattice $D_n^*$ has the following theta series:

$$\Theta_{D_n^*}(z) = \Theta_2^n(z) + \Theta_3^n(z) \quad (80)$$

The last lattice, which is also important in the context of wavelet-based image compression, is the laminated lattice $\Lambda_{16}$. The theta series of this lattice is

$$\Theta_{\Lambda_{16}}(z) = \frac{1}{2}\left(\Theta_2^{16}(2z) + \Theta_3^{16}(2z) + \Theta_4^{16}(2z) + 30\Theta_2^8(2z)\Theta_3^8(2z)\right) \quad (81)$$

with the coefficients given in Table 4.

### C. Relationship Between Lattices and Codes

As already mentioned, the laminated lattice $\lambda_{16}$ can be generated from the first-order Reed-Muller code. This is an example of the relationships between lattices and codes that are examined in this section (MacWilliams and Sloane, 1977; Conway and Sloane, 1993; Leech and Sloane, 1971; Sloane, 1977).

Let $F_2$ denote a finite field with two elements: $F_2 = \{0, 1\}$. A *binary linear code* $C$ over the field $F_2$ of length $n$ is a subset of $F_2^n$. In other words, the

TABLE 3
NUMBERS OF POINTS ON THE SURFACE OF A SPHERE OF A GIVEN RADIUS
FOR THE LATTICE $D_4$

| $m$ | 0 | 2 | 4 | 6 | 8 | 10 | 12 | 14 | 16 | 18 | 20 | 22 | 24 | 26 | 28 |
|---|---|---|---|---|---|---|---|---|---|---|---|---|---|---|---|
| $A_m$ | 1 | 24 | 24 | 96 | 24 | 144 | 96 | 192 | 24 | 312 | 144 | 288 | 96 | 336 | 192 |

## TABLE 4
### Numbers of Points on the Surface of a Sphere of a Given Radius for the Lattice $\Lambda_{16}$

| $m$ | 0 | 2 | 4 | 6 | 8 | 10 | 12 | 14 | 16 |
|---|---|---|---|---|---|---|---|---|---|
| $A_m$ | 1 | 0 | 4,320 | 61,440 | 522,720 | 2,211,840 | 8,960,640 | 23,224,320 | 67,154,400 |

code $C$ is a set of binary vectors of length $n$. The *Hamming distance* between two vecsors $c_1$ and $c_2$ is equal to the number of positions where these two vectors differ and is denoted by dist $(c_1, c_2)$. For instance, if $c_1 = 10111$ and $c_2 = 00101$, then dist $(c_1, c_2) = 2$. The *Hamming weight* of a vector is the number of nonzero components in it and is denoted by wt($c$). For the $c_1$ specified here, we have wt($c_1$) = 4. It is easy to derive the following relation between the Hamming distance and Hamming weight:

$$\text{dist}(c_1, c_2) = \text{wt}(c_1 - c_2)$$

A code $C$ can also be characterized by its minimum distance between any two codewords $C$:

$$d(C) = \min \text{dist}(c_1, c_2), \qquad \forall c_1, c_2 \in C \quad \text{and} \quad c_1 \neq c_2.$$

A code with a minimum distance $d$ can correct $\lceil (d-1)/2 \rceil$ errors, where $\lceil a \rceil$ denotes the usual ceiling function, which gives the greatest integer less than or equal to $a$. Typically, an $[n, k, d]$ binary linear code refers to a code with $2^k$ codewords of length $n$ that differ in at least $d$ places. The first-order Reed-Muller code related to the $\Lambda_{16}$ laminated lattice is denoted by [16, 5, 8].

For an $[n, k, d]$ linear code $C$, let $A_i$ be the number of codewords with the Hamming weight of $i$. Thus, $\sum_{i=0}^{n} A_i = 2^k$. The numbers $A_i$ are called *weight distribution* of $C$. Now, we can associate with the code $C$ a homogeneous (containing terms that are all of the same degree) polynomial in the following way:

$$\begin{aligned} W_C(x, y) &= \sum_{c \in C} x^{n-\text{wt}(c)} y^{\text{wt}(c)} \\ &= \sum_{i=0}^{n} A_i x^{n-i} y^i \\ &= x^n + A_1 x^{n-1} y^1 + \cdots + y^n \end{aligned} \qquad (82)$$

This polynomial is called the *weight enumerator* of the code $C$. In a weight enumerator, the variable $x$ effectively indicates the number of 0s in a codeword, whereas $y$ indicates 1s.

With the preceding fundamentals in mind, it is now possible to draw a parallel between the weight enumerators of codes and the theta series of lattices. Consider the coefficients $A_m$ of the theta series given by Eq. (74) and the coefficients $A_i$ of the weight enumerators in Eq. (82). In both cases they specify the number of points, or codewords, at a certain distance from the origin. Therefore, these polynomials contain the essential information about the distribution of vectors in the subspace under consideration.

Extending the connection established here between lattices and codes, we can observe that there exist a number of methods for constructing sphere packings from codes. It can be shown that a sphere packing constructed from a code is a lattice packing if and only if the code is linear (MacWilliams and Sloane, 1977). Therefore, by choosing a linear code as a basis for any of the available constructions, we obtain a lattice corresponding to this code. Three of the most popular construction methods, referred to as constructions A, B, and C, are outlined next.

### 1. Construction A

This construction yields the simplest way to associate a lattice with a code. Let $C$ be a binary code. The centers $x$ of spheres from the sphere packing corresponding to the code $C$ in $\mathbb{R}^n$ are those congruent (modulo 2) to codewords of $C$:

$$\sqrt{2}x \bmod a \in C \tag{83}$$

In this way the centers of spheres are obtained by adding even numbers to the codewords of $C$ and then dividing the result by $\sqrt{2}$. The multiplication with $\sqrt{2}$ is simply a scaling operation, and, as we shall see later, it does not alter enumeration of lattice points and is often omitted.

As we have already mentioned, the code must be linear in order to obtain a corresponding lattice packing, $L(C)$. Let us consider a codeword $c = (c_1, c_2, \ldots, c_n)$ in a linear code $C$. Following Eq. (83) the corresponding lattice points, $l = (l_1, l_2, \ldots, l_n) \in L(C)$, can be expressed as

$$L(c) = (l_1, l_2, \ldots, l_n) \quad \text{with} \quad l_i \in \frac{c_i + 2Z}{\sqrt{2}} = \frac{1}{\sqrt{2}} c_i + \sqrt{2}Z \tag{84}$$

where $L(c) \in L(C)$ denotes a lattice point constructed from a codeword $c \in C$. Because $c_i$ is defined on $F_2 = \{0, 1\}$, we need to consider two cases, namely, $c_i = 0$ and $c_i = 1$. From the previous subsection we recall that $\Theta_{Z_n}(z) = \Theta_3^n(z)$. Thus, $\Theta_Z(z) = \Theta_3(z)$, and we can deduce the form of $\Theta_{\sqrt{2}Z}$ as follows:

$$\Theta_{\sqrt{2}Z}(z) = \sum_{m=-\infty}^{\infty} q^{(\sqrt{2}m)^2} = \sum_{m=-\infty}^{\infty} q^{2m^2} = \Theta_3(2z), \quad \text{because} \quad q = e^{\pi i z}$$

Similarly, for $\Theta_{1/\sqrt{2}+\sqrt{2}z}$ we have

$$\Theta_{1/\sqrt{2}+\sqrt{2}z}(z) = \sum_{m=-\infty}^{\infty} q^{(1/\sqrt{2}+\sqrt{2}m)^2} = \sum_{m=-\infty}^{\infty} q^{2(1/2+m)^2} = \Theta_2(2z)$$

Now we can write

$$\Theta_{L(c)}(z) = \prod_{i=1}^{n-\text{wt}(c)} \Theta_3(2z) \prod_{i=1}^{\text{wt}(c)} \Theta_2(2z) = \Theta_3^{n-\text{wt}(c)}(2z) \Theta_2^{\text{wt}(c)}(2z)$$

where wt($c$) is the number of 1s and $n - \text{wt}(c)$ is the number of 0s in $c$. Consequently,

$$\Theta_{L(c)}(z) = \sum_{c \in C} \Theta_{L(c)}(z) = W_C(\Theta_3(2z), \Theta_2(2z)) \tag{85}$$

In summary, if we assume that $C$ is a linear code with weight enumerator $W_C(x, y)$ given by Eq. (82), then the theta function of the corresponding lattice $L(C)$ is given by Eq. (85). For example, using construction A, we can generate lattice $E_8$ from the [8, 4, 4] extended Hamming code $H_8$ (Sloane, 1977) with weight enumerator

$$W_{H_8}(x, y) = x^8 + 14x^4 y^4 + y^8$$

## 2. Construction B

Consider an [$n, k, 8$] binary linear code $C$ with codeword weights divisible by 4. The centers $x$ of the corresponding sphere packing are points that satisfy the following properties

1. $\sqrt{2}x \pmod 2 \in C$ \hfill (86)

2. $4 | \sqrt{2} \sum_{i=1}^{n} x_i$ \hfill (87)

It can be shown [38] that the lattice sphere packing $L(C)$ obtained by the following construction B has a theta series as follows:

$$\Theta_{L(c)}(z) = \frac{1}{2} W_C(\Theta_3(2z), \Theta_2(2z)) + \frac{1}{2}\Theta_4^n(2z).$$

For example, by applying construction B, the lattice $E_8$ can be generated from the repetition code [8, 1, 8] consisting of the codewords $\{\{0\}^8, \{1\}^8\}$.

The procedures given by constructions A and B can be simplified by introducing a coordinate array for each lattice point defined as follows. The number of columns in this array equals the dimensionality of the generated lattice. Each column is the binary representation of the corresponding coordinate of the point. For a negative number, 2's complement notation is used. For example, the coordinate array of $x = (3, 2, 1, 0, -1, -2, -3)$ is

| 3 | 2 | 1 | 0 | −1 | −2 | −3 |
|---|---|---|---|----|----|----|
| 1 | 0 | 1 | 0 | 1  | 0  | 1  |
| 1 | 1 | 0 | 0 | 1  | 1  | 0  |
| 0 | 0 | 0 | 0 | 1  | 1  | 1  |
| 0 | 0 | 0 | 0 | 1  | 1  | 1  |

Now, if the coefficient $\sqrt{2}$ in constructions A and B specified by Eqs. (83), (86), and (87) is omitted, we have the following simplified algorithms. Construction A is effectively reduced to finding the centers (points) for which the top rows of their coordinate arrays are codewords of code $C$. Construction B can be redefined as a method of finding the centers that have the top rows being codewords of code $C$ and the second top rows with even weights.

## 3. Construction C

This construction is based on a linear code $C$ with codewords of length $n$. The corresponding lattice packing consists of points for which the $2^i$'s rows ($i = 0, \ldots, n$) of their coordinate arrays are in $C$.

In our discussion of lattice quantization, we have covered definitions from the lattice theory (Section IV), types of lattices (Sections IV.A and IV.C), and construction of lattices (Sections IV.B and IV.C). Fast quantization algorithms for lattices can be found in Conway and Sloane (1982, 1983). From the point of view of lattice quantization, another problem remains open. We need a scaling algorithm that can be applied to wavelet coefficients as a preprocessing step before utilization of the fast quantization algorithms. Such a scaling algorithm is discussed in the next section.

## VII. Scaling Algorithm

Although the probability distribution function of the wavelet coefficients studied in Section II.C.2 has similar shapes for each block, e.g., (*LL*), (*LH*), (*HL*), and (*HH*), some statistical parameters, such as minimum and maximum values, standard deviation, and so on, are expected to vary from block to block. Therefore, for the purpose of quantization, each block of the wavelet coefficient matrix should be treated separately.

According to the results presented in Section III.C an optimal quantization of wavelet coefficients can be achieved by uniform placement of the codevectors on the surfaces of concentric spheres with centers at the origin. It was shown in the previous section that each lattice actually provides a set of points located on the surfaces of spheres centered at the origin. The number of such points for each sphere is defined by the theta series of Eq. (74). Assuming that each lattice point is a codeword, we need to know how many codewords are required for quantization of a given block, or, alternatively, what the size of the codebook should be. One possible solution to the problem of determining the number of codewords was presented in Section II.C in the from of an optimal bit-allocation routine. The output of this routine is the number of bits per pixel, or, in other words, the compression ratio for each block. From the target compression ratio we derive the size of the goal codebook (the number of codewords required for quantization), $N$. By summing the coefficients of the theta series, the number of the lattice points corresponding to that size can be obtained. Because the theta series gives only the numbers of lattice points (see Tables 2, 3, and 4) lying on surfaces of the spheres around the origin, the size of the codebook actually used is often an approximation of the one generated by the optimal bit-allocation procedure.

The available size of the codebook consisting of the lattice points is given by $N = \sum_{i=0}^{m} A_i$. Equivalently, given the appropriately chosen radius $\sqrt{m}$ of the largest sphere that accommodates vectors of the squared norm (or energy) $m$, we obtain a set of codevectors belonging to the selected lattice. Now we require that the wavelet coefficients (collected into vectors, because we use vector quantization) with a certain norm $E$ be scaled to the norm $m$ of the surface of the outer sphere. We select $E$ so that a prescribed number of vectors, say, 70 to 80%, have their norm not greater than $E$ (Fig. 20). The scaling factor between the $m$ and $E$ shells is given by

$$s = \sqrt{\frac{E}{m}} \qquad (88)$$

After quantization this scaling factor is included in the compressed data stream (Barlaud et al., 1994).

There are a number of possible ways of handling vectors with squared norm greater than $E$. The simplest method is to truncate them to the surface of the outer sphere $E$. However, it results in the introduction of some additional distortion of the high-frequency edges in an image. The approach we have adopted and tested in this work is to set a separate scaling factor $s_i$ for each such vector, given by

$$s_i = \sqrt{\frac{E_i}{m}}, \qquad \forall \, E_i > E \qquad (89)$$

where $E_i$ is the squared norm of the $i$th vector. Thus, the high-frequency

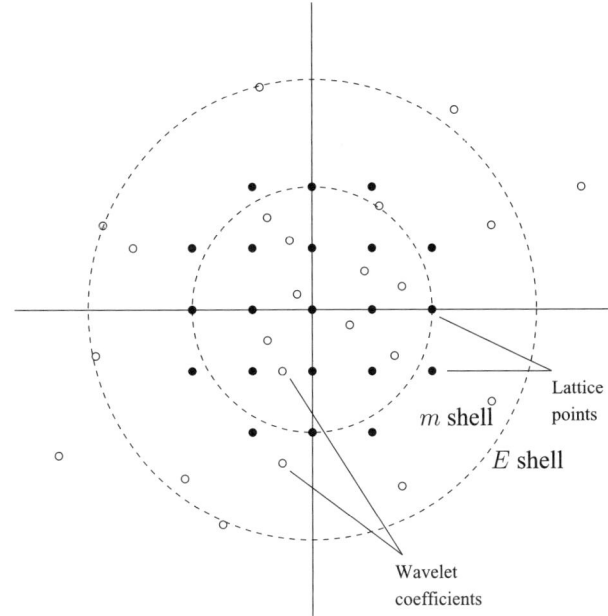

FIGURE 20. Lattice quantization scheme.

information remains preserved at the expense of a slight increase in the bit rate, which depends on the value of the threshold $E_i$. The scaling factors $s_i$ are to be transmitted together with the corresponding vectors. Thus, at the receiving end, dequantization is followed by scaling each vector by the factor $1/s$ for the vectors with norm not exceeding $E$ and the factor $1/s_i$ for the vectors with norm greater than $E$.

Let us summarize the considerations presented in this section in the form of the following scaling algorithm to be used as a preprocessing step in lattice quantization of wavelet coefficients:

1. Compute the squared norm of each vector of wavelet coefficients and determine the norm $E$.
2. Determine the size of the codebook for a chosen lattice as well as the corresponding norm $m$ (see Section II.C).
3. Determine the scaling factor as in (88) and include $s$ in the bit stream.
4. For each vector (norm $E_i$):
   (a) If $E_i < E$, scale the vector by a factor of $s$; otherwise, determine $s_i$ as in Eq. (89) and scale the vector by a factor of $s_i$.
   (b) Encode the vector with a fast quantization algorithm (see Section V).
   (c) If $E_i > E$, include $s_i$ in the bit stream.

In this section, we have examined the problem of lattice quantization. As shown previously, there exist a variety of lattices that can be used for this purpose. The final choice depends on a particular application and relies on the evaluation of a number of criteria, such as the fidelity of the system, the time of encoding, or software or hardware implementation. This issue, namely the choice of a suitable lattice, is addressed in the following section.

## VIII. Selecting a Lattice for Quantization

From Section IV we know that every lattice is determined by its generator matrix $M$ and the points that comprise a lattice are specified by Eq. (30). We can choose these points as codevectors for quantization of an $n$-dimensional input sequence.

Let $w_i$ denote an $i$th codevector and $V_i$ be a quantization region around it. As a result of quantization, every entry in the input sequence located within the region $V_i$ is represented by the codeword $w_i$. The distortion introduced during quantization can be measured as the mean-squared error (MSE) between the input and output sequences of the quantizer. Here we use a dimensionless quantity $D$ known as MSE per dimension, given by

$$D = \frac{1}{n} \sum_{i=1}^{N} \int_{V_i} \|x - w_i\|^2 \, p(x) \, dx \tag{90}$$

where $x$ is the input sequence and $p(x)$ is the probability distribution function of $x$. For many applications, an optimization of quantization means an appropriate selection of codevectors $w_i$ with the aim to minimize the distortion. Thus, it is important to determine the minimum distortion $D^*$ subject to the set of codevectors $w_i$ $(0 \leq i \leq N-1)$,

$$D^* = \inf_{w_i} D. \tag{91}$$

Solving the preceding equation for each lattice will enable the choice of an appropriate lattice for quantization of the source $x$. It is evident that the distortion depends not only on the position of the codevectors in the quantization space but also on their quantity. Therefore, to allow a correct comparison of the performance of various lattices, the number of lattice points $N$ used for quantization should be fixed for every lattice.

We know that the quantization regions $V_i$ formed by a lattice are congruent to its fundamental region $\bar{V}$. Often, in order to fit $N$ quantization regions in a quantization space $\Omega$, quantization regions are scaled:

$$\text{vol}(\Omega) = N \, \text{vol}(q \bar{V}) \tag{92}$$

## STILL IMAGE COMPRESSION WITH LATTICE QUANTIZATION

where $q$ is the scaling factor. It was shown by Zador (1982) that for large $N$, the following equality is satisfied:

$$D^* = G_P(n, r) N^{-r/n} \left( \int_{R^n} p(x)^{n/(n+r)} dx \right)^{(n+r)/n} \qquad (93)$$

where $r$ is set to 2 for the MSE measure and $G_P(n, r)$ is known as the dimensionless second moment of a polytope $P$. In the case of lattice quantization, a polytope $P$ becomes a quantization region $\bar{V}$ of a lattice.

After setting $r$ to 2, we have that the moment $G_{\bar{V}}(n, 2)$ depends only on $n$, which is the dimensionality of the input $x$. Assume that $w_i$ are selected so that the distortion $D^*$ is minimized. Thus, $G_{\bar{V}}(n, 2)$ can be calculated as

$$G_{\bar{V}}(n, 2) = \frac{1}{n} \frac{N^{2/n} \sum_{i=1}^{N} \int_{V_i} \|x - w_i\|^2 p(x) dx}{\left( \int_{R^n} p(x)^{n/(n+2)} dx \right)^{(n+2)/n}}$$

$$= \frac{1}{n} \frac{\frac{1}{N} \sum_{i=1}^{N} \int_{V_i} \|x - w_i\|^2 p(x) dx}{\left( \frac{1}{N} \sum_{i=1}^{N} \int_{V_i} p(x)^{n/(n+2)} dx \right)^{(n+2)/n}} \qquad (94)$$

In this equation $\int_{R^n} p(x)^{n/(n+2)} dx$ has been replaced by $\sum_{i=1}^{N} \int_{V_i} p(x)^{n/(n+2)} dx$, because the quantization space is entirely covered with the quantization regions. The values of the second moment $G_{\bar{V}}(n, 2)$ can be tabulated for each lattice by letting $x$ be uniformly distributed over the quantization space $\Omega$, i.e., $p(x) = $ const. After setting $p(x)$ to a constant, Eq. (94) becomes

$$G_{\bar{V}}(n, 2) = \frac{1}{n} \frac{\frac{1}{N} \sum_{i=1}^{N} \int_{V_i} \|x - w_i\|^2 dx}{\left( \frac{1}{N} \sum_{i=1}^{N} \int_{V_i} dx \right)^{(n+2)/n}} \qquad (95)$$

Because all quantization regions $V_i$ are congruent to the fundamental region of a lattice, we have

$$\int_{V_i} dx = \text{vol}(V_i) = \text{vol}(\bar{V}) \qquad (96)$$

where $\bar{V}$ is the fundamental region of a lattice. Also, provided that $w_i$ are located in the centroids of the corresponding quantization regions, the distortions produced by quantization of the uniformly distributed source $x$ within all quantization regions are equivalent. Finally, assuming that the centroid $w$ of the fundamental region is located in the origin, Eq. (95) can be simplified as follows:

$$G_{\bar{V}}(n, 2) = \frac{1}{n} \frac{\int_{\bar{V}} x^T x \, dx}{\text{vol}(\bar{V})^{(n+2)/n}} \qquad (97)$$

## TABLE 5
### Dimensionless Second Moment $G_P(n, 2)$ for Selected Lattices

| $n$ | 1 | 2 | 3 | | 4 | | |
|---|---|---|---|---|---|---|---|
| Lattice | $Z_1$ | $A_2, A_2^*$ | $A_3, D_3$ | $A_3^*, D_3^*$ | $A_4$ | $A_4^*$ | $D_4, D_4^*$ |
| $G_P(n, 2)$ | 0.083333 | 0.080188 | 0.078745 | 0.078543 | 0.078020 | 0.077559 | 0.076603 |

| 5 | | | | 6 | | 7 | |
|---|---|---|---|---|---|---|---|
| $A_5$ | $A_5^*$ | $D_5$ | $D_5^*$ | $E_6$ | $E_6^*$ | $E_7$ | $E_7^*$ |
| 0.077647 | 0.076922 | 0.075786 | 0.075625 | 0.074347 | 0.074244 | 0.073231 | 0.073116 |

| 8 | | | | | 12 | 16 | 24 |
|---|---|---|---|---|---|---|---|
| $A_8$ | $A_8^*$ | $D_8$ | $D_8^*$ | $E_8, E_8^*$ | $K_{12}$ | $\Lambda_{16}, \Lambda_{16}^*$ | $\Lambda_{24}, \Lambda_{24}^*$ |
| 0.077391 | 0.075972 | 0.075914 | 0.074735 | 0.071682 | 0.070100 | 0.068299 | 0.065771 |

Table 5, reproduced from Conway and Sloane (1993), shows the dimensionless second moment $G_V(n, 2)$ for various popular lattices.

Equation (93) can be now rewritten as

$$D^* = G_{\tilde{V}}(n, 2) N^{-2/n} \left( \sum_{i=1}^{N} \int_{V_i} p(\mathbf{x})^{n/(n+2)} d\mathbf{x} \right)^{(n+2)/n} \tag{98}$$

According to the scaling algorithm presented in Section 7, every input vector, $\mathbf{x}$ is multiplied by the scaling factor $1/s$ before it is fed into the quantizer. Thus, for the minimum distortion, $D^*$, we have

$$D^* = G_{\tilde{V}}(n, 2) N^{-2/n} \left( s^n \sum_{i=1}^{N} \int_{V_i} p(s\mathbf{x})^{n/(n+2)} d\mathbf{x} \right)^{(n+2)/n} \tag{99}$$

Suppose that the lattice points cover the quantization space densely enough to assume that the PDF of the input is uniform within each quantization region. With this assumption, Eq. (99) can be simplified further as

$$D^* = G_V(n, 2) N^{-2/n} \left( s^n \sum_{i=1}^{N} p(s\mathbf{w}_i)^{n/(n+2)} \int_{V_i} d\mathbf{x} \right)^{(n+2)/n} \tag{100}$$

where $p(\mathbf{w}_i)$ is the probability of the source at the centroid of the $i$-quantization region.

# STILL IMAGE COMPRESSION WITH LATTICE QUANTIZATION

In this derivation we have assumed that all quantization regions are congruent and the PDF within each of them is constant; thus, Eq. (96) is satisfied. Using Eq. (96), we can rewrite Eq. (100) as

$$D^* = G_V(n,2) N^{-2/n} (s^n \mathrm{vol}(\bar{V}))^{(n+2)/n} \left( \sum_{i=1}^{N} p(s\mathbf{w}_i)^{n/(n+2)} \right)^{(n+2)/n} \quad (101)$$

It is easy to see that the value of $s^n \mathrm{vol}(\bar{V})$ in this expression is, effectively, the normalized volume of the fundamental region of a lattice, and the scaling factor $s$ corresponds to the scaling factor $q$ in Eq. (92). An appropriate selection of the factor $s$ will ensure that the normalized volumes of the fundamental regions of all lattices are equal to each other, and a collection of $N$ lattice points covers completely the quantization space $\Omega$. A connection between $s$ and $N$ can be also seen from the fact that $s$ is given by the scaling equation, Eq. (88), and $m$ in Eq. (88) is chosen so that $N = \sum_{j=0}^{m} A_j$. $N$ denotes the number of points that belong to a lattice. For quantization, lattice points become codevectors. In (101) the codevectors are denoted by $\mathbf{w}_i$, with $i$ between 0 and $N-1$. As mentioned in Section VI the number of lattice points is given by the theta series (74). Therefore, Eq. (101) can be modified in the following way:

$$D^* = G_{\bar{V}}(n,2) N^{-2/n} (s^n \mathrm{vol}(\bar{V}))^{(n+2)/n} \left( \sum_{j=0}^{m} \sum_{k=1}^{A_j} p(s\mathbf{w}_k)^{n/(n+2)} \right)^{(n+2)/n} \quad (102)$$

with $\mathbf{w}_k^T \mathbf{w}_k = j$, $A_j$ being the coefficients of the theta series. Expression (102) gives a good approximation of the distortion measure provided that the following assumptions are fulfilled:

- Quantization codevectors are densely distributed in the quantization space.
- The probability distribution within each quantization region is almost uniform.

In order to select an optimal lattice for quantization of a source, one should fix the value of $N$ and then solve Eq. (101) for every lattice, with $p(\cdot)$ set to the PDF of the source. The final selection of a lattice for quantization does not depend on the actual values of the distortion measure $D^*$ across the range of lattices but rather on which lattice attains the smallest $D^*$. In this context, it appears valuable to note that after fixing $N$ for some $n$-dimensional quantization space $\Omega$, the term $N^{-2/n} (s^n \mathrm{vol}(\bar{V}))^{(n+2)/n}$ becomes constant:

$$C = N^{-2/n} (s^n \mathrm{vol}(\bar{V}))^{(n+2)/n} \quad (103)$$

Thus, we have

$$D^*(N, \Omega) = CG_{\bar{V}}(n, 2) \left( \sum_{j=0}^{m} \sum_{k=1}^{A_j} p(s\mathbf{w}_k)^{n/(n+2)} \right)^{(n+2)/n} \quad (104)$$

Because our primary concern in this paper is quantization of wavelet coefficients, the PDF that must be used in Eq. (104) is the PDF of wavelet coefficients. Recall from Section II that the PDF of wavelet coefficients can be modeled by the Gaussian-type function. Naturally, we can use this model in Eq. (104).

Consider the following example of selecting a two-dimensional lattice for quantization of the lowest-frequency band of the wavelet coefficients. Assume that we have a bank of two lattices, namely, lattices $Z_2$ and $A_2$. As shown in Eq. (7), for the lowest frequency band of the wavelet coefficient matrix (texture), the generalized Gaussian function is reduced to the Gaussian PDF. Assuming that the DPCM predictor used in the lowest frequency band removes correlation between neighboring samples, we have

$$p(\mathbf{x}) = \frac{1}{(2\pi)^{n/2}} \exp\left(-\frac{1}{2}\mathbf{x}^T \mathbf{x}\right) \quad (105)$$

given that the standard deviation is $\sigma = 1$. For this Gaussian PDF, the distortion measure can be expressed in the following way:

$$D^*(N, \Omega) = CG_{\bar{V}}(n, 2) \left( \sum_{j=0}^{m} A_j p(s\sqrt{j})^{n/(n+2)} \right)^{(n+2)/n} \quad (106)$$

where $p(s\sqrt{j})$ is the probability of the source vector having the squared norm of $s^2 j$:

$$p(s\mathbf{x}) = \frac{1}{(2\pi)^{n/2}} \exp\left(-\frac{1}{2}s^2\mathbf{x}^T \mathbf{x}\right) = \frac{1}{(2\pi)^{n/2}} \exp\left(-\frac{1}{2}s^2 j\right) = p(s\sqrt{j}) \quad (107)$$

where $j$ is the squared norm of $\mathbf{x}$. For an estimation of the distortion measure of Eq. (106), we must select the scaling factor $s$. In Section VII, $s$ was defined as a scaling factor for mapping the outer sphere of the quantization space $\Omega$ onto the outer sphere comprising the lattice points to be used for quantization. After testing, it was found that a good choice of the outer sphere of the quantization space for wavelet coefficients is when it includes between 70 and 80% of wavelet coefficients. This corresponds to selecting the value of $E$ in Eq. (88) to be approximately $(1.3\sigma)^2$, where $\sigma$ is the standard deviation. Recall that $\sigma$ was chosen to be 1 in Eq. (105).

TABLE 6
NUMBERS OF POINTS OF THE LATTICES $Z_2$ AND $A_2$

| j | 0 | 1 | 2 | 3 | 4 | 5 | 6 | 7 | 8 | 9 | 10 | 11 | 12 | 13 | 14 | 15 | 16 |
|---|---|---|---|---|---|---|---|---|---|---|----|----|----|----|----|----|----|
| $Z_2 : A_j$ | 1 | 4 | 4 | 0 | 4 | 8 | 0 | 0 | 4 | 4 | 8 | 0 | 0 | 8 | 0 | 0 | 4 |
| $A_2 : A_j$ | 1 | 6 | 0 | 6 | 6 | 0 | 0 | 12 | 0 | 6 | 0 | 0 | 6 | 12 | 0 | 0 | 6 |

In order to calculate $s$ specified in Eq. (88), it is required to select the number of lattice points/codevectors used for quantization; this number is denoted by $N$. Because $N = \sum_{j=0}^{m} A_j$, from $N$ we can find $m$, which is needed in Eq. (88). As mentioned before, $A_j$ are coefficients of the theta series, which, for the lattices $Z_2$ and $A_2$ used in this example, are given in Table 6.

Let us set $N = 45$, for example. Thus, we have $\sum_{j=0}^{13} A_j = 45$ for the lattice $Z_2$ and $\sum_{j=0}^{12} A_j = 43 \approx 45$ for the lattice $A_2$. Using this, the scaling factors for the lattices $Z_2$ and $A_2$ become $s = 1.3/\sqrt{13}$ and $s = 1.3/\sqrt{12}$, respectively.

Now, Eq. (106) can be tabulated for the test lattices. The results are

$$\frac{D_{Z_2}^*(45, \Omega)}{C} = 17.2136 \quad \text{and} \quad \frac{D_{A_2}^*(\approx 45, \Omega)}{C} = 15.7095$$

In conclusion, by quantization of the wavelet coefficients with the $A_2$ lattice, one may expect to obtain approximately 10% improvement in the mean-squared error, compared with quantization of the same source with the $Z_2$ lattice. Not surprisingly, this result coincides with the result obtained by Conway and Sloane (1993) for an uniformly distributed source. Although for an uniformly distributed source, the benefit of using $A_2$ over $Z_2$ is approximately 4%.

## IX. ENTROPY CODING OF LATTICE VECTORS

In an image-compression system, lattice quantization is typically followed by an entropy coder that exploits an unevenness of the PDF of lattice vectors. Therefore, an entropy coder requires the knowledge of the probability of each codevector before encoding. In many systems, the codevectors, together with their probabilities, are included in the output bitstream. This results in undesirable increase in the bit rate.

Because the lattice quantizer has a regular structure, all codevectors can be generated by the decoder from the fundamental region of the lattice. What remains to be included into the bitstream is the probability for each vector. However, since we know that the probability distribution of the wavelet coefficients has a form of the generalized Gaussian function, the probability of a codevector $w_i$ can be approximated as

$$P(w_i) = s^n \text{vol}(\bar{V}) p(w_i) \quad \forall 1 \leq i \leq N \tag{108}$$

where $N$ is the number of lattice points used for quantization and $p(\cdot)$ is the generalized Gaussian function with the mean of 0 and the standard deviation equals the standard deviation of the wavelet coefficients. This estimation can be done independently at the encoder and decoder using the standard deviation of the coefficients, which can be included into the bitstream without noticable overhead.

Although the preceding approach for tabulating the probability of each codevector is correct, it appears rather demanding to enumerate all lattice points with their probabilities. It was shown in the previous sections that the generalized Gaussian function with an appropriately selected parameter $\xi$ closely resembles the PDF of the wavelet coefficients. Obviously, the shape of the surfaces of constant probabilities of the GGF function depend on the parameter $\xi$. The two-dimensional equiprobable surfaces for some choices of $\xi$ are depicted in Figure 21. When $\xi = 2$, the generalized Gaussian becomes the well-known Gaussian; consequently, the regions of equal probabilities are located on spheres centered around the origin, which is 0 in the case of wavelet coefficients. If $\xi$ equals 0.7 the equiprobable surfaces become somehow distorted spheres. The further the value of $\xi$ departs from being 2, the more distant from a spherical shape the equiprobable surfaces become. For instance, when $\xi$ is 1, i.e., the Laplacian PDF, the equiprobable surfaces are pyramids (Fischer, 1986).

The number of lattice points on each equiprobable surface is given by the theta series specified in Eq. (74). Provided that we have a theta series for each lattice, instead of enumeration of the lattice points, as before, we can enumerate spheres and calculate the probability of any vector, say, a representative, located on each sphere. The remaining vectors have the same probability as their representatives. In order to accomplish this enumeration, we need a theta series for each lattice. Let us define a $\xi$-norm of an $M$-dimensional vector $x$ as

$$\|x\|_\xi = \left( \sum_{i=1}^{M} |x_i|^\xi \right)^{1/\xi} \tag{109}$$

Employing this definition, a general form of the theta series can be written as

$$\Theta_L(z, \xi) = \sum_{x \in L} q^{\|x\|_\xi^\xi} = \sum_{m=0}^{\infty} A_m q^m \tag{110}$$

By setting $\xi$ to 2, we obtain the theta series discussed in Section VI.B. For

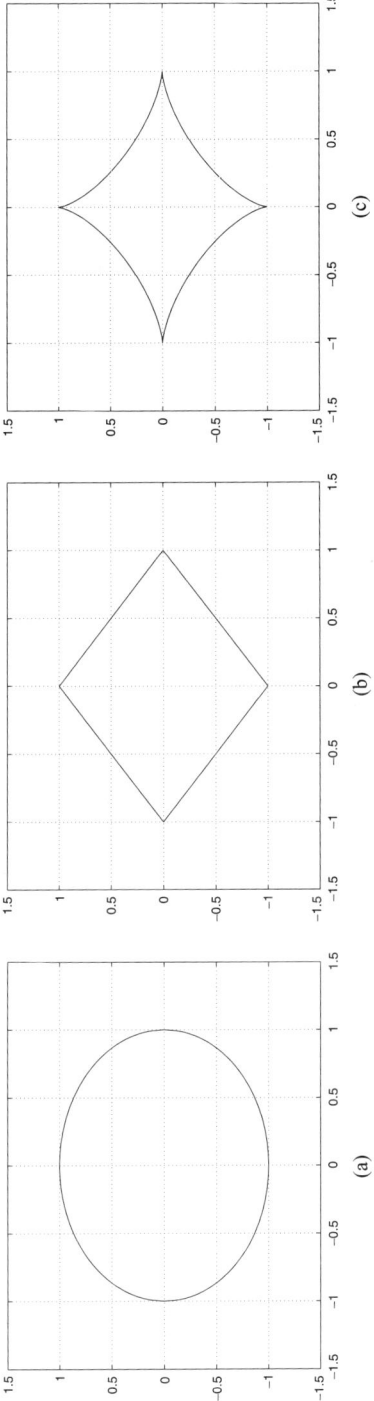

FIGURE 21. Constant-probability surfaces for GGF with various $\xi$: (a) $\xi = 2$; (b) $\xi = 1$; (c) $\xi = 7/10$.

$\xi = 0.7$ a new set of theta series must be tabulated. The theta series for the 1D lattice $Z_1$ with the norm 0.7 is

$$\Theta_{Z_1}(z, 0.7) = \sum_{m=-\infty}^{\infty} q^{|m|^{0.7}} = 1 + 2 \sum_{m=1}^{\infty} q^{m^{0.7}} \quad (111)$$

Equation (111) corresponds to the theta series $\Theta_3(z)$ introduced in Section VI.B. We will denote the new theta series by $\Theta_3(z, 0.7)$.

Another important theta series with the norm $\xi = 0.7$ corresponds to $\Theta_2(z)$ from Section VI.B:

$$\Theta_2(z, 0.7) = \sum_{m=-\infty}^{\infty} q^{|m+1/2|^{0.7}} \quad (112)$$

The $n$-dimensional lattice $Z_n$ has the following theta series derived from Eq. (111):

$$\Theta_{Z_n}(z, 0.7) = \Theta_{Z_1}^n(z, 0.7) \quad (113)$$

As an example, $\Theta_{Z_2}(z, 0.7)$ of the lattice $Z_2$ is enumerated in Table 7.

For each norm of $m^{1/\eta}$, the number of lattice points $A_m$ of this norm and the coordinates of one of the points located on the surface with this norm are shown. The points $w_m$ can be referred as representatives of the norm $m^{1/\eta}$.

Similarly to spheres, more complex lattices can be obtained for the norm of 0.7 by utilization of construction A discussed in Section VI.C.

Because $c_i$ is defined on $F_2 = \{0, 1\}$, we need to consider two cases: $c_i = 0$ and $c_i = 1$. Note that the factor $\sqrt{2}$ is used only for scaling and can be ignored in further discussion. For $c_i = 0$, the form of $\Theta_{2Z}$ can be deduced as follows:

$$\Theta_{2Z}(z, 0.7) = \sum_{r=-\infty}^{\infty} q^{|2r|^{0.7}} = 1 + 2 \sum_{r=1}^{\infty} q^{(2r)^{0.7}} = \Theta_3(2^{0.7}z, 0.7) \quad (114)$$

because $\Theta_Z(z, 0.7) = \Theta_3(z, 0.7)$.

TABLE 7

NUMBERS OF POINTS ON THE EQUIPROBABLE SURFACES OF THE LATTICE $Z_2$ WITH NORM OF 0.7

| $m$ | 0 | 1 | 1.6245 | 2 | 2.1577 | 2.6245 | 2.6390 | 3.0852 | 3.1577 |
|---|---|---|---|---|---|---|---|---|---|
| $w_m$ | [0 0] | [1 0] | [2 0] | [1 1] | [3 0] | [2 1] | [4 0] | [5 0] | [3 1] |
| $A_m$ | 1 | 4 | 4 | 4 | 4 | 8 | 4 | 4 | 8 |
| | 3.2490 | 3.5051 | 3.6390 | 3.7822 | 3.9045 | 4.0852 | ... |
| | [2 2] | [6 0] | [4 1] | [2 3] | [7 0] | [5 1] | ... |
| | 4 | 4 | 8 | 8 | 4 | 8 | ... |

Similarly, for $\Theta_{1+2Z}$ that corresponds to $c_i = 1$, we have

$$\Theta_{1+2Z}(z, 0.7) = \sum_{r=-\infty}^{\infty} q^{|1+2r|^{0.7}} = 2\sum_{r=0}^{\infty} q^{(2r+1)^{0.7}} = \Theta_2(2^{0.7}z, 0.7) \quad (115)$$

Now, by analogy to Eq. (85), we can write

$$\Theta_{L(C)}(z, 0.7) = W_C(\Theta_3(2^{0.7}z, 0.7), \Theta_2(2^{0.7}z, 0.7)) \quad (116)$$

***Example IX.1*** Consider an example of using Eq. (84). The binary code $[n, 2^n, 1]$, which is known as the universal code $F_2^n$, has the weight enumerator $W_{F_2^n}(x, y) = (x + y)^n$. The lattice $Z_n$ that can be generated from this code has the following theta series:

$$\Theta_{Z_n}(z, 0.7) = \Theta_{L(F_2^n)}(z, 0.7) = (\Theta_3(2^{0.7}z, 0.7) + \Theta_2(2^{0.7}z, 0.7))^n$$

which coincides with Eq. (113).

***Example IX.2*** Another example of constructing lattices from codes is the lattice $D_n$, which can be obtained from the parity-check code with the weight enumerator $W_{P_2^n}(x, y) = \frac{1}{2}((x + y)^n + (x - y)^n)$. Then, following Eq. (116), the theta series of the lattice $D_n$ is

$$\Theta_{D_n}(z, 0.7) = \frac{1}{2}((\Theta_3(2^{0.7}z, 0.7) + \Theta_2(2^{0.7}z, 0.7))^n$$
$$+ (\Theta_3(2^{0.7}z, 0.7) - \Theta_2(2^{0.7}z, 0.7))^n) \quad (117)$$

The theta series of $D_2$ is given in Table 8.

By using this theta series, the numbers of points on equiprobable surfaces can be tabulated and then used in Eq. (108) to tabulate probabilities of the codevectors. This approach greatly simplifies the compilation of probabilities of the codevectors for entropy coding of the generalized Gaussian sources with $\xi$ equal 2 or 0.7.

TABLE 8

NUMBERS OF POINTS ON THE EQUIPROBABLE SURFACES OF THE LATTICE $D_2$ WITH NORM OF 0.7

| $m$ | 0 | 1.6245 | 2 | 2.6390 | 3.1577 | 3.2490 | 3.5051 | 4.0852 | 4.2635 |
|---|---|---|---|---|---|---|---|---|---|
| $w_m$ | [0 0] | [2 0] | [1 1] | [4 0] | [3 1] | [2 2] | [6 0] | [5 1] | [4 2] |
| $A_m$ | 1 | 4 | 4 | 4 | 8 | 4 | 4 | 8 | 8 |

| | 4.2871 | 4.9045 | 5.0119 | 5.1296 | 5.2428 | 5.6555 | ... |
|---|---|---|---|---|---|---|---|
| | [8 0] | [7 1] | [10 0] | [2 6] | [3 5] | [9 1] | ... |
| | 4 | 8 | 4 | 8 | 8 | 8 | ... |

An alternative method for estimation of the probabilities was suggested in Shnaider and Paplinski (1997). Consider a random two-dimensional sample $x = \{x_1, x_2\}$ from a generalized Gaussian distribution with standard deviation 1 and mean 0. Let $y$ be defined as

$$y = \sum_{i=1}^{2} |x_i|^\xi \tag{118}$$

The probability distribution function of $y$ is given by

$$G(y) = \Pr\left(\sum_{i=1}^{2} |x_i|^\xi \leq y\right) \tag{119}$$

Then, provided that $x_1$ and $x_2$ are independent, the probability distribution, $G(y)$, can written in the following form:

$$G(y) = \begin{cases} \iint_S a^2 \exp\left(-b^\xi \sum_{i=1}^{2} |x_i|^\xi\right), & \text{if } y \geq 0 \\ 0, & \text{otherwise} \end{cases} \tag{120}$$

where $S$ is the collection of random samples of size 2 located within the space bounded by $y^{1/\xi}$ and centered at $(0, 0)$. The parameter $b$ is assumed positive. Introducing a new variable $c = b^\xi$, we have, for $y \geq 0$,

$$G(y) = \iint_S a^2 \exp\left(-c \sum_{i=1}^{2} |x_i|^\xi\right) dx_0 \, dx_1 \quad \text{and} \quad a = \frac{c^{1/\xi} \xi}{2\Gamma\left(\frac{1}{\xi}\right)} \tag{121}$$

The change to the spherical coordinates leads to

$$\sum_{i=1}^{2} |x_i|^\xi = \rho^\xi, \qquad x_0 = \rho \cos\phi, \quad \text{and} \quad x_1 = \rho \sin\phi$$

with $\rho$ between 0 and $y^{1/\xi}$ and $\phi$ between 0 and $2\pi$. Upon substitution of the boundry conditions, Eq. (121) becomes

$$G(y) = a^2 \int_0^{y^{1/\xi}} \int_0^{2\pi} \rho \exp(-c\rho^\xi) \, d\phi \, d\rho$$

$$= 2\pi a^2 \int_0^{y^{1/xi}} \rho \exp(-c\rho^\xi) \, d\rho \tag{122}$$

Let $\rho = w^{1/\xi}$; then we have

$$G(y) = 2\pi a^2 \int_0^y \frac{1}{\xi} w^{2/\xi - 1} \exp(-cw) \, dw \tag{123}$$

From Eq. (123), the PDF of $y$ is

$$g(y) = G'(y) = \frac{2\pi}{\xi} a^2 y^{2/\xi - 1} \exp(-cy) \qquad (124)$$

After substitution of $a$ specified in Eq. (121), the last equation can be written in the following form:

$$g(y) = \frac{\pi \xi c}{2\Gamma^2(1/\xi)} (cy)^{2/\xi - 1} \exp(-cy)$$

Finally,

$$g(y) = \frac{\pi \xi c}{2B\Gamma\left(\frac{2}{\xi}\right)} (cy)^{2/\xi - 1} \exp(-cy) \qquad (125)$$

where $B = \int_0^1 (t - t^2)^{1/\xi - 1} dt$.

The last part of Eq. (125), namely,

$$\frac{1}{\Gamma\left(\frac{2}{\xi}\right)} (cy)^{2/\xi - 1} \exp(-cy)$$

is the gamma distribution. In general, the PDF of $y$ follows the gamma distribution. In other words, $g(y)$ is proportional to

$$\frac{1}{\Gamma\left(\frac{r}{\xi}\right)} (cy)^{r/\xi - 1} \exp(-cy) \quad \text{where} \quad y = \sum_{i=1}^{r} |x_i|^{\xi}$$

When $\xi$ is set to 2, i.e., in the case of the Gaussian PDF, $g(y)$ becomes the well-known chi-square distribution with $r$ degrees of freedom.

In conclusion, we can say that using this approach, the probabilities of codevectors located on the equiprobable surfaces can be closely estimated. The close form solution for these probabilities depends on the dimensionality $r$ of the quantized vectors and the norm $\xi$ and is always in the form of the gamma distribution.

## X. Experimental Results

As discussed previously, the basic one-dimensional wavelet transform can be performed as an expansion using a pair of filters, of which one is a low-pass and the other high-pass. The typical extension to the two-dimensional cases is based on the consecutive application of the algorithm to the columns and rows of the input matrix, as depicted in Figures 1, 2, and 3. The depth of the transforms is limited only by the size of input images. In our experiments we have used the three-level wavelet expansion tree.

Most of the information in real image data is concentrated in the low- to medium-frequency components. Therefore, we increase the dimension of the quantization vectors moving from low to high frequencies; as a result, we obtain fine quantization at low frequencies and coarse quantization at higher frequencies. In the examples that follow, we use $1 \times 1$ blocks, or scalar quantization, for the wavelet coefficients of level 3, $2 \times 2$ blocks for level 2, and $4 \times 4$ blocks for level 1. Thus, we consider three possible dimensions of the quantization vectors, namely, one-dimensional, four-dimensional, and sixteen-dimensional vectors.

For simulation purposes we use three test images: "Lena" (Fig. 4), "Mandrill" (Fig. 22), and "Coast" (Fig. 5), each of size $512 \times 512$ pixels. The test images have been expanded by means of the three-level wavelet transform. This tree structure was examined in a number of studies (Antonini, *et al.;* Shnaider and Paplinski, 1994;) and was found to be well suited for image coding applications. The biorthogonal filters (4, 4) (Daubechies, 1992) used in the experiments are specified in Table 9 and Eq. (126).

$$\tilde{g}_n = (-1)^{n+1} h_{-n+1} \quad \text{and} \quad g_n = (-1)^{n+1} \tilde{h}_{-n+1} \qquad (126)$$

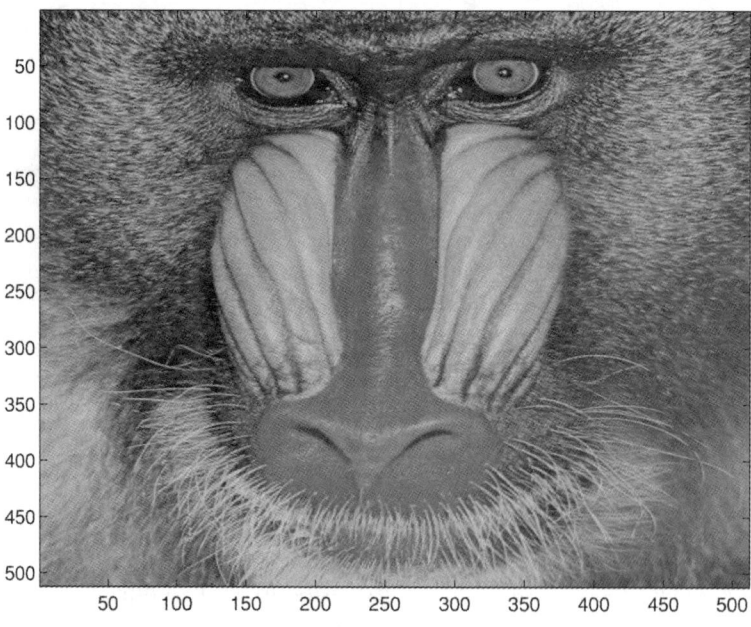

FIGURE 22. Image "Mandrill": original.

TABLE 9
BIORTHOGONAL FILTERS (4, 4) USED IN SIMULATION

| $n$ | 0 | ±1 | ±2 | ±3 | ±4 |
|---|---|---|---|---|---|
| $h_n$ | 0.6029 | 0.2669 | −0.0782 | −0.0169 | 0.0267 |
| $\tilde{h}_n$ | 0.5575 | 0.2956 | −0.0288 | −0.0456 | 0 |

The matrix of wavelet coefficients consists of 7 blocks, as listed in Table 10, each block representing certain spatial frequencies. The texture block denoted by 3 in Table 10 and located in the upper left corner of the matrix of wavelet coefficients has been encoded with scalar quantization using the $Z_1$ lattice. The coefficient blocks of level 2 (2H, 2V, 2D) have been quantized with 4D vectors using the $D_4$ lattice. To quantize the highest-frequency blocks of level 1 (1H, 1V, 1D), we have used 16D vectors coded using the $\Lambda_{16}$ lattice. Before quantization, approximately three-quarters of the wavelet coefficients within each block were rescaled with the same scaling factors $s$, following the scaling algorithm presented in Section VII. The rest of the coefficients were scaled individually. The optimal size of the codebook for each quantizer has been determined using the routine developed in Section 2 with the primary bit per pixel parameter set to 0.6. The bit allocation for each block of the test images is given in Table 10. It can be seen from Table 10 that the number of bits per pixel required for optimal quantization decreases when moving from low- to high-frequency blocks. This is consistent with our preliminary assumption that the energy of real-image data is concentrated mostly in low and medium frequencies. Due to the nature of lattice quantization, the bit rates obtained were rounded up to the nearest codebook length available for the corresponding lattice. For example, the number of bits per coefficient for the block 2H of the image "Coast" is 2.04 (Table 10). Because each block is 2 × 2 coefficients, the number of bits per block is 8.16, with a corresponding codebook size of approximately 286. From Table 3 we have that the nearest number of lattice points that comprise the codebook is $\sum_{m=0}^{10} A_m = 313$.

TABLE 10
BIT ALLOCATIONS FOR IMAGES "LENA," "MANDRILL," AND "COAST" WITH THE AVERAGE BIT RATE OF 0.6 BPP

| Image | 3 | 2H | 2V | 2D | 1H | 1V | 1D |
|---|---|---|---|---|---|---|---|
| "Lena" | 3.66 | 1.77 | 0.93 | 0.57 | 0.67 | 0 | 0 |
| "Mandrill" | 2.95 | 0.98 | 1.55 | 0.69 | 0.12 | 0.74 | 0 |
| "Coast" | 4.07 | 2.04 | 1.75 | 0.25 | 0 | 0.38 | 0 |

Finally, the quantized coefficients were encoded with the Huffman coder. The resulting compressed images were of the following sizes: 10,908 bytes for the "Lena" image, 18,566 bytes for "Mandrill," and 14,674 bytes for "Coast." These correspond to compression ratios of 24.0:1, 14.1:1, and 17.9:1, respectively. The specified compression ratios are based on the size of the file containing the encoded image and all additional information required to decode it.

The reconstructed images are shown in Figures 23, 24, and 25. The fidelity of compression was measured using the peak signal-to-noise ratio (PSNR). The results obtained are collected in Table 11. As might have been expected, the PSNR for the image "Mandrill" is lower than for the other test images due to the high complexity of this image.

In addition to the PSNR, in image compression another factor of general concern is the speed of coding. It is possible to observe that lattice quantization is significantly faster than most of the existing quantization methods due to the use of the fast algorithms for calculating the closest lattice point (or codevector) and the elimination of the need to generate the codebook. For lattice quantization it is possible to reduce the time for encoding further by choosing lattices with low-complexity quantization algorithms. One of the most advantageous possibilities is to use $D$-type lattices, which deliver

FIGURE 23. Image "Lena" encoded with the $\Lambda$-type lattices.

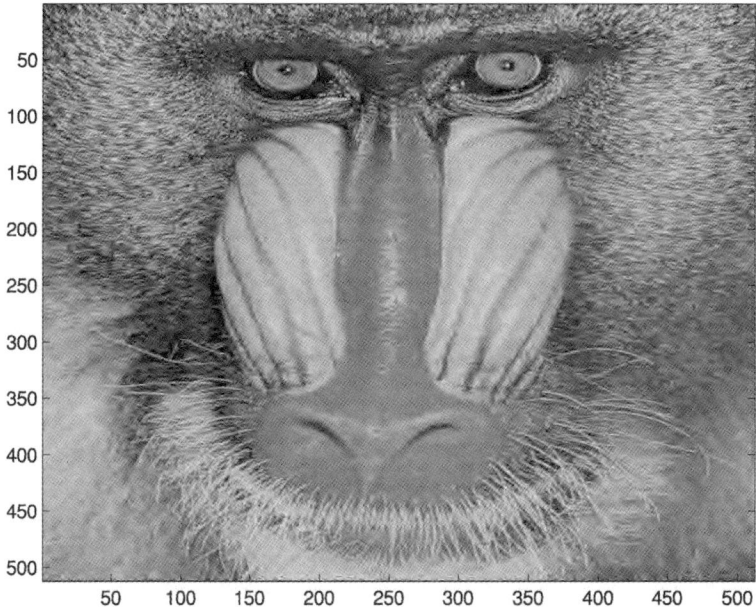

FIGURE 24. Image "Mandrill" encoded with the $\Lambda$-type lattices.

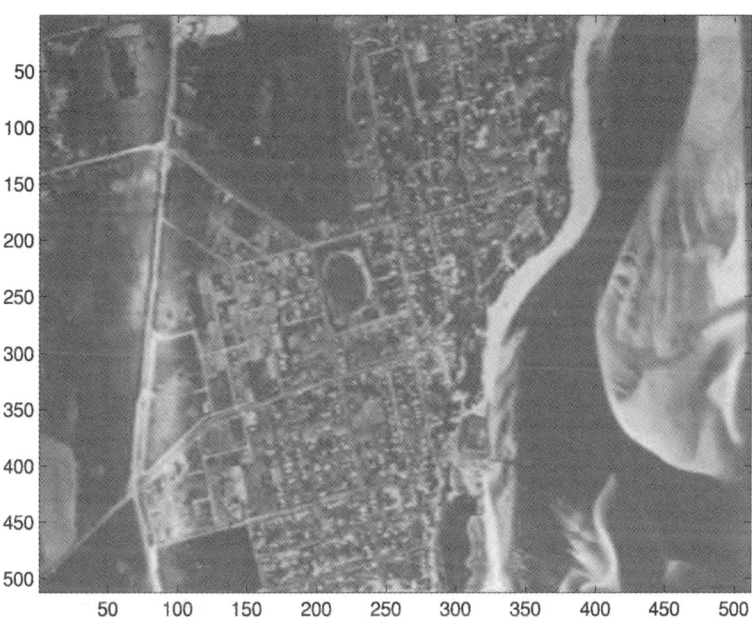

FIGURE 25. Image "Coast" encoded with the $\Lambda$-type lattices.

TABLE 11
COMPRESSION RESULTS USING THE "OPTIMAL"
LATTICE QUANTIZERS

| Image | Compression ratio | PSNR (dB) |
|---|---|---|
| "Lena" | 24.0:1 | 32.06 |
| "Mandrill" | 14.1:1 | 23.82 |
| "Coast" | 17.9:1 | 34.23 |

high-speed encoding algorithms. The combination of wavelet transform with $D$-type lattices was examined in Shnaider and Paplinski (1997).

## XI. CONCLUSIONS

One of the most prominent directions in the area of coding of still images is based on the utilization of spatial correlation between neighboring pixels in an image. Such correlation can be exploited by mapping the image data into a frequency plane using one of the transforms, the wavelet transforms being the best option contender. The resulting set of transform coefficients represents the original image. The coefficients are subsequently quantized and coded to achieve the desired compression ratio. The original image is recovered by the inverse procedure. The quality of an image-compression system is assessed by the fidelity of the output images, the compression ratio, and the time required for encoding and decoding, as well as by a number of secondary criteria. Typically, it is not easy to satisfy all the criteria for a high-quality image-coding system, and therefore a trade-off between various requirements must be found, with emphasis on the prime objectives, which are the compression ratio, quality of encoding, and the time for encoding and decoding.

As has been shown in this work, an image-coding system based on the wavelet transform and lattice quantization supported by appropriate bit allocation procedure and entropy coding gives an excellent fidelity for a given compression ratio, maintains high speed in both encoding and decoding operations, and is not very demanding in terms of the cost of implementation.

Although the combination of the wavelet transform and different methods of quantization has been investigated for image compression in a number of recent studies, the introduction of lattice quantization in such an image compression system seems a rather novel direction of research. It has been gaining popularity following a number of recent developments in the lattice theory. The study of lattice theory presented in this text is a summary of current developments in the context of the utilization of lattices for quantization. We have shown that the lattices constitute optimal of near-optimal quantizer systems for encoding the wavelet coefficients. We have also examined two of the most interesting types of lattices, namely, root and laminated lattices. Another problem addressed in this work has been the number of lattice points available for quantization. We have presented methods for calculation of an approximate solution, as well as the exact solution to this problem.

In order to minimize the distortion introduced by the procedure of quantizing the wavelet coefficients, we have developed an optimal bit-allocation routine. This routine allows distribution of the available bits so that the fidelity of the compression system is maximized. By using a model of the probability distribution function of coefficients instead of the actual PDF, we have significantly reduced the amount of computation required to obtain the solution of the distortion function and, as a result, lowered the overall time of encoding.

In the experimental part of this work we have tested the performance of our coding system based on the discrete wavelet transform and lattice quantization.

In conclusion we can state that in this study we have both investigated the theoretical foundation for an image coding system based on the wavelet transform and lattice quantization and developed and tested such a system. Through a number of simulations, this system has demonstrated its capacity to encode and decode gray-scale images at low to medium bit rates while maintaining excellent fidelity and a high coding speed. Optimization of the algorithm that has been developed has resulted in both a reduction of its complexity and, consecutively, simplicity of its implementation.

We also note that the high speed of coding makes the system that has been developed particularly advantageous in the context of real-time image compression.

## XII. CARTAN MATRICES OF SOME ROOT SYSTEMS

The Cartan matrices for the root systems, $A_n$, $B_n$, $C_n$, $D_n$, $E_6$, $E_7$, $E_8$, $F_4$ and $G_2$ are

$$A_n = \begin{bmatrix} 2 & -1 & 0 & \cdot & \cdot & \cdot & \cdot & \cdot & 0 \\ -1 & 2 & -1 & 0 & \cdot & \cdot & \cdot & \cdot & 0 \\ 0 & -1 & 2 & -1 & 0 & \cdot & \cdot & \cdot & 0 \\ \cdot & & & & & & & & \cdot \\ 0 & \cdot & & \cdot & & \cdot & 0 & -1 & 2 \end{bmatrix} \quad (127)$$

$$B_n = \begin{bmatrix} 2 & -1 & 0 & \cdot & \cdot & \cdot & & \cdot & 0 \\ -1 & 2 & -1 & 0 & \cdot & \cdot & & \cdot & 0 \\ \cdot & \cdot & & & & & & & \cdot \\ 0 & \cdot & & \cdot & \cdot & \cdot & -1 & 2 & -2 \\ 0 & \cdot & & \cdot & \cdot & \cdot & 0 & -1 & 2 \end{bmatrix} \quad (128)$$

$$C_n = \begin{bmatrix} 2 & -1 & 0 & \cdot & \cdot & \cdot & \cdot & \cdot & \cdot & 0 \\ -1 & 2 & -1 & \cdot & \cdot & \cdot & \cdot & \cdot & \cdot & 0 \\ 0 & -1 & 2 & -1 & \cdot & \cdot & \cdot & \cdot & \cdot & 0 \\ \cdot & \cdot & \cdot & & & & & & & \cdot \\ 0 & \cdot & & \cdot & \cdot & \cdot & -1 & 2 & -1 \\ 0 & \cdot & & \cdot & \cdot & \cdot & 0 & -2 & 2 \end{bmatrix} \qquad (129)$$

$$D_n = \begin{bmatrix} 2 & -1 & 0 & \cdot & \cdot & & \cdot & \cdot & \cdot & 0 \\ -1 & 2 & -1 & \cdot & \cdot & & \cdot & \cdot & \cdot & 0 \\ \cdot & & & & & & & & & \cdot \\ 0 & \cdot & & \cdot & \cdot & -1 & 2 & -1 & 0 & 0 \\ 0 & \cdot & & \cdot & \cdot & \cdot & -1 & 2 & -1 & -1 \\ 0 & \cdot & & \cdot & \cdot & \cdot & 0 & -1 & 2 & 0 \\ 0 & \cdot & & \cdot & \cdot & \cdot & 0 & -1 & 0 & 2 \end{bmatrix} \qquad (130)$$

$$E_6 = \begin{bmatrix} 2 & 0 & -1 & 0 & 0 & 0 \\ 0 & 2 & 0 & -1 & 0 & 0 \\ -1 & 0 & 2 & -1 & 0 & 0 \\ 0 & -1 & -1 & 2 & -1 & 0 \\ 0 & 0 & 0 & -1 & 2 & -1 \\ 0 & 0 & 0 & 0 & -1 & 2 \end{bmatrix} \qquad (131)$$

$$E_7 = \begin{bmatrix} 2 & 0 & -1 & 0 & 0 & 0 & 0 \\ 0 & 2 & 0 & -1 & 0 & 0 & 0 \\ -1 & 0 & 2 & -1 & 0 & 0 & 0 \\ 0 & -1 & -1 & 2 & -1 & 0 & 0 \\ 0 & 0 & 0 & -1 & 2 & -1 & 0 \\ 0 & 0 & 0 & 0 & -1 & 2 & -1 \\ 0 & 0 & 0 & 0 & 0 & -1 & 2 \end{bmatrix} \qquad (132)$$

$$E_8 = \begin{bmatrix} 2 & 0 & -1 & 0 & 0 & 0 & 0 & 0 \\ 0 & 2 & 0 & -1 & 0 & 0 & 0 & 0 \\ -1 & 0 & 2 & -1 & 0 & 0 & 0 & 0 \\ 0 & -1 & -1 & 2 & -1 & 0 & 0 & 0 \\ 0 & 0 & 0 & -1 & 2 & -1 & 0 & 0 \\ 0 & 0 & 0 & 0 & -1 & 2 & -1 & 0 \\ 0 & 0 & 0 & 0 & 0 & -1 & 2 & -1 \\ 0 & 0 & 0 & 0 & 0 & 0 & -1 & 2 \end{bmatrix} \qquad (133)$$

$$F_4 = \begin{bmatrix} 2 & -1 & 0 & 0 \\ -1 & 2 & -2 & 0 \\ 0 & -1 & 2 & -1 \\ 0 & 0 & -1 & 2 \end{bmatrix} \qquad (134)$$

$$G_2 = \begin{bmatrix} 2 & -1 \\ -3 & 2 \end{bmatrix} \qquad (135)$$

## References

Antonini, M., Barlaud, M., Mathieu, P., and Daubechies, I. (1992). Image coding using wavelet transform, *IEEE Trans. Image Proc.* **1,** 205–220.
Barlaud, M., Sole, P., Moureaux, J., Antonini, M., and Gauthier, P. (1993). Elliptical codebook for lattice vector quantization, *Proc. of ICASSP* **5,** 590–593.
Barlaud, M., Sole, P., Gaidon, T., Antonini, M., and Mathieu, P. (1994). Pyramidal lattice vector quantization for multiscale image coding, *IEEE Trans. Image Proc.* **3,** 367–381.
Berger, T. (1971). *Rate Distortion Theory; A mathematical Basis for Data Compression.* Upper Saddle River, N.J.: Prentice Hall.
Birkhoff, G. *Lattice theory,* 3rd ed., Providence: Amer. Math. Soc., 1967.
Campbell, F., and Robson, J. (1968). Application of Fourier analysis to the visibility of gratings, *J. Physiology* **197,** 551–566.
Charavi, H., and Tabatabai, A. (1988). Subband coding of monochrome and colour images, *IEEE Trans. Circuits and Syst.* **35,** 207–214.
Chen, F., Gao, Z., and Villasenor, J. (1997). Lattice vector quantization of generalized gaussian sources, *IEEE Trans. Information Theory* **43,** 92–103.
Conway, J., and Sloane, N. (1982). Voronoi regions of lattices, second moments of polytopes, and quantization, *IEEE Trans. Inform. Theory* **28,** 211–226.
Conway, J., and Sloane, N. (1982). Fast quantizing and decoding algorithms for lattice quantizers and codes, *IEEE Trans. Inform. Theory* **28,** 227–232.
Conway, J., and Sloane, N. (1983). A fast encoding method for lattice codes and quantizers, *IEEE Trans. Inform. Theory* **29,** 820–824.
Conway, J., and Sloane, N. *Sphere Packings, Lattices and Groups.* Springer-Verlag, Second Ed., New York, 1993.
Cosman, P., Gray, R., and Vetterli, M. (1996). Vector quantization of image subbands: A survey, *IEEE Trans. Image Proc.* **5,** 202–225.
Coxeter, H. *Regular Polytopes.* 3rd ed., Dover, NY., 1973.
Daubechies, I. *Ten Lectures on Wavelets.* SIAM, Philadelphia: 1992.
DeVore, R., Jawerth, B., and Lucier, B. (1992). Image compression through wavelet transform coding, *IEEE Trans. Inform. Theory* **38,** 719–746.
Ebeling, W. *Lattices and Codes.* Wiesbaden: Vieweg, 1994.
Fischer, T. R. (1986). A pyramid vector quantizer, *IEEE Trans. Information Theory* **32,** 568–583.
Gersho, A. (1979). Asymptotically optimal block quantization, *IEEE Trans. Inform. Theory* **25,** 373–380.
Gersho, A., and Gray, R. *Vector Quantization and Signal Compression.* Kluwer Academic Publishers, Norway: 1992.
Gibson, J., and Sayood, K. (1988). Lattice quantization, *Adv. Electronics and Electron Physics,* **72,** 259–330.
Gilmore, R. *Lie Groups, Lie Algebras, and Some of Their Applications.* John Wiley and Sons, Inc., New York, 1974.
Grosswald, E. *Representations of Integers as Sums of Squares.* Springer-Verlag, NY, 1985.
Grove, L., and Benson, C. *Finite Reflection Groups.* Springer-Verlag, New York, Second Ed., 1972.
Humphreys, J. *Introduction to Lie Algebras and Representation Theory.* Springer-Verlag, New York, 1972.
Igusa, J. *Theta Functions.* Springer-Verlag, NY, 1972.
Krätzel, E. *Lattice Points.* VEB Deutscher Verlag der Wissenschaften, Berlin, 1998.
Lang, S. *Introduction to Modular Forms.* Springer-Verlag, Berlin, 1976.

Leech, J., and Sloane, N. (1971). Sphere packings and error-correcting codes, *Canadian Journal of Math.* **23**, 718–745.

Lewis, A., and Knowles, G. (1992). Image compression using the 2-D wavelet transform, *IEEE Trans. Image Proc.* **1**, 244–250.

Li, W., and Zhang, Y. (1994). A study of vector transform coding of subband decomposed images, *IEEE Trans. Circuits Sys. Video Tech.* **4**, 383–391.

MacWilliams, F., and Sloane, N. *The Theory of Error-Correcting Codes*. North-Holland, NY, 1977.

Mallat, S. (1989). Multifrequency channel decompositions of images and wavelet models, *IEEE Trans. ASSP* **37**, 2091–2110.

*MATLAB Reference Guide,* The Math Works, Inc., 1992.

Meyer, Y. *Wavelets: Algorithms and Applications,* SIAM, Philadelphia, 1993.

Rankin, R. *Modular Forms and Functions.* Cambridge University Press, NY, 1977.

Riskin, E. (1991). Optimal bit allocation via the generalized BFOS algorithm, *IEEE Trans. Inform. Theory* **37**, 400–402.

Sakrison, D. (1968). A geometric treatment of the source encoding of a gaussian random variable, *IEEE Trans. Inform. Theory* **14**, 481–486.

Sakrison, D. Image coding applications of vision models, in *Image Transmission Techniques,* (K. Pratt, ed.), pp. 21–71, Academic Press, 1979.

Sampson, D., da Silva, E., and Ghanbari, M. (1994). Wavelet transform image coding using lattice vector quantization, *Electron. Letters* **30**, 1477–1478.

Sayood, J., Gibson, K., and Rost, M. (1984). An algorithm for uniform vector quantizer design, *IEEE Trans. Inform. Theory* **30**, 805–814.

Senoo, T., and Girod, B. (1992). Vector quantization for entropy coding of image subbands, *IEEE Trans. Image Proc.* **1**, 526–532.

Shannon, C. (1948). A mathematical theory of communication, *Bell System Tech. J.* **27**, 379–423 and 623–656.

Shapiro, J. (1993). Embedded image coding using zerotrees of wavelet coefficients, *IEEE Trans. Signal Processing* **41**, 3445–3462.

Shnaider, M., and Paplinski, A. (1995). Compression of fingerprint images using wavelet transform and vector quantization, in *Proc. of Int. Symp. Sig. Proc. Appl., ISSPA'96* (Gold Coast, Australia).

Shnaider, M. (1997). *A Study on an Image Coding System Based on the Wavelet Transform and Lattice Vector Quantization,* PhD thesis, Monash University, Australia.

Shnaider, M., and Paplinski, A. (1994). Wavelet transform for image coding, *Technical Report* 94-11, Monash University, Department of Robotics and Digital Technology.

Shnaider, M., and Paplinski, A. (1995). A novel wavelet toolbox with optimal vector quantizer, *Proc. of Dig. Image Comp. Tech. Appl., DICTA'95* (Brisbane, Australia).

Shnaider, M., and Paplinski, A. (1997). Image coding through D lattice quantization of wavelet coefficients, *Graphical Models and Image Processing* **59**, 193–204.

Shnaider, M., and Paplinski, A. (1999). Lattice vector quantization for wavelet based image coding, in *Advances in Imaging and Electron Physics* (P. Hawkes, ed.), New York: Academic Press.

Shnaider, M., and Paplinski, A. (2000). Selecting lattices for quantization of wavelet coefficients of images, *Opt. Eng.* **39**, 1327–1337.

Sloane, N. Binary codes, lattices and sphere packings, in *Combinatorial Surveys: Proceedings of The Sixth British Combinatorial Conference* (P. Cameron, ed.), pp. 117–164. Academic Press, 1977.

Vetterli, M. (1984). Multi-dimensional sub-band coding: some theory and algorithms, *Signal Proc.* **6**, 97–112.

Wang, X., Chan, E., Mandal, M., and Panchanathan, S. (1996). Wavelet-based image coding using nonlinear interpolative vector quantization, *IEEE Trans. Image Proc.* **5,** 518–522.

Woods, J., and O'Neil, S. (1986). Subband coding of images, *IEEE Trans. Acoust. Speech Signal Proc.* **34,** 1278–1288.

Zador, P. (1982). Asymptotic quantization error of continuous signals and their quantization dimension, *IEEE Trans. Information Theory* **28,** 139–149.

Zettler, W., Huffman, J., and Linden, D. (1990). Application of compactly supported wavelets to image compression, *Proc. of SPIE Conf. Image Proc. Algorithms, Techn.* pp. 150–160.

# Morphological Scale-Spaces

## PAUL T. JACKWAY

*Cooperative Research Centre for Sensor Signal and Information Processing*
*School of Computer Science and Electrical Engineering*
*The University of Queensland, Brisbane, Queensland 4072, Australia*

|   |   |
|---|---|
| I. Introduction . . . . . . . . . . . . . . . . . . . . . . . . . . . . . . . . . | 124 |
|    A. Gaussian Scale-Space . . . . . . . . . . . . . . . . . . . . . . . . . | 125 |
|    B. Related Work and Extensions . . . . . . . . . . . . . . . . . . . . . | 128 |
| II. Multiscale Morphology . . . . . . . . . . . . . . . . . . . . . . . . . . . . | 131 |
|    A. Scale-Dependent Morphology . . . . . . . . . . . . . . . . . . . . . . | 134 |
|    B. Semigroup and General Properties of the Structuring Function . . . . . | 138 |
| III. Multiscale Dilation-Erosion Scale-Space . . . . . . . . . . . . . . . . . . . | 139 |
|    A. Continuity and Order Properties of the Scale-Space Image . . . . . . . | 141 |
|    B. Signal Extrema in Scale-Space . . . . . . . . . . . . . . . . . . . . | 143 |
| IV. Multiscale Closing-Opening Scale-Space . . . . . . . . . . . . . . . . . . . . | 146 |
|    A. Properties of the Multiscale Closing-Opening . . . . . . . . . . . . . | 147 |
|    B. Monotone Theorem for the Multiscale Closing-Opening . . . . . . . . . . | 151 |
| V. Fingerprints in Morphological Scale-Space . . . . . . . . . . . . . . . . . . . | 153 |
|    A. Equivalence of Fingerprints . . . . . . . . . . . . . . . . . . . . . | 154 |
|    B. Reduced Fingerprints . . . . . . . . . . . . . . . . . . . . . . . . . | 156 |
|    C. Computation of the Reduced Fingerprint . . . . . . . . . . . . . . . . | 157 |
| VI. Structuring Functions for Scale-Space . . . . . . . . . . . . . . . . . . . . | 161 |
|    A. Semigroup Properties . . . . . . . . . . . . . . . . . . . . . . . . . | 161 |
|    B. A More General Umbra . . . . . . . . . . . . . . . . . . . . . . . . . | 162 |
|    C. Dimensionality . . . . . . . . . . . . . . . . . . . . . . . . . . . . | 164 |
|    D. The Poweroid Structuring Functions . . . . . . . . . . . . . . . . . . | 168 |
| VII. A Scale-Space for Regions . . . . . . . . . . . . . . . . . . . . . . . . . . | 170 |
|    A. The Watershed Transform . . . . . . . . . . . . . . . . . . . . . . . | 171 |
|    B. Homotopy Modification of Gradient Functions . . . . . . . . . . . . . . | 173 |
|    C. A Scale-Space Gradient Watershed Region . . . . . . . . . . . . . . . . | 176 |
| VIII. Summary, Limitations, and Future Work . . . . . . . . . . . . . . . . . . . | 179 |
|    A. Summary . . . . . . . . . . . . . . . . . . . . . . . . . . . . . . . | 179 |
|    B. Limitations . . . . . . . . . . . . . . . . . . . . . . . . . . . . . | 180 |
|    C. Future Work . . . . . . . . . . . . . . . . . . . . . . . . . . . . . | 180 |
| IX. Appendix . . . . . . . . . . . . . . . . . . . . . . . . . . . . . . . . . . | 181 |
|    A. Proof of Mathematical Results . . . . . . . . . . . . . . . . . . . . | 181 |
|       1. Proof of Proposition II.1 . . . . . . . . . . . . . . . . . . . . . | 181 |
|       2. Proof of Proposition II.2 . . . . . . . . . . . . . . . . . . . . . | 182 |
|       3. Proof of Corollary III.1.2 . . . . . . . . . . . . . . . . . . . . | 182 |
|       4. Proof of Proposition IV.1 . . . . . . . . . . . . . . . . . . . . . | 183 |
|       5. Proof of Proposition IV.3 . . . . . . . . . . . . . . . . . . . . . | 183 |
|       6. Proof of Proposition V.1 . . . . . . . . . . . . . . . . . . . . . | 183 |

7. Proof of Proposition V.1 . . . . . . . . . . . . . . . . . . . . . . . . 184
8. Proof of Proposition XII.1 . . . . . . . . . . . . . . . . . . . . . . . 184
B. Computer Code . . . . . . . . . . . . . . . . . . . . . . . . . . . . . 184
References . . . . . . . . . . . . . . . . . . . . . . . . . . . . . . . . . 186

## I. Introduction

In the 3 years since the current paper was published in *Advances in Imaging and Electron Physics* (Jackway, 1998), several findings on scale-space have emerged. Firstly, it is now clear (Weickert *et al.*, 1999) that the ideas and fundamental theorems underlying linear scale-space were first discovered and reported in Japan in the late 1950s Taizo Iijima and his students (1959, 1962, 1971). Secondly, the field continues to grow. There are now several books devoted to this and related topics (Weickert, 1998; Florack, 1997; Sporring *et al.*, 1997; Lindeberg, 1994), and an ongoing string of successful conferences (Haar Romeny *et al.*, 1997; Nielsen *et al.*, 1999). Interestingly, the conference held July 2001 in Vancouver was titled "Scale-Space and Morphology in Computer Vision," which sees the two major themes of this current paper come together at one meeting.

The term *scale-space* has grown from its modest launch in the title of a 4-page conference paper by Witkin (1983) to denote a whole subfield of study and a raft of generalizations, extensions, and theories (unfortunately, not all of them compatible!). As a indication of the enduring nature of Witkin's contribution, we note that at the present time (nearly two decades since its publication), Witkin's paper still receives nearly 50 citations per annum in the *Science Citation Index*.

Except for this revised introduction, I have left the remainder of the paper as originally published, although I am now aware that many of the mathematical results herein may be based on known results from Convex Analysis (Hiriart-Urruty and Lemarechal, 1993; van den Boomgaard and Heijmans, 2000) and may be able to be presented with much greater clarity, elegance, and generality by those skilled in that art.

What then is scale-space about? Let's follow Witkin's introduction:

> Any sophisticated signal understanding task must rely on a description of the signal which extracts meaningful objects or events. The problem of "scale" has emerged consistently as a fundamental source of difficulty in finding a good signal descriptor, as we need to separate events at different scales arising from distinct physical processes (Marr, 1982). It is possible to introduce a "parameter of scale" by smoothing the signal with a mask of variable size, but every setting of the scale parameter yields a different description! How can we decide which if any of this continuum of descriptions is "right"?

For many tasks it has become apparent that no one scale of description is categorically correct so there has been considerable interest in multi-scale descriptions (Ballard and Brown, 1982; Rosenfeld and Thurston, 1971; Marr and Poggio, 1979; Marr and Hildreth, 1980). However, merely computing signal descriptions at multiple scales does not solve the problem; if anything it exacerbates it by increasing the volume of data. Some means must be found to organize the description by relating one scale to another. *Scale-space filtering* provides a means for managing the ambiguity of scale in an organized and natural way. (Witkin, 1983)

From this passage we see that scale-space filtering concerns signals, in particular, signal understanding or analysis. It proceeds by dealing with descriptions of the signal smoothed by masks of varying sizes corresponding to multiple scales. Important too is the idea of dealing with all the resulting descriptions as a whole—we are not trying to determine which is the single "best" scale for analysis.

As we will see in the next section, the stack of signal descriptions is organized by relying on continuity properties of signal features across the scale dimension. Indeed, scale-space is most useful if we demand that the signal representation gets simpler with increasing smoothing. This turns out to lead to very interesting theoretical questions, such as: For what signal features, and for which smoothers, and which class of signals do the required properties exist? And given the signal/smoother/feature combination, what are the stability, uniqueness, invertability, and differential properties of the resulting signal representation.

These theoretical questions have exercised the minds and pens of many researchers, starting from Witkin (1983) who speculated that (under some restrictions) the Gaussian filter was the unique filter in 1D that possessed the required properties, and there is quite a large body of work on these topics (for examples, see Yuille and Poggio, 1985; Babaud *et al.*, 1986; Hummel, 1986; Hummel and Moniot, 1989; Wu and Xie, 1990; ter Haar Romeny *et al.*, 1991; Alvarez and Morel, 1994; Jackway and Deriche, 1996).

## A. Gaussian Scale-Space

To start, we need to review Witkin's (1983) approach. Suppose we have a signal, $f(x): R^n \to R$ and a smoothing kernel $g(\mathbf{x}, \sigma): R^n \times R \to R$. The *scale-space image* $F(\mathbf{x}, \sigma): R^n \times R \to R$ of the signal is obtained by smoothing the signal at all possible scales and is a function on the $(n+1)$-dimensional space called *scale-space:*

$$F(\mathbf{x}, \sigma) = f(\mathbf{x}) * g(\mathbf{x}, \sigma), \tag{1}$$

where ∗ denotes a smoothing operation. $F$ is known as the *scale-space image* of the signal.

The ideas behind scale-space first appeared in a report on expert systems by Stansfield (1980) who was looking at ways to extract features from graps of commodity prices. The scale-space concept was named, formalized, and brought to image analysis by Witkin (1984). Both these authors used the linear convolution as the smoothing operation:

$$F(\mathbf{x}, \sigma) = \int f(\xi) g(\mathbf{x} - \xi, \sigma) \, d\xi \qquad (2)$$

and Gaussian functions as the scale-dependent smoothing kernel:

$$g(\mathbf{x}, \sigma) = (2\pi\sigma^2)^{-n/2} \exp\left(-\frac{1}{2\sigma^2} \mathbf{x}^\mathsf{T} \mathbf{x}\right) \qquad (3)$$

With this smoother, the Marr-Hildreth edge detector (Marr and Hildreth, 1980) (zero-crossings of the second derivative of the signal) is the appropriate feature detector.

Witkin's idea is elegant: If scale is considered as a *continuous* variable rather than a parameter, then a signal feature at one scale is identified with that at another scale if they lie on the same feature path in the resulting scale-space.

A central idea in Witkin's work is that important signal features would persist through to relatively coarse scales even though their location may be distorted by the filtering process. However, by *coarse-to-fine tracking* they could be tracked back down a path in scale-space to zero-scale to be located exactly on the original signal. In this way the benefit of large smoothing to detect the major features could be comined with precise localization. In a way these linkages across scale are used to overcome the *uncertainty principle,* which states that spatial localization and frequency domain localization are conflicting requirements (Wilson and Granlund, 1984).

A defining feature of scale-space theory, in contrast to other multiscale approaches, is the property that a signal feature, once present at some scale, must persist all the way through scale-space to zero-scale (otherwise the feature would be spurious: being caused by the filter and not the original signal). This is called a *monotone property* since the number of features must necessarily be a monotone decreasing function of scale. If $\mathbf{Z}_{f(\mathbf{x})}$ denotes the point set of the positions of features in a signal $f(x)$, and if $C[\mathbf{Z}]$ denotes the number of features in the set, then we require

$$C\left[\mathbf{Z}_{F(\mathbf{x},\sigma_1)}\right] \leq C\left[\mathbf{Z}_{F(\mathbf{x},\sigma_2)}\right] \qquad \text{for all } \sigma_1 > \sigma_2 > 0 \qquad (4)$$

A *continuity property* is also implied since the feature paths should be continuous across scale to enable tracking. The plot of signal features positions versus

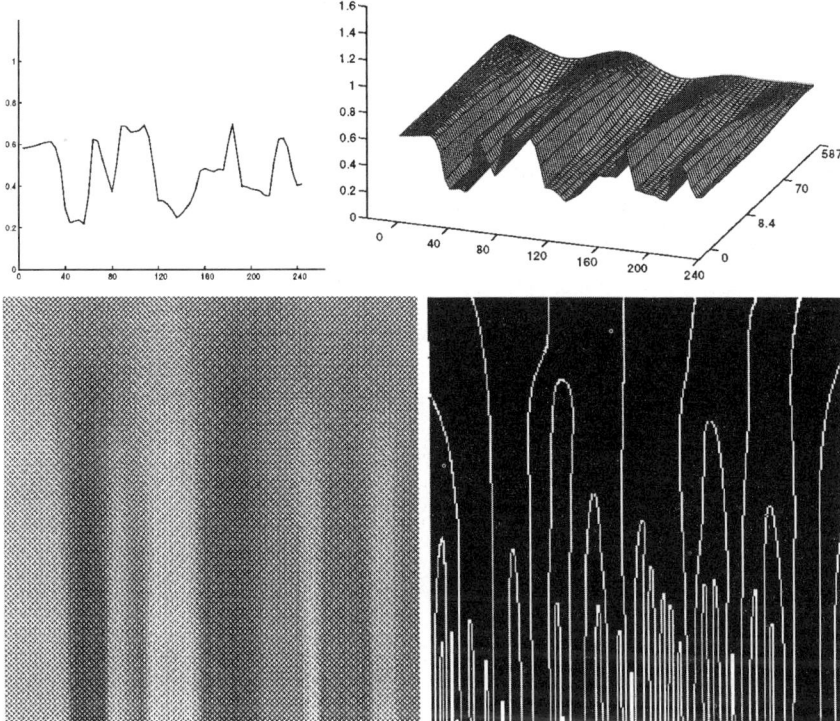

FIGURE 1. The Gaussian scale-space analysis of a 1D signal (a scan line from the "Lena" image). From left to right, top to bottom: the signal; the scale-space image as a surface plot; the scale-space image as a gray-scale image; the fingerprint (the plot of zero-crossings of the second spatial derivative).

scale has been termed the *fingerprint* of the signal (Yuille and Poggio, 1985). Figure 1 shows the Gaussian scale-space analysis of a 1D signal. Note that the monotone and continuity properties ensure that all the fingerprint lines of a signal form continuous paths in scale-space. All the paths start at zero-scale and continue upward until they stop at some scale (possibly infinite) which is characteristic for that feature. It was Witkin's plan to understand the signal by using its fingerprint.

Because of problems with signals in higher dimensions, the chief applications of Gaussian scale-space have been those involving 1D signals, for example, the description and recognition of planar curves (Asada and Brady, 1986; Mokhtarian and Mackworth, 1986), histogram analysis (Carlotto, 1987), signal matching (Witkin *et al.*, 1987), ECG signal analysis (Tsui *et al.*, 1988), the pattern matching of 2D shapes (Morita *et al.*, 1991), boundary contour refinement in images (Raman *et al.,* 1991), the analysis of facial profiles (Campos *et al.,* 1993), and the matching of motion trajectories (Rangarajan *et al.,* 1993).

To motivate the use of scale-space and to illustrate the general flavor of the above scale-space applications, consider the following example. Suppose we want to represent the shape of an object in a binary image. First, we express the boundary curve of the shape $C$ as a pair of functions of path length $t$ along the curve:

$$C = \{x(t), y(t)\} \tag{5}$$

Then the curvature at each point can be computed:

$$\kappa(t) = \frac{\dot{x}\ddot{y} - \dot{y}\ddot{x}}{(\dot{x}^2 + \dot{y}^2)^{3/2}} \tag{6}$$

where $\dot{x}$ and $\ddot{x}$ denote the first and second derivatives with respect to $t$. Now we smooth the curvature function with a scaled Gaussian [Eqs. (2) and (3)], directly (Asada and Brady, 1986) or via smoothing $x(t)$ and $y(t)$ (Mokhtarian and Mackworth, 1986) to give a scale-space curvature image $\kappa(t, \sigma)$, and then plot the zero-crossings of this image (to detect points of inflection of the curve) to give a fingerprint diagram representing the shape.

Now, to recognize or match this shape we can use the fingerprints. The idea is that noise and minor features are confined to small scales in the fingerprint while the most important features persist through to larger scales in the representation. Thus, it makes sense and is very efficient to match fingerprints in a hierarchical fashion starting at the larger scales (Mokhtarian and Mackworth, 1986; Jackway *et al.,* 1994). Other ways of dealing with fingerprints are also possible; for example, they can be represented as a ternary tree (Witkin, 1984) and the stability of various branches of this tree considered (Bischof and Caelli, 1988).

## B. Related Work and Extensions

Unfortunately, it is generally impossible to find smoothing filters which would satisfy all the desired properties on images and higher dimensional signals. Therefore, various authors, in an effort to extend and generalize the theory to higher dimensions and nonlinear smoothing operators, have emphasized certain properties and sacrificed others.

Koenderink (1984) emphasized the differential structure of scale-space; that is, what are the laws governing the shape of the surface surrounding an arbitrary point $F(\mathbf{x}_0, \sigma_0)$ in scale-space? Koenderink showed that Gaussian filtering is the Green's function (DuChateau and Zachmann, 1986) of the differential equation known as the *heat equation*. That is, the Gaussian scale-space image $F(\mathbf{x}, \sigma)$ given by Eqs. (2) and (3) is a solution of:

$$\nabla^2 F = -k \frac{\partial F}{\partial \sigma} \tag{7}$$

With this approach, the original signal is the initial condition $\sigma = 0$, which propagates into scale-space under control of Eq. (7).

The solutions to Eq. (7) obey a *maximum principle* (Protter and Weinberger, 1967), which states that if $F$ is a solution to Eq. (7) on the open and bounded region with $F$ of class $C^2$ and continuous on the closure of the region, then $F$ assumes its maximum at some point on the boundary of the region or for $\sigma = 0$. It has been shown that the maximum principle implies an evolution property for zero-crossings of the solution of the heat equation (Hummel and Moniot, 1989): Let $C$ be a connected component of the set of zero-crossings in the domain: $\{(\mathbf{x}, \sigma) : \mathbf{x} \in R^n, S_1 \leq \sigma \leq S_2\}$, where $0 \leq S_1 < S_2$. Then $C \cap \{(\mathbf{x}, \sigma) : \sigma = S_1\} \neq \emptyset$. This property ensures that a new zero-crossing component cannot begin at nonzero-scale, and that all zero-crossing components can be traced to features on the original signal.

This evolution property, called *causality* (Koenderink, 1984), under certain conditions leads uniquely to the scaled Laplacian-of-Gaussian filter (Babaud et al., 1986). Important later work generalizing the heat equation has shown that space variant anisotropic operators can also satisfy causality while not degrading image edges with increasing scale (Perona and Malik, 1990). The maximum principle and its evolution or causality properties are indeed one way to extend Witkin's 1D results to images and higher dimensions. However, part of the elegance of the 1D result is lost since, "a closed zero-crossing contour can split into two as the scale increases, just as the trunk of a tree may split into two branches" (Yuille and Poggio, 1986). This may be a problem as two separate contours at a coarse scale may in fact be caused by the same signal feature (see, for example, the diagrams in Lifshitz and Pizer (1990)). The monotone property in the sense of our Eq. (4) is therefore not valid, which is a disadvantage with the linear scale-space formulations using zero-crossings in 2D and higher dimensions.

The recently developed field of mathematical morphology (Serra, 1982; Haralick et al., 1987) deals in its own right with the analysis of images but also provides quite general nonlinear operators, which can be used to remove structure from a signal. Therefore, scaled morphological operations have been used as scale-space smoothers. Chen and Yan (1989) have used a scaled disk for the morphological opening of objects in binary images to create a scale-space theorem for zero-crossings of object boundary curvature. These results have since been extended to general compact and convex structuring elements (Jang and Chin, 1991). Unfortunately, these results only apply to zero-crossings of boundary curvature of objects in binary images, although they can also be applied to 1D functions, through the use of umbras (Sternberg, 1986). However, the extension to higher dimensions seems problematic (Jackway, 1995a).

Recent work has also considered to construction of a scale-space through scaled morphological operations (van den Boomgaard and Smeulders, 1994). Van den Boomgaard's approach proceeds by considering the nonlinear differential equation, which governs the propagation of points on a signal into scale-space under morphological operations with a scaled convex structuring function. However, this work mainly examined the differential structure of the scale-space itself and did not explicitly emphasize a monotone property. All the points on a signal propagate into scale-space and van den Boomgaard did not explicitly consider special signal points (features) except for singular points which do not obey a monotone property.

Alvarez and Morel have recently presented an excellent theoretical unification and axiomization of many multiscale image analysis theories including most of those mentioned above (Alvarez and Morel, 1994). Once again this approach emphasizes the partial differential equations governing the propagation of the image into scale-space. Here the causality principle is essentially the maximum principle already discussed. The importance of image features and the monotone property is not stressed in this work.

The way we have chosen to extend Witkin's work is to seek to return to first principles. If the aim is to examine the deep structure of images, then we should seek to relate signal features across differing scales of image blurring. To be a scale-space theory, we require a monotone property that ensures that increasing scale removes features from the image.

In Witkin's original work, extrema of the signal and its first derivative are seen as fundamental signal features (Witkin, 1984). However, as discussed in Lifshitz and Pizer (1990), there is no convolution kernel with the property that it does not introduce new extrema with increasing scale in 2D, so the monotone property does not hold for linear filters and signal extrema. Therefore, we must turn to nonlinear filters.

We have found that scaled operations from mathematical morphology can act as signal smoothers and allow a monotone property for signal extrema, and indeed this result holds for signals on arbitrary dimensional space (Jackway and Deriche, 1996). We have also found a way to combine the dilation and erosion to give meaning to negative values of the scale parameter, thereby creating a full-plane ($\sigma \in R$) scale-space. The emphasis in our work is on the monotone property for signal features. We use signal extrema rather than zero-crossings as the signal feature of importance. We no longer use a linear smoothing operator, and we do not restrict the scale parameter to nonnegative values. In the remainder of this article we discuss these developments.

Multiscale dilation-erosion is introduced and its scale-space properties are discussed in Section II, followed in Section III by a discussion of the multiscale closing-opening scale-space and its relation to the dilation-erosion. Section IV considers *dimensionality* and the selection of the structuring function, and Section V extends the scale-space theory to regions via the watershed transform. Finally, Section VI is a summary and conclusion.

Where possible, the reader is directed to previously published papers for proofs of the various mathematical results, proofs of the new material are placed in an appendix. Parts of this work can be found in earlier conference papers (Jackway, 1992; Jackway *et al.,* 1994) and more fully later (Jackway, 1995a, 1995b, 1996; Jackway and Deriche, 1996).

## II. Multiscale Morphology

Mathematical morphology grew out of theoretical investigations of a geometrical or probabilistic nature needed in the analysis of spatial data from geology. The work was carried out by a team at the Fontainebleau research center of the Paris School of Mines from 1964. This theoretical work was first released widely with the publication of the book by Matheron (1975). A more practical book related to image analysis was later published by Serra (1982), followed by a second volume on theoretical advances (Serra, 1988).

Mathematical morphology, developed originally for sets, can be applied to numerical functions either via umbras (Sternberg, 1986) or directly and preferably via the complete lattice approach (Heijmans and Ronse, 1990). However, because we will consider only functions, we can skip the preliminaries and merely *define* the required operations directly on functions. Also, since notation varies between sources, we need to state that we will follow that of (Haralick *et al.,* 1987).

Denoting the functions $f : D \subset R^n \to R$ and $g : G \subset R^n \to R$, the two fundamental operations of gray-scale morphology are as follows.

**Definition II.1** *(Dilation):* The *dilation* of the function $f(\mathbf{x})$ by the function $g(\mathbf{x})$ is denoted by $(f \oplus g)(\mathbf{x}) : D \subset R^n \to R$ and is defined by

$$(f \oplus g)(\mathbf{x}) = \bigvee_{\mathbf{t} \in G \cap \check{D}_{-\mathbf{x}}} \{f(\mathbf{x} - \mathbf{t}) + g(\mathbf{t})\}$$

**Definition II.2** *(Erosion):* The *erosion* of the function $f(\mathbf{x})$ by the function $g(\mathbf{x})$ is denoted by $(f \ominus g)(\mathbf{x}) : D \subset R^n \to R$ and is defined by

$$(f \ominus g)(\mathbf{x}) = \bigwedge_{\mathbf{t} \in G \cap D_{-\mathbf{x}}} \{f(\mathbf{x} + \mathbf{t}) - g(\mathbf{t})\}$$

Where $D_\mathbf{x}$ is the translate of $D$, $D_\mathbf{x} = \{\mathbf{x} + \mathbf{t} : \mathbf{t} \in D\}$, $\check{D}$ is the reflection of $D$, $\check{D}, = \{x : -x \neq D\}$, and $\bigvee\{f\}$ and $\bigwedge\{f\}$ refer to the *supremum* (least upper bound) and *infimum* (greatest lower bound) of $f$ (DePree and Swartz, 1988).

In the discrete case (and for computation) where the function is a countable set of points, max$\{f\}$ and min$\{f\}$ are used for $\bigvee\{f\}$ and $\bigwedge\{f\}$. These definitions are general; in practice one function, say $f$, denotes the signal and the other $g$ is a compact shape called the *structuring function*. Note also that we have taken particular care with the *edge effects* by intersecting the supports of the two functions.

It is well known that the composition of the above two operations possesses a new property, that of *idempotence* (Serra, 1982), which is so important that we define the following operations:

**Definition II.3** *(Opening):* The *opening* of the function $f(\mathbf{x})$ by the function $g(\mathbf{x})$ is denoted by $(f \circ g)(\mathbf{x}) : D \subset R^n \to R$, and is defined by

$$(f \circ g)(\mathbf{x}) = ((f \ominus g) \oplus g)(\mathbf{x})$$

**Definition II.4** *(Closing):* The *closing* of the function $f(\mathbf{x})$ by the function $g(\mathbf{x})$ is denoted by $(f \bullet g)(\mathbf{x}) : D \subset R^n \to R$, and is defined by

$$(f \ominus g)(\mathbf{x}) = ((f \oplus g) \ominus g)(\mathbf{x})$$

These four basic morphological operations possess many interesting relationships and properties, which can be found in the standard sources (Matheron, 1975; Serra, 1982, 1988) and in the tutorial (Haralick et al., 1987). We will list here just those few used in this article.

1. All the morphological operations are *nonlinear,*

$$\psi(af + bh) \neq \psi(af) + \psi(bh) \qquad \text{for all functions } f, h \qquad (8)$$

   where $\psi(f) = (f \oplus g), (f \ominus g), (f \bullet g),$ or $(f \circ g)$.
2. Dilation and erosion are *duals:*

$$(f \oplus g) = -((-f) \ominus \check{g}) \qquad (9)$$

   where the reflection is

$$\check{g}(\mathbf{x}) = g(-\mathbf{x}) \qquad (10)$$

3. Closing and opening are *duals:*

$$(f \bullet g) = -((-f) \circ \check{g}) \qquad (11)$$

4. Closing and opening are *idempotent:*

$$((f \bullet g) \bullet g) = (f \bullet g) \qquad (12)$$

$$((f \circ g) \circ g) = (f \circ g) \qquad (13)$$

5. Dilation is *commutative* and *associative* and erosion admits a *chain rule*,
$$(f \oplus g) = (g \oplus f) \tag{14}$$
$$(f \oplus g) \oplus h = f \oplus (g \oplus h) \tag{15}$$
$$(f \ominus g) \ominus h = f \ominus (g \oplus h) \tag{16}$$

6. Some further identities also prove useful:
$$f \oplus g = (f \oplus g) \circ g = (f \bullet g) \oplus g \tag{17}$$
$$f \oplus g = (f \ominus g) \bullet g = (f \circ g) \ominus g \tag{18}$$

Now we need to define the following partial order relation on the functions $f, h : D \subset R^n \to R : f \le h$ is defined as $f(\mathbf{x}) \le h(\mathbf{x})$ for all $\mathbf{x} \in D$.

7. All the morphological operations are *increasing*:
$$f \le h \Rightarrow \psi(f) \le \psi(h) \tag{19}$$
where $\psi(f) = (f \oplus g), (f \ominus g), (f \bullet g),$ or $(f \circ g)$.

8. If the structuring function contains the origin $g(\mathbf{0}) \ge 0$, the dilation is *extensive* and erosion is *antiextensive*:
$$g(\mathbf{0}) \ge 0 \Rightarrow f \le (f \oplus g) \tag{20}$$
$$g(\mathbf{0}) \ge 0 \Rightarrow (f \ominus g) \le f \tag{21}$$

9. Closing is *extensive* and opening is *antiextensive*:
$$f \le (f \bullet g) \tag{22}$$
$$(f \circ g) \le f \tag{23}$$

Putting the previous two properties together, we have the order relation:

10. If the structuring function contains the origin $g(\mathbf{0}) \ge 0$, then:
$$(f \ominus g) \le (f \circ g) \le f \le (f \bullet g) \le (f \oplus g) \tag{24}$$

Finally, we say that function $g$ is *open* with respect to $h$ if
$$(g \circ h) = g \tag{25}$$

11. The opening and closing obey a sieving property, for if $g$ is open with respect to $h$, then
$$(f \circ g) \circ h = (f \circ h) \circ g = (f \circ g) \tag{26}$$
$$(f \bullet g) \bullet h = (f \bullet h) \bullet g = (f \bullet g) \tag{27}$$

12. They also obey an ordering property,
$$(f \circ g) \le (f \circ h) \le f \le (f \bullet h) \le (f \bullet g) \tag{28}$$

## A. Scale-Dependent Morphology

We will now proceed to review a scale parameterized signal smoothing operation we have developed for the purpose of scale-space using the basic morphological operations of dilation and erosion (Jackway and Deriche, 1996). Note, from the aspect of morphology, there is little new in the following sections, multiscale morphology has been used since the beginning, for example, in the "Granulometries" as introduced in Matheron (1975). Granulometries enable the distribution of particle sizes in an image to be found by measuring the residue following morphological openings with increasingly large structuring elements. The emphasis in the following work is not on the morphological operations but on the resulting scale-space properties; the fact that the operations used come from morphology is incidental—(but fortunate!) due to the large volume of well-developed theory on which to draw.

The simplest structuring function might be that representing an $(n+1)$-dimensional ball of radius $r$, this can be written as:

$$g_{\text{ball}}(\mathbf{x}) = \sqrt{r^2 - \|\mathbf{x}\|^2} \qquad \|\mathbf{x}\| \le r \tag{29}$$

where $\|\cdot\| = \sqrt{x_1^2 + x_2^2 + \cdots + x_n^2}$ is the norm in the $n$-dimensional Euclidean space.

In general, the result of dilation or erosion depends on the position of the *origin* $g(\mathbf{0})$ of the structuring function. We can see this by performing the dilation and erosion of the constant function $f(\mathbf{x}) = 0, \forall \mathbf{x} \in \mathbf{R}^n$ by the structuring function given by (29),

$$(f \oplus g_{\text{ball}})(\mathbf{x}) = \bigvee_{\mathbf{t} \in G} \{g_{\text{ball}}(\mathbf{t})\} = g_{\text{ball}}(\mathbf{0}) = r \qquad \forall \mathbf{x} \in \mathbf{R}^n \tag{30}$$

$$(f \ominus g_{\text{ball}})(\mathbf{x}) = \bigwedge_{\mathbf{t} \in G} \{-g_{\text{ball}}(\mathbf{t})\} = -g_{\text{ball}}(\mathbf{0}) = -r \qquad \forall \mathbf{x} \in \mathbf{R}^n \tag{31}$$

These level-shifting effects are easy to fix; we simply require:

$$\bigvee_{\mathbf{t} \in G} \{g(\mathbf{t})\} = 0 \tag{32}$$

Hence, the structuring function should be everywhere nonpositive with a global maximum value of zero: $g(\mathbf{t}) \le 0$ for all $\mathbf{t} \in G$. Additionally, to avoid horizontal translation effects, we require this maximum to occur at the origin,

$$g(\mathbf{0}) = \bigvee_{\mathbf{t} \in G} \{g(\mathbf{t})\} \tag{33}$$

We can make the $(n+1)$-dimensional ball function satisfy the foregoing conditions by shifting down by the radius:

$$g_{\text{ball2}}(\mathbf{x}) = \sqrt{r^2 - \|\mathbf{x}\|^2} - r \qquad \|\mathbf{x}\| \leq r \qquad (34)$$

We henceforth assume that all structuring functions discussed in this article satisfy conditions (32) and (33).

The morphological dilation, erosion, closing, and opening can be made scale dependent by the use of a *scaled structuring function*, $g_\sigma : \mathbf{G}_\sigma \subset \mathrm{R}^n \to \mathrm{R}$. The most natural way is to equate scale to the radius of the ball in Eq. (34). For reasons that will become clear in the next section, we will in fact equate radius to the *magnitude* of the scale parameter $\sigma \neq 0$, so the scaled ball becomes:

$$g_\sigma(\mathbf{x}) = \sqrt{|\sigma|^2 - \|\mathbf{x}\|^2} - |\sigma| \qquad \|\mathbf{x}\| \leq |\sigma| \qquad (35)$$

We can see how this relates to a prototype ball of unit radius by noting:

$$\sqrt{|\sigma|^2 - \|\mathbf{x}\|^2} - |\sigma| = |\sigma|(\sqrt{1 - \||\sigma|^{-1}\mathbf{x}\|^2} - 1)$$
$$= |\sigma| g_{\text{unitball}}(|\sigma|^{-1}\mathbf{x}) \qquad \sigma \neq 0 \qquad (36)$$

and for the support region,

$$\|\mathbf{x}\| \leq |\sigma| \Rightarrow \||\sigma|^{-1}\mathbf{x}\| \leq 1 \qquad (37)$$

This suggests that given any prototype structuring function $g : \mathbf{G} \subset \mathrm{R}^n \to \mathrm{R}$, we use the scaled structuring functions given by

$$g_\sigma(\mathbf{x}) = |\sigma| g(|\sigma|^{-1}\mathbf{x}) \qquad \mathbf{x} \in \mathbf{G}_\sigma, \qquad \sigma \neq 0, \qquad (38)$$
$$\mathbf{G}_\sigma = \{\mathbf{x} : |\sigma|^{-1}\mathbf{x} \in \mathbf{G}\}. \qquad (39)$$

Now, if $\mathbf{G}$ is bounded, say, $\mathbf{G} \subseteq \{\mathbf{x} : \|\mathbf{x}\| < R\}$ for some $R$, then $g_\sigma$ is defined on $\mathbf{G}_\sigma \subseteq \{\mathbf{x} : \|\mathbf{x}\| < |\sigma|R\}$ so the support region of the structuring function scales correctly, in particular:

$$\mathbf{G}_\sigma \to \{\mathbf{0}\} \qquad \text{as } |\sigma| \to 0 \qquad (40)$$

However, further conditions need to be imposed on the structuring function to ensure reasonable scaling behavior in all cases. Consider the *threshold set* of the structuring function $\mathcal{G}_\sigma(t) = \{\mathbf{x} : g_\sigma(\mathbf{x}) \geq t\}$ for any $t < 0$. To incorporate the idea of scaling, we wish to ensure that for all $t < 0$:

$$|\sigma| \to 0 \Rightarrow \mathcal{G}_\sigma(t) \to \{\mathbf{0}\} \qquad (41)$$
$$|\sigma_1| < |\sigma_2| \Rightarrow \mathcal{G}_{\sigma_1}(t) \subset \mathcal{G}_{\sigma_2}(t) \qquad (42)$$
$$|\sigma| \to \infty \Rightarrow \mathcal{G}_\sigma(t) \supset \{\mathbf{x} : \|\mathbf{x}\| \leq R\} \qquad \text{for all } R > 0 \qquad (43)$$

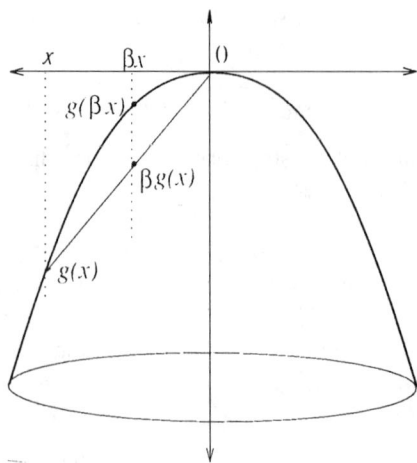

FIGURE 2. A chord from the origin to any point on the structuring function should lie on or below the structuring function.

In terms of the functions involved, this is equivalent to

$$|\sigma| \to 0 \Rightarrow g_\sigma(\mathbf{x}) \to \begin{cases} 0 & \text{if } \mathbf{x} = \mathbf{0} \\ -\infty & \text{if } \mathbf{x} \neq \mathbf{0} \end{cases} \quad (44)$$

$$0 < |\sigma_1| < |\sigma_2| \Rightarrow g_{\sigma_1}(\mathbf{x}) \leq g_{\sigma_2}(\mathbf{x}) \quad \mathbf{x} \in G_{\sigma_1} \quad (45)$$

$$|\sigma| \to \infty \Rightarrow g_\sigma(\mathbf{x}) \to 0 \quad \text{for all } \mathbf{x} \quad (46)$$

Expanding Eq. (45) we have

$$0 < |\sigma_1| < |\sigma_2| \Rightarrow |\sigma_1|g(|\sigma_1|^{-1}\mathbf{x}) \leq |\sigma_2|g(|\sigma_2|^{-1}\mathbf{x}) \quad \mathbf{x} \in G \quad (47)$$

Divide by $|\sigma_2|$ and let $\beta = |\sigma_1|/|\sigma_2|$, then,

$$0 < \beta < 1 \Rightarrow \beta g(\mathbf{x}) \leq g(\beta \mathbf{x}) \quad \mathbf{x} \in G. \quad (48)$$

This relation indicates that a chord from the origin to any point on the structuring function should lie on or below the structuring function (see Fig. 2). This, together with the nonpositivity condition (32), means that the structuring function should be monotone decreasing along any radial direction from the origin. It also means the function is *convex*.*

In fact in almost all practical cases we use functions from the slightly smaller class of continuous convex structuring functions. Convex shapes are

---

*By convention, in mathematical morphology, a *convex* function has any chord entirely on or below the function. Note, this would be called a *concave* function in analysis.

widely used as structuring elements in morphology as they possess many useful properties (Serra, 1982).

In image analysis it is often desirable to ensure isotropic properties in any filtering; this translates directly to the morphological structuring functions being circularly symmetric. A useful family of isotropic structuring functions is given by power functions of the vector norm $\|\mathbf{x}\| = \sqrt{\mathbf{x}^\top \mathbf{A} \mathbf{x}} = \sqrt{x_1^2 + x_2^2 + \cdots + x_n^2}$. So we define the *poweroid* family of scaled structuring functions:

***Definition II.5*** *(Poweroid Structuring Functions):* The scaled poweroid structuring functions are given by,

$$g_\sigma(\mathbf{x}) = -|\sigma|(\|\mathbf{x}\|/|\sigma|)^\alpha \qquad \alpha \geq 0, \qquad \sigma \neq 0$$

Some commonly used structuring functions are presented in Figure 3 including representative members of the 2D poweroid functions

$$g(x, y) = -(\sqrt{x^2 + y^2})^\alpha \tag{49}$$

This family includes cones ($\alpha = 1$), paraboloids ($\alpha = 2$), and cylinders ($\alpha = \infty$).

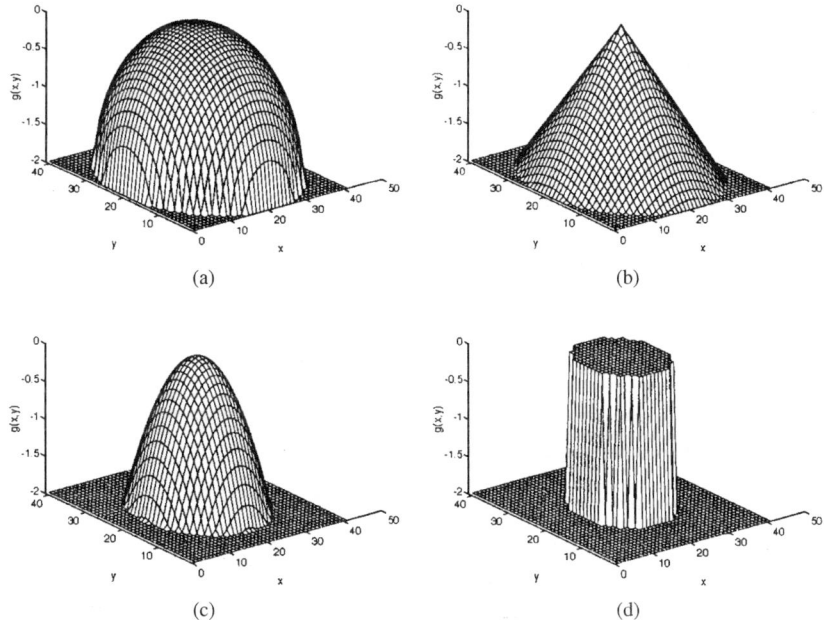

FIGURE 3. Various structuring functions on 2D: (a) sphere; (b) cone ($\alpha = 1$); (c) paraboloid ($\alpha = 2$); (d) cylinder ($\alpha = \infty$). Note: (b) to (d) belong to the circular poweroid family, $g(x, y) = -(x^2 + y^2)^{\alpha/2}$.

To cater for nonisotropic (directional) filtering we can define a *elliptic poweroid* family of scaled structuring functions:

$$g_\sigma(\mathbf{x}) = -|\sigma|(\sqrt{\mathbf{x}^T \mathbf{A} \mathbf{x}}/|\sigma|)^\alpha \qquad \alpha > 0, \qquad \sigma \neq 0 \qquad (50)$$

where $\mathbf{A}$ is a symmetric positive definite matrix. In $R^2$ the contours $g_\sigma(\mathbf{x}) =$ *constant* are ellipses of various orientation and eccentricity. If $\mathbf{A}$ is the unit matrix then Eq. (50) reduces to the isotropic case (Definition II.5). This idea is an extension of van den Boomgaard's (1992) nonisotropic *quadratic structuring function*.

The elliptic poweroid structuring functions have some special properties and are discussed further in Section VI.

## B. Semigroup and General Properties of the Structuring Function

Serra (1982) has presented the following proposition:

> A family $\boldsymbol{B}_\lambda$ ($\lambda \geq 0$) of nonempty compact sets is a one-parameter continuous semigroup (i.e., $\boldsymbol{B}_\lambda \oplus \boldsymbol{B}_\mu = \boldsymbol{B}_{\lambda+\mu}$, $\lambda, \mu \geq 0$) if and only if $\boldsymbol{B}_\lambda = \lambda \boldsymbol{B}$ where $\boldsymbol{B}$ is a convex compact set.

We will present a related result for our scaled structuring function, which helps to explain our choice of the scaling Eq. (38) and why the convex property of structuring functions is necessary. We could obtain the umbras of our functions (Sternberg, 1986) and directly use Serra's result, but it is more informative to work from first principles to show how the definition of convexity comes into play. We introduce the following proposition:

**Proposition II.1** *A family $g_\sigma (\sigma \geq 0)$ of scaled structuring functions given by Eq. (38), which is convex, is a one-parameter continuous semigroup. That is, $g_\sigma \oplus g_\mu = g_{\sigma+\mu}$ for $\sigma, \mu \geq 0$.*

*Proof.* A proof of this proposition can be found in Section X. ∎

As our scaled structuring functions are dependent only on the magnitude of the scale parameter, we have the further result,

$$g_\sigma \oplus g_\mu = g_{|\sigma|+|\mu|}, \qquad \sigma, \mu \in R \qquad (51)$$

The concept of being morphologically open, Eq. (25), places an order on the structuring functions. We have the following proposition:

**Proposition II.2** *If $g_\sigma(\mathbf{x})$ denotes a convex scaled structuring function given by Eq. (38), and if $|\sigma_2| \geq |\sigma_1|$, then the scaled structuring function $g_{\sigma_2}(\mathbf{x})$ is morphologically open with respect to $g_{\sigma_1}(\mathbf{x})$.*

*Proof.* A proof of this proposition can be found in Section X. ∎

## III. MULTISCALE DILATION-EROSION SCALE-SPACE

Using the scaled structuring functions just defined, we can join dilation and erosion at zero-scale to form a single multiscale operation, which unifies the two morphological operations as follows:

**Definition III.1** *(Multiscale Dilation-Erosion):* The multiscale dilation-erosion of the signal $f(\mathbf{x})$ by the scaled structuring function $g_\sigma(\mathbf{x})$ is denoted* by $f \circledast g_\sigma$ and is defined by

$$(f \circledast g_\sigma)(\mathbf{x}) = \begin{cases} (f \oplus g_\sigma)(\mathbf{x}) & \text{if } \sigma > 0; \\ f(\mathbf{x}) & \text{if } \sigma = 0; \\ (f \ominus g_\sigma)(\mathbf{x}) & \text{if } \sigma < 0. \end{cases}$$

That is, for positive scales we perform a dilation, for negative scales an erosion. With this method, scale may be negative; it is $|\sigma|$ which corresponds to the intuitive notion of scale. Unlike linear operators, dilation and erosion and "non-self-dual" (Serra, 1988); therefore, positive and negative scales in scale-space contain differing aspects of the information in a signal. As we shall see, positive scales pertain to local maxima in the signal, whereas negative scales pertain to local minima.

Other authors (e.g., van den Boomgaard, 1992) have considered scaled dilations and erosions separately, and it is well known from mathematical morphology that if $f$ is sufficiently smooth, then both $\lim_{\sigma \to 0}(f \oplus g_\sigma) \to f$, and, $\lim_{\sigma \to 0}(f \ominus g_\sigma) \to f$, but we explicitly combine these operations into a *single* operation. We specifically wish to consider the scale-space fingerprint for positive and negative scales as a whole since the information content of a signal is expanded into this entire region. This approach is consistent with the scale-space philosophy of treating the scale-space image as a whole for the purposes of analysis.

Having defined a suitable operator, we now define the associated scale-space image $F : \boldsymbol{D} \subseteq \mathbf{R}^n \times \mathbf{R} \to \mathbf{R}$ defined by (cf. Eq. (1)):

$$F(\mathbf{x}, \sigma) = (f \circledast g_\sigma)(\mathbf{x}) \tag{52}$$

where the $(n+1)$-dimensional space given by $\boldsymbol{D} \times \mathbf{R}$ is known as the *multiscale dilation-erosion scale-space*.

Since we are using the operations of mathematical morphology to smooth a signal, the well-known geometric visualizations of dilation and erosion are intuitively helpful: For the moment, take the scaled structuring function to be a ball with the radius as a scale parameter with a positive radius corresponding to rolling the ball along the top of the "surface" of the signal, and a negative

---

*The symbol $\circledast$ has previously been used by Serra (1982) to refer to the *hit or miss transform*, which does not appear in this article.

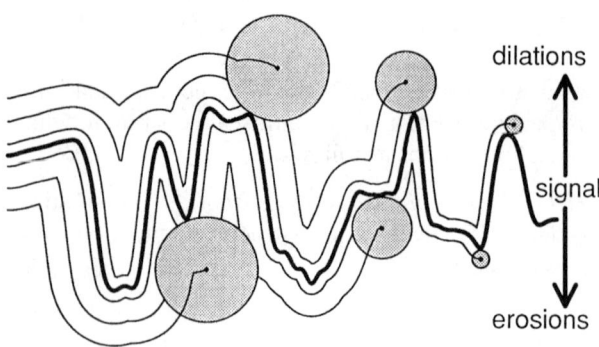

FIGURE 4. Smoothing of a 1D signal by multiscale morphological dilation-erosion.

radius to rolling the ball along the underneath. The smoothed signal can be visualized as the surface traced out by the center of the ball when it is traced over the top (dilation) or underneath (erosion) of the surface of the signal. We illustrate this operation for a 1D signal in Fig. 4. The multiscale dilation-erosion smoothing of the "Lena" image is shown in Fig. 5 for several scales, positive and negative.

Intuitively, this new surface is smoother (in the sense of having smoother and less hills) than the original signal, and furthermore the larger the radius the smoother the filtered surface becomes. In the limits, as the radius approaches zero the original image is recovered, and as the radius approaches infinity the output becomes flat. It should be apparent that if the ball touches the top of a hill (local maxima), then a hill will appear on the output at exactly that point. If, however, the radius is such that the ball is prevented form touching that hill by nearby hills, then no hill will appear at that point on the output, and more importantly that hill cannot reappear for any increased value of radius, $r$. Thus, it would seem that the number of local maxima may be a monotone decreasing function of $r$.

This turns out to be so—but it is not quite so easy to prove! The scale-space properties of the multiscale dilation-erosion will be reviewed formally in the form of propositions building to a theorem and corollaries expressing the scale-space monotone principle. The proofs of Propositions III.1 to III.5 will not be presented here because they can be found in Jackway and Deriche (1996).

In outlining the results of this section, we will often make use of the following duality between morphological dilation and erosion, Eq. (9). Therefore, many results on $f \oplus g_\sigma$ for $\sigma > 0$, which correspond to dilation, can be immediately applied to $\sigma < 0$, corresponding to erosion. In practice most structuring functions are symmetrical about the origin ($g(\mathbf{x}) = g(-\mathbf{x})$) so that $\check{g}(\mathbf{x}) = g(\mathbf{x})$.

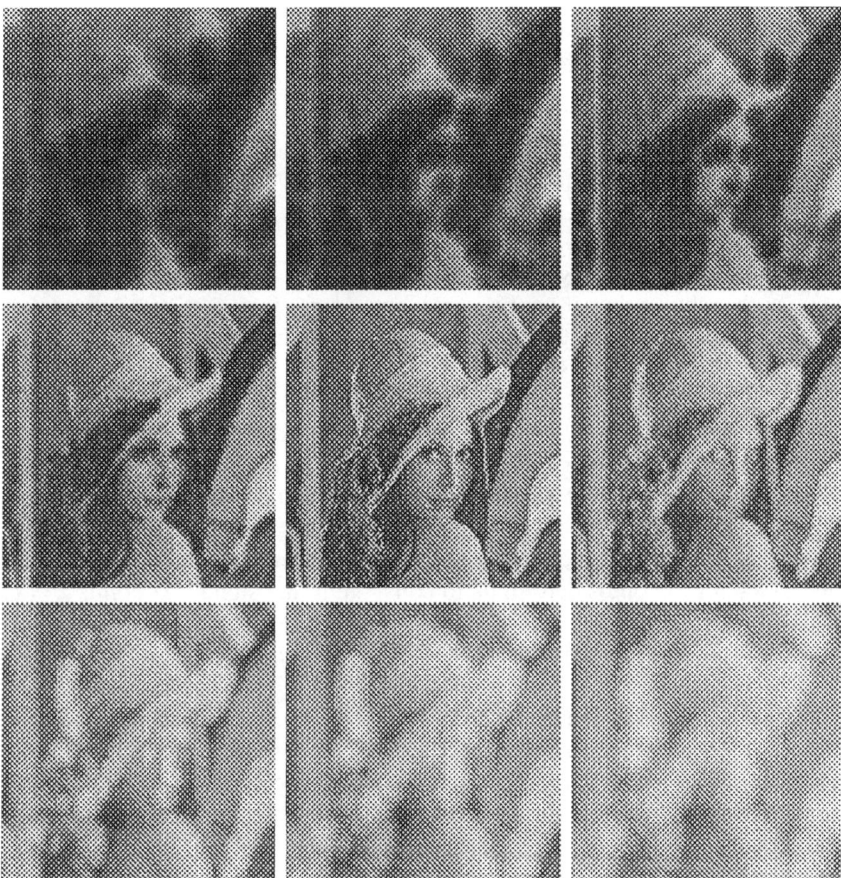

FIGURE 5. The "Lena" image smoothed by multiscale dilation-erosion. From left to right, top to bottom, the scales are: −1.6, −0.9, −0.4, −0.1, 0.0, 0.1, 0.4, 0.9, 1.6. A circular paraboloid structuring function was used.

In this case the erosion of a signal by a structuring function may be obtained by negating the signal, performing a dilation with that same structuring function, and negating the result.

### A. Continuity and Order Properties of the Scale-Space Image

The definition of the multiscale dilation-erosion (Definition III.1) consists of three parts corresponding to positive, zero, or negative values of the scale parameter. It is therefore natural to consider the behavior of the scale-space image $F(\mathbf{x}, \sigma)$ across this seam at $\sigma = 0$ in scale-space.

Because the structuring function is zero at the origin [see Eq. (33)], the dilation is *extensive* and the erosion *antiextensive* [Eqs. (20) and (21)]; therefore, we have the result

$$(f \circledast g_\sigma)(\mathbf{x})_{\sigma<0} \leq f(\mathbf{x}) \leq (f \circledast g_\sigma)(\mathbf{x})_{\sigma>0} \quad \forall \mathbf{x} \in D \quad (53)$$

This order property looks promising and indeed a further continuity property applies that shows that at continuous points of $f(\mathbf{x})$, $(f \circledast g_\sigma)(\mathbf{x})$ approaches $f(\mathbf{x})$ as the scale parameter approaches zero from either above or below. We have the following proposition.

**Proposition III.1** *If the bounded signal $f(\mathbf{x})$ is continuous at some $\mathbf{x} \in D$, then the scale-space image $F(\mathbf{x}, \sigma)$ is continuous with respect to $\sigma$ at $\sigma = 0$. That is, at points $\mathbf{x}$ where $f(\mathbf{x})$ is continuous, $F(\mathbf{x}, \sigma) \to f(\mathbf{x})$ as $\sigma \to 0$.*

In fact a slightly stronger result holds; full continuity of the signal is not necessary for the one-sided limits to converge to the signal. At points $\mathbf{x} = \mathbf{x}_u$ where $f(\mathbf{x}_u)$ is *upper semicontinuous* (u.s.c.) $F(\mathbf{x}_u, \sigma) \to f(\mathbf{x}_u)$ as $\sigma \to 0^+$, and at points $\mathbf{x} = \mathbf{x}_l$ where $f(\mathbf{x}_l)$ is *lower semicontinuous* (l.s.c) $F(\mathbf{x}_l, \sigma) \to f(\mathbf{x}_l)$ as $\sigma \to 0^-$.

We recall that a function $f(\mathbf{x})$ is said to be upper semicontinuous at $\mathbf{x}_u$ if the *nondeleted limit superior* $\operatorname{Lim\,sup}_{\mathbf{x} \to \mathbf{x}_u} f(\mathbf{x}) = f(\mathbf{x}_u)$, see, for example, Bartle (1964). Further $f$ is lower semicontinuous at $\mathbf{c}$ iff $-f(\mathbf{c})$ is u.s.c. Upper semicontinuous functions are often used to model pictures because their threshold sets, $\mathcal{F}(t) = \{\mathbf{x} : f(\mathbf{x}) \geq t\}$, are *closed* sets (Serra, 1982).

If the structuring function is sufficiently smooth, this property transfers to the scale-space image and we have the following proposition:

**Proposition III.2** *If the structuring function $g(\mathbf{t})$ is a continuous function on $R^n$, then the scale-space image of the bounded signal $f(\mathbf{x})$ is continuous on $R^n \times R$ for all $\mathbf{x} \in D$, $\sigma \neq 0$.*

This applies for any signal $f(\mathbf{x})$ (as long as it is bounded) and shows that the scale-space image is much better behaved than the signal itself. This is to be expected since the signal has been *smoothed* to give the scale-space image.

The following pointwise order properties of the scale-space image follow directly from the extensivity, and increasing properties of the morphological dilation and erosion [Eqs. (19)–(21)], and the order properties of the scaled structuring function [Eqs. (44)–(46)].

**Proposition III.3** *The scale-space image $F(\mathbf{x}, \sigma) = (f \otimes g_\sigma)(\mathbf{x})$ possesses the following properties:*

$$F(\mathbf{x}, 0) = f(\mathbf{x}) \quad \text{for all } \mathbf{x} \in D$$

$$F(\mathbf{x}, \infty) = \bigvee_{\mathbf{t} \in D} \{f(\mathbf{t})\} \quad \text{for all } \mathbf{x} \in D$$

$$F(\mathbf{x}, -\infty) = \bigwedge_{\mathbf{t} \in D} \{f(\mathbf{t})\} \quad \text{for all } \mathbf{x} \in D$$

$$\sigma_q < \sigma_p \Rightarrow F(\mathbf{x}, \sigma_q) \leq F(\mathbf{x}, \sigma_p) \quad \text{for all } \sigma_p, \sigma_q \in \mathrm{R}; \quad \mathbf{x} \in D$$

### B. Signal Extrema in Scale-Space

Propositions III.1 to III.3 show that the scale-space image has good continuity and order properties but we have yet to show the essential scale-space monotone property. The major result of this section is a theorem, which shows in a precise way how $f \otimes g_\sigma$ becomes smoother with increasing $|\sigma|$. Furthermore, we show that the monotone property holds for local extrema of the signal so this is the signal feature appropriate to the multiscale dilation-erosion scale space. Prior to presenting this theorem some necessary partial results are obtained.

The first result relates the position and amplitude of a local maximum (or minimum) in the filtered signal to that in the original signal.

**Proposition III.4** *Let the structuring function have a single maximum at the origin; that is, $g(\mathbf{x})$ a local maximum implies $\mathbf{x} = \mathbf{0}$, then:*

(a) *If $\sigma > 0$ and $(f \otimes g_\sigma)(\mathbf{x}_{\max})$ is a local maximum, then, $f(\mathbf{x}_{\max})$ is a local maximum of $f(\mathbf{x})$ and $(f \otimes g_\sigma)(\mathbf{x}_{\max}) = f(\mathbf{x}_{\max})$;*
(b) *If $\sigma < 0$ and $(f \otimes g_\sigma)(\mathbf{x}_{\min})$ is a local minimum, then, $f(\mathbf{x}_{\min})$ is a local minimum of $f(\mathbf{x})$ and $(f \otimes g_\sigma)(\mathbf{x}_{\min}) = f(\mathbf{x}_{\min})$.*

We are now able to relate a signal feature at nonzero-scale to the original signal (zero-scale). However, to obtain a monotone result we need the next proposition.

**Proposition III.5** *Let the structuring function have a single local maximum at the origin; that is, $g(\mathbf{x})$ is a local maximum implies $\mathbf{x} = \mathbf{0}$, then:*

(a) *If $\sigma_0 > \sigma > 0$ and $(f \otimes g_{\sigma_0})(\mathbf{x}_{\max})$ is a local maximum, then, $(f \otimes g_\sigma)(\mathbf{x}_{\max})$ is a local maximum and, $(f \otimes g_\sigma)(\mathbf{x}_{\max}) = (f \otimes g_{\sigma_0})(\mathbf{x}_{\max})$;*
(b) *If $\sigma_0 < \sigma < 0$ and $(f \otimes g_{\sigma_0})(\mathbf{x}_{\min})$ is a local minimum, then, $(f \otimes g_\sigma)(\mathbf{x}_{\min})$ is a local minimum and, $(f \otimes g_\sigma)(\mathbf{x}_{\min}) = (f \otimes g_{\sigma_0})(\mathbf{x}_{\min})$.*

These propositions provide very important scale-space results because they enable coarse-to-fine tracking in the scale-space image. If a signal feature (extrema) appears at some scale $\sigma_0$, it also appears at zero-scale and all scales in between. Stated as a monotone property, we can state that the number of features may not decrease as scale approaches zero. This property is now encapsulated in a theorem.

**Theorem III.1** (Scale-Space Monotone Property for Extrema): *Let $f : D \subseteq \mathbb{R}^n \to \mathbb{R}$ denote a bounded function, $g_\sigma : G \subseteq \mathbb{R}^n \to \mathbb{R}$ a scaled structuring function satisfying the conditions of Proposition III.5, and the point sets, $\boldsymbol{E}_{\max}(f) = \{\mathbf{x} : f \text{ is a local maximum}\}$, and, $\boldsymbol{E}_{\min}(f) = \{\mathbf{x} : f \text{ is a local minimum}\}$ denote the local extrema of $f$. Then, for any scales $\sigma_1 < \sigma_2 < 0 < \sigma_3 < \sigma_4$,*

(a) $\boldsymbol{E}_{\min}(f \circledast g_{\sigma_1}) \subseteq \boldsymbol{E}_{\min}(f \circledast g_{\sigma_2}) \subseteq \boldsymbol{E}_{\min}(f)$;

*and*

(b) $\boldsymbol{E}_{\max}(f \circledast g_{\sigma_4}) \subseteq \boldsymbol{E}_{\max}(f \circledast g_{\sigma_3}) \subseteq \boldsymbol{E}_{\max}(f)$.

*Proof.* Suppose the theorem is false and $\boldsymbol{E}_{\max}(f \circledast g_{\sigma_4}) \not\subseteq \boldsymbol{E}_{\max}(f \circledast g_{\sigma_3})$ for some $0 < \sigma_3 < \sigma_4$; then there exists some $\mathbf{x}_{\max} \in D$ such that $F(\mathbf{x}_{\max}, \sigma_4)$ is a local maximum but $F(\mathbf{x}_{\max}, \sigma_3)$ is not, which contradicts Proposition III.4(a). The case for $\boldsymbol{E}_{\min}$ is proved similarly using Proposition III.4(b). ∎

This theorem is actually stronger than required since it governs the positions of the extrema as well as their number. To obtain a monotone property of the form of Eq. (4), we need some functional $\#: \mathbb{R}^n \to \mathbb{R}$ such that

$$E_1 \subseteq E_2 \Rightarrow \#(E_1) \leq \#(E_2) \qquad \text{for all } E_1, E_2 \subset \mathbb{R}^n \qquad (54)$$

For the practical case where $E \subset \mathbb{Z}^n$ we simply choose $\#[E] =$ *the number of points in $E$*. We have the following corollary to Theorem III.1.

**Corollary III.1.1** (Scale-Space Monotone Property for the Number of Local Extrema): *For $\#: \mathbb{R}^n \to \mathbb{R}$, such that $E_1 \subseteq E_2 \subset \mathbb{R}^n \Rightarrow \#(E_1) \leq \#(E_2)$ then, for any $\sigma_1 < \sigma_2 < 0 < \sigma_3 < \sigma_4$,*

(a) $\#[\boldsymbol{E}_{\min}(f \circledast g_{\sigma_1})] \leq \#[\boldsymbol{E}_{\min}(f \circledast g_{\sigma_2})] \leq \#[\boldsymbol{E}_{\min}(f)]$;

*and*

(b) $\#[\boldsymbol{E}_{\max}(f \circledast g_{\sigma_4})] \leq \#[\boldsymbol{E}_{\max}(f \circledast g_{\sigma_3})] \leq \#[\boldsymbol{E}_{\max}(f)]$.

*Proof.* The proof follows from the direct substitution of Eq. (54) in Theorem III.1. ∎

We can further extend Theorem III.1 from local to regional extrema, which makes it more useful when dealing with operations that rely on the number of regional extrema, such as the watershed transform (to be discussed later). First, we recall the definitions of the various types of extrema:

1. $f$ is said to have a *strict local maximum* at $\mathbf{x} = \mathbf{x}_0$ if there exists a neighborhood $N(\mathbf{x}_0)$ such that $f(\mathbf{x}) < f(\mathbf{x}_0)$ for all $\mathbf{x} \in N(\mathbf{x}_0)$.
2. $f$ is said to have a local maximum at $\mathbf{x} = \mathbf{x}_0$ if there exists a neighborhood $N(\mathbf{x}_0)$ such that $f(\mathbf{x}) \leq f(\mathbf{x}_0)$ for all $\mathbf{x} \in N(\mathbf{x}_0)$.
3. $f$ is said to have a regional maximum of value $h$ on the connected component $M$ if there exists neighborhood of $M$, $N(M)$, such that $f(\mathbf{x}) = h$ for all $\mathbf{x} \in M$, and $f(\mathbf{x}) < h$ for all $\mathbf{x} \in N(M)$.
4. The corresponding definitions for minima follow directly with the inequalities reversed.

Because we will be counting regional extrema, we will assume that $f$ has a finite number of connected components in all upper and lower thresholds.

We can now present the following corollary to Theorem III.1.

**Corollary III.1.2** (Scale-Space Monotone Property for the Number of Regional Extrema): *Let $C[R_{\max}(f)]$ and $C[R_{\min}(f)]$ denote the number of connected components in the point sets of the regional extrema of a signal $f$, then for any scales, $\sigma_1 < \sigma_2 < 0 < \sigma_3 < \sigma_4$,*

(a) $C[R_{\min}(f \otimes g_{\sigma_1})] \leq C[R_{\min}(f \otimes g_{\sigma_2})] \leq C[R_{\min}(f)];$

*and*

(b) $C[R_{\max}(f \otimes g_{\sigma_4})] \leq C[R_{\max}(f \otimes g_{\sigma_3})] \leq C[R_{\max}(f)].$

*Proof.* A proof of this corollary is given in Jackway (1996). ∎

Corollaries III.1.1 and III.1.2 are monotone properties of the form of Eq. (4) and we can therefore claim the production of a scale-space.

This scale-space allows all input signals in any dimensionality as long as they are bounded (infinite amplitudes upset the morphological operations!). The signal is *expanded* into a scale-space image by smoothing with the multiscale morphological dilation-erosion. The features in this scale-space are the signal local extrema (maxima for positive scales, minima for negative scales). We have given a meaning to the concept of *negative scale* through the use of the morphological erosion.

We have shown that the number of features may not increase with increasing scale but we have not shown that they decrease! However, if a signal contains information at different scales, this will generally be reflected as a decrease in the number of features with increasing scale magnitude. If the signal has a single unique global maximum (minimum), then for sufficiently large positive (negative) scale, there remains only a single feature in the scale-space image.

## IV. Multiscale Closing-Opening Scale-Space

We have developed a scale-space theory based on the morphological dilation-erosion, but to some readers it may seem strange that we did not use the opening or closing operations.

First, the morphological dilation and erosion are not true *morphological filters* (as they are are not idempotent) like the opening and closing (Serra, 1988). Second, Chen and Yan (1989) have published a well-known paper titled "A Multiscaling Approach Based on Morphological Filtering," in which they demonstrate a scale-space causality property for the zero-crossings of curvature on the boundaries of objects in binary images when opend by multiscale disks. This work has since been generalized by Jang and Chin (1991) to show that *convexity* and *compactness* of the structuring element are the necessary and sufficient conditions for the monotonic property of the multiscale morphological opening filter. Their theorem is:

**Theorem IV.1** (Monotonic Property of the Multiscale Opening: Jang and Chin, 1991): *Suppose $X$ is a compact set in $R^2$. $Z[\partial X]$ denotes the finite number of zero-crossings of curvature function along the contour $\partial X$, and $CN[X]$ is the number of connected components of $X$. For any $r > 0$,*

$$X \circ B(r) \neq \emptyset \quad \text{and} \quad CN[X] = CN[X \circ B(r)] = 1$$

*we have*

$$Z(\partial[X \circ B(r)]) \leq Z[\partial X]$$

*and $Z(\partial[X \circ B(r)])$ is monotonic decreasing as r increasing if and only if $B(r)$ is a compact convex set.*

A review and comparison of Gaussian and morphological opening scale-spaces for shape analysis have recently appeared in the literature (Jang and Chin, 1992). Interestingly this review stresses the *signal feature–smoothing filter* aspects of scale-space and the importance of the scale-space causality or monotone property as we do here.

Note some technical problems have been found in the approaches of both Chen and Yan (1989) and Jang and Chin (1991), which limit the generality of their results (Nacken, 1994; Jackway, 1995a). However, the paper by Chen and Yan (1989) is noteworthy in being the first attempt to use nonlinear operations to create a scale-space.

An advantage of the preceding approach is that in using zero-crossings of boundary curvature as the feature, there is an obvious close connection with the Gaussian approach that uses zero-crossings of the second derivative.

When applied to functions, zero-crossings of curvature are equivalent to zero-crossings of the second derivative, because

$$\kappa_f(x) = \frac{f''(x)}{(1+f'^2(x))^{3/2}} > 0 \Rightarrow f''(x) > 0; \tag{55}$$

$$\kappa_f(x) < 0 \Rightarrow f''(x) < 0. \tag{56}$$

In unpublished work we have extended Theorem III.2 to functions and obtained a scale-space monotone property for zero-crossings of the second derivative of 1D functions smoothed with a multiscale closing-opening operation (Jackway, 1995a). In common with all the other zero-crossing approaches, including Gaussian scale-space, the big disadvantage with using zero-crossings as the scale-space feature is that there is no obvious extension to functions on higher dimensional spaces (e.g., images).

Our previous success with using signal extrema as the feature with morphological dilation-erosion scale-space suggests that using extrema rather than zero-crossings with the morphological opening may be a way to achieve a scale-space for higher dimensions. This is indeed so, and we will now present the development of a closing-opening scale-space and then we will examine its relation to the dilation-erosion scale-space.

### A. Properties of the Multiscale Closing-Opening

Throughout this section we will work with the multiscale closing-opening, which can be defined in terms of the closing and openings with scaled structuring functions:

**Definition IV.1** (*Multiscale Closing-Opening*): The multiscale closing-opening of the signal $f(\mathbf{x})$ by the scaled structuring function $g_\sigma(\mathbf{x})$ is denoted* by $f \odot g_\sigma$, and is defined by

$$(f \odot g_\sigma)(\mathbf{x}) = \begin{cases} (f \bullet g_\sigma)(\mathbf{x}) & \text{if } \sigma > 0; \\ f(\mathbf{x}) & \text{if } \sigma = 0; \\ (f \circ g_\sigma)(\mathbf{x}) & \text{if } \sigma < 0. \end{cases}$$

Note: A similar multiscale operation has also been defined by van den Boomgaard (1992). The multiscale dilation-erosion smoothing of the "Lena" image is shown in Figure 6 for several scales, positive and negative.

---

*The symbol $\odot$ has previously been used by Serra (1982) to refer to the *thickening*, which does not appear in this article.

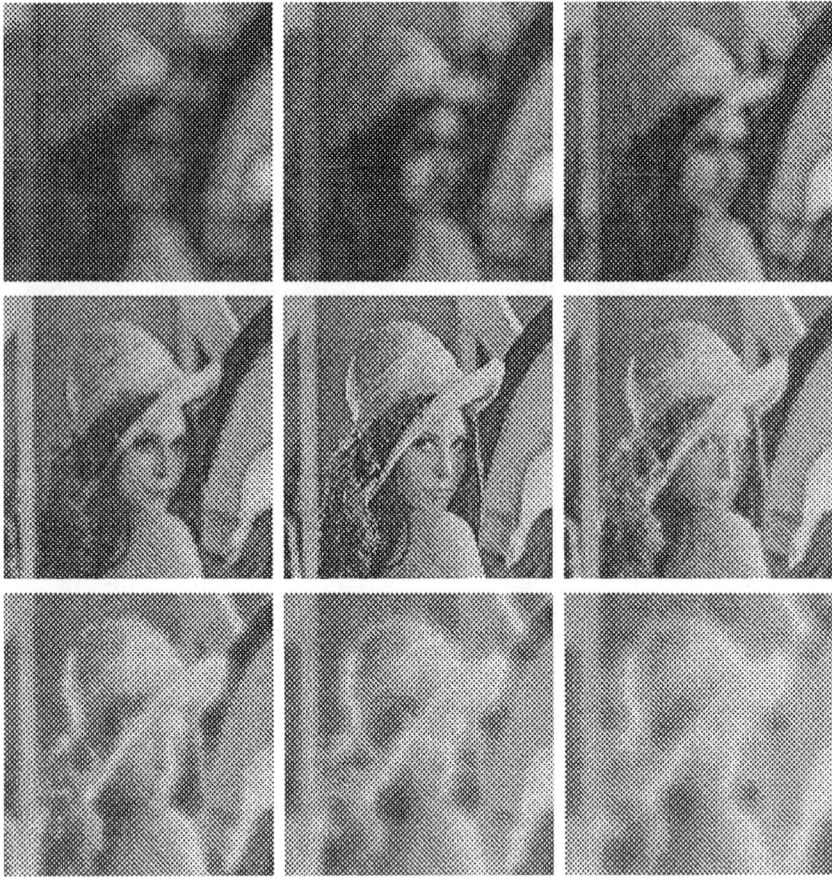

FIGURE 6. The "Lena" image smoothed by multiscale closing-opening. From left to right, top to bottom, the scales are: $-1.6, -0.9, -0.4, -0.1, 0.0, 0.1, 0.4, 0.9, 1.6$. A circular paraboloid structuring function was used.

In this section we will often obtain results for scale $\sigma < 0$ because this corresponds to the opening operation used in the literature. In these cases we then appeal to the duality principle of the opening and closing (Eq. (11)) to extend the results to the morphological closing and hence to the combined closing-opening operation.

Because both the closing and opening are idempotent, Eqs. (12) and (13), the closing-opening is as well:

$$(f \odot g_\sigma) \odot g_\sigma = f \odot g_\sigma \qquad (57)$$

Because the closing is *extensive* and the opening *antiextensive* [Eqs. (22) and (23)], we have the following result:

$$(f \odot g_\sigma)(\mathbf{x})_{\sigma<0} \le f(\mathbf{x}) \le (f \odot g_\sigma)(\mathbf{x})_{\sigma>0} \qquad \forall \mathbf{x} \in \mathbf{D} \tag{58}$$

and we have the following order properties with respect to scale.

**Proposition IV.1** *If $\sigma_1 < \sigma_2 < 0 < \sigma_3 < \sigma_4$, then*

$$(f \odot g_{\sigma_1}) \le (f \odot g_{\sigma_2}) \le f \le (f \odot g_{\sigma_3}) \le (f \odot g_{\sigma_4}). \tag{59}$$

*Proof.* A proof of this proposition is given in Section IX. ∎

The multiscale closing-opening filter also satisfies the following scale-related conditions (cf. Chen and Yan, 1989):

1. It is scale invariant, i.e.,

$$f(t) \odot g_\sigma(t) = \sigma\left(\frac{1}{\sigma} f(\sigma t) \odot g_1(t)\right) \tag{60}$$

2. The filter recovers the input signal for zero-scale (by definition!),

$$(f \odot g_0)(t) = f(t) \tag{61}$$

3. As scale approaches positive (negative) infinity, the output approaches the global maximum (minimum) of the input signal, i.e.,

$$\lim_{\sigma \to \infty} \{(f \odot g_\sigma)(t)\} = \bigvee_{t \in D} \{f(t)\} \tag{62}$$

$$\lim_{\sigma \to -\infty} \{(f \odot g_\sigma)(t)\} = \bigwedge_{t \in D} \{f(t)\} \tag{63}$$

Therefore, the multiscale closing-opening appears suited to the formation of a scale-space, similar to that formed by the multiscale dilation-erosion. This is, we should consider the *multiscale closing-opening scale-space* $F: \mathbf{D} \subseteq \mathbf{R}^n \times \mathbf{R} \to \mathbf{R}$ defined by

$$F(\mathbf{x}, \sigma) = (f \odot g_\sigma)(\mathbf{x}) \tag{64}$$

Now we need to obtain monotonic properties for signal features within this scale-space.

The results of Chen and Yan (1989) and Jang and Chin (1991) depend on partitioning the result of the opening operation into arcs of the original set and arcs of the translated structuring element; we will extend this partitioning idea to work with functions on multidimensional spaces and, therefore, with multidimensional arcs or *patches* of the structuring function. The first step is a basic morphological result, outlined in Haralick *et al.* (1987), which provides a geometrical interpretation to the opening and closing:

To obtain the opening of $f$ by a paraboloid structuring element, for example, take the paraboloid, apex up, and slide it under all the surface of $f$ pushing it hard up against the surface. The apex of the paraboloid may not be able to touch all points of $f$. For example, if $f$ has a spike narrower than the paraboloid, the top of the apex may only reach as far as the mouth of the spike. The opening is the surface of the highest points reached by any part of the paraboloid as it slides under all the surface of $f$. (...) To close $f$ with a paraboloid structuring element, we take the reflection of the paraboloid in the sense of (Eq. (10)), turn it upside down (apex down), and slide it all over the top of the surface of $f$. The closing is the surface of all the lowest points reached by the sliding paraboloid.

In terms of the opening we have the following proposition.

**Proposition IV.2**

$$f \circ g = T\left[\bigcup_{\{z:U[g]\subseteq U[f]\}} U[g]_z\right]$$

where $U[g]$ is the umbra of $g$, i.e., $U[g] = \{(\mathbf{x}, y) : y \leq g(\mathbf{x})\}$. $T[U[g]] : \mathbb{R}^n \to \mathbb{R}$ is the "top surface" of the umbra; i.e., $T[U[g]](\mathbf{x}) = \max\{y : (\mathbf{x}, y) \in U[g]\}$. $U[g]_\mathbf{z}$ indicates the translate of $U[g]$ by $\mathbf{z} \in \mathbb{R}^n \times \mathbb{R}$, $U[g]_\mathbf{z} = \{\mathbf{u} + \mathbf{z} : \mathbf{u} \in U[g]\}$.

*Proof.* This result is proved in Proposition 71 of Haralick *et al.* (1987). ∎

From this geometrical interpretation of the opening (or closing), we see that the output signal can be partitioned; i.e., with $f : D \in \mathbb{R}^n \to \mathbb{R}$,

$$(f \circ g_\sigma)(\mathbf{x}) = \begin{cases} f(\mathbf{x}) & \text{if } \mathbf{x} = S'(\sigma); \\ s(\mathbf{x}) & \text{if } \mathbf{x} = S''(\sigma). \end{cases} \quad (65)$$

with

$$S'(\sigma) \cup S''(\sigma) = D \quad (66)$$

$$S'(\sigma) \cap S''(\sigma) = \emptyset \quad (67)$$

$$s(\mathbf{x}) < f(\mathbf{x}) \quad \mathbf{x} = S''(\sigma) \quad (68)$$

$$s(\mathbf{x}) = \bigcup_{i \in I} \text{PATCH}[(g_\sigma)_{z_i}] \quad (69)$$

$$\text{PATCH}[(g_\sigma)_{z_i}] \cap \text{PATCH}[(g_\sigma)_{z_j}] = \emptyset \quad \text{for } i \neq j \quad (70)$$

where PATCH $[(g_\sigma)_{z_i}]$ is a patch on the structuring function $g_\sigma$, which has the origin translated to $z_i \in U[f]$. Note that $I$ is a *finite* index family (Jang and Chin, 1991). In words, the opening of a signal consists of patches of the original signal combined with patches of translated structuring functions. By duality,

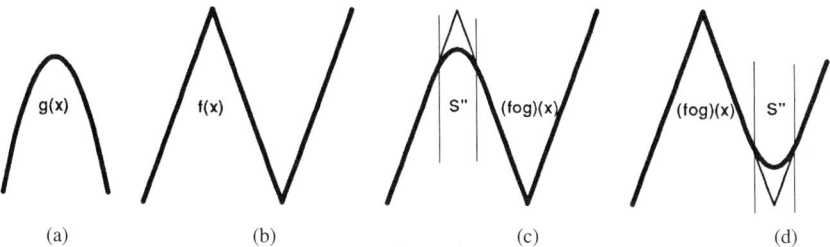

FIGURE 7. Geometrical interpretation of the opening and closing with partitioning. (a) parabolic structuring function $g(x)$; (b) signal $f(x)$; (c) opening $f \circ g$; (d) closing $f \bullet g$.

the closing can be partitioned in a similar way. This geometrical interpretation of the opening and closing (on a 1D function) is illustrated in Figure 7.

Now we examine how this partitioning varies with scale and we obtain the following proposition:

**Proposition IV.3**  *Given that $\sigma_1 < \sigma_2 < 0$, then*

$$f \circ g_{\sigma_1} = f(\mathbf{x})|_{\mathbf{x} \in S'(\sigma_1)} \cup \bigcup_{i \in I} \text{PATCH}[(g_{\sigma_1})_{z_i}] \tag{71}$$

$$f \circ g_{\sigma_2} = f(\mathbf{x})|_{\mathbf{x} \in S'(\sigma_2)} \cup \bigcup_{i \in I} \text{PATCH}[(g_{\sigma_2})_{z_i}] \tag{72}$$

*and*

$$S'(\sigma_1) \subseteq S'(\sigma_2) \subseteq D \tag{73}$$

*Proof.* A proof of this proposition is given in Section X. ∎

This proposition states that with increasing scale, the opening replaces more and more of the original function with patches from the structuring element. Likewise, from the duality of opening and closing, we see that a corresponding result applies for the closing, with the function being replaced with patches from the inverted structuring function.

## B. Monotone Theorem for the Multiscale Closing-Opening

The patches of the upright (inverted) structuring function possess several smoothness properties, namely, that they are convex (concave), cannot contain a local minimum (local maximum), and contain at most one local maximum (local minimum). Therefore, we have the following scale-space monotone theorem.

**Theorem IV.2** (Monotone Property of Local Extrema of the Multiscale Closing-Opening): *Let $f: D \subseteq R^n \to R$ denote a bounded function, $g_\sigma : G \subseteq R^n \to R$ a scaled structuring function satisfying the conditions of Proposition 10, and the point sets, $\mathbf{E}_{\max}(f) = \{\mathbf{x} : f \text{ is a local maximum}\}$, and, $\mathbf{E}_{\min}(f) = \{\mathbf{x} : f \text{ is a local minimum}\}$, denote the local extrema of $f$. Then, for any scales $\sigma_1 < \sigma_2 < 0 < \sigma_3 < \sigma_4$,*

(a) $\mathbf{E}_{\min}(f \odot g_{\sigma_1}) \subseteq \mathbf{E}_{\min}(f \odot g_{\sigma_2}) \subseteq \mathbf{E}_{\min}(f)$

*and,*

(b) $\mathbf{E}_{\max}(f \odot g_{\sigma_4}) \subseteq \mathbf{E}_{\max}(f \odot g_{\sigma_3}) \subseteq \mathbf{E}_{\max}(f)$

*Proof.* The patches for the opening contain no local minima,

$$\mathbf{E}_{\min}\left(\bigcup_{i \in I} \text{PATCH}[(g_{\sigma_1})_{z_i}]\right) = \emptyset \tag{74}$$

$$\mathbf{E}_{\min}\left(\bigcup_{i \in I} \text{PATCH}[(g_{\sigma_2})_{z_i}]\right) = \emptyset \tag{75}$$

The occurrence of a local minimum is a *local* property of a function, therefore, the replacement of part of $f$ with a patch from $g_\sigma$ cannot affect the existence of local extrema outside the patch except possibly at the patch boundary. However, from Eq. (68), the patch is everywhere less than the function it replaces, and no new minima can be created at the patch boundaries; therefore,

$$\mathbf{E}_{\min}(f \odot g_{\sigma_1}) = \mathbf{E}_{\min}\left(f(\mathbf{x})|_{\mathbf{x} \in S'(\sigma_1)}\right) \tag{76}$$

$$\mathbf{E}_{\min}(f \odot g_{\sigma_2}) = \mathbf{E}_{\min}\left(f(\mathbf{x})|_{\mathbf{x} \in S'(\sigma_2)}\right) \tag{77}$$

But, by relation (73), $S'(\sigma_1) \subseteq S'(\sigma_2) \subseteq D$, so

$$\mathbf{E}_{\min}\left(f(\mathbf{x})|_{\mathbf{x} \in S'(\sigma_1)}\right) \subseteq \mathbf{E}_{\min}\left(f(\mathbf{x})|_{\mathbf{x} \in S'(\sigma_2)}\right) \subseteq \mathbf{E}_{\min}(f(\mathbf{x})|_{\mathbf{x} \in D}) \tag{78}$$

and, therefore,

$$\mathbf{E}_{\min}(f \odot g_{\sigma_1}) \subseteq \mathbf{E}_{\min}(f \odot g_{\sigma_2}) \subseteq \mathbf{E}_{\min}(f) \tag{79}$$

And, as usual, by duality,

$$\mathbf{E}_{\max}(f \odot g_{\sigma_4}) \subseteq \mathbf{E}_{\max}(f \odot g_{\sigma_3}) \subseteq \mathbf{E}_{\max}(f) \tag{80}$$

This theorem is the direct analog of Theorem III.1. Corollaries III.1.1 and III.1.2 can also be applied to Theorem IV.2 and we will state them for completeness without proof. ∎

**Corollary IV.2.1** (Scale-Space Monotone Property for the Number of Local Extrema): *For* $\#\colon \mathbf{R}^n \to \mathbf{R}$, *such that* $\mathbf{E}_1 \subseteq \mathbf{E}_2 \subset \mathbf{R}^n \Rightarrow \#(\mathbf{E}_1) \leq \#(\mathbf{E}_2)$ *then, for any* $\sigma_1 < \sigma_2 < 0 < \sigma_3 < \sigma_4$,

(a) $\#[\mathbf{E}_{\min}(f \odot g_{\sigma_1})] \leq \#[\mathbf{E}_{\min}(f \odot g_{\sigma_2})] \leq \#[\mathbf{E}_{\min}(f)]$

and

(b) $\#[\mathbf{E}_{\max}(f \odot g_{\sigma_4})] \leq \#[\mathbf{E}_{\max}(f \odot g_{\sigma_3})] \leq \#[\mathbf{E}_{\max}(f)]$

**Corollary IV.2.2** (Scale-Space Monotone Property for the Number of Regional Extrema): *Let* $\mathbf{C}[\mathbf{R}_{\max}(f)]$ *and* $\mathbf{C}[\mathbf{R}_{\min}(f)]$ *denote the number of connected components in the point sets of the regional extrema of a signal f, then for any scales,* $\sigma_1 < \sigma_2 < 0 < \sigma_3 < \sigma_4$,

(a) $\mathbf{C}[\mathbf{R}_{\min}(f \odot g_{\sigma_1})] \leq \mathbf{C}[\mathbf{R}_{\min}(f \odot g_{\sigma_2})] \leq \mathbf{C}[\mathbf{R}_{\min}(f)]$

and

(b) $\mathbf{C}[\mathbf{R}_{\max}(f \odot g_{\sigma_4})] \leq \mathbf{C}[\mathbf{R}_{\max}(f \odot g_{\sigma_3})] \leq \mathbf{C}[\mathbf{R}_{\max}(f)]$

Having presented another scale-space, the question now is how are the two scale spaces, $F_{\circledast}(\mathbf{x}, \sigma) = (f \circledast g_\sigma)(\mathbf{x})$ and $F_{\odot}(\mathbf{x}, \sigma) = (f \odot g_\sigma)(\mathbf{x})$ related? We can answer this in the next section using fingerprints.

## V. Fingerprints in Morphological Scale-Space

As mentioned in the introduction, the central idea of scale-space filtering is that we trace the paths of signal features through the scale-space. These paths are termed the *scale-space fingerprint* and we saw in Figure 1 a typical Gaussian scale-space fingerprint. We have already defined the sets of feature positions, $\mathbf{E}_{\max}$ and $\mathbf{E}_{\min}$ for use in Theorems III.1 and IV.2, now we start with some further definitions and notations concerning fingerprints.

**Definition V.1** (*Multiscale Dilation-Erosion Scale-Space Fingerprint*): The *multiscale dilation-erosion scale-space fingerprint* is a plot, versus scale, of the scale-dependent point-set:

$$\mathbf{E}^{\circledast}(\sigma) = \mathbf{E}_{\max}(f \circledast g_\sigma) \cup \mathbf{E}_{\min}(f \circledast g_\sigma). \tag{81}$$

Likewise, for the multiscale closing-opening, we define:

**Definition V.2** (*Multiscale Closing-Opening Scale-Space Fingerprint*): The *multiscale closing-opening scale-space fingerprint* is a plot, versus scale, of the scale-dependent point-set:

$$\mathbf{E}^{\odot}(\sigma) = \mathbf{E}_{\max}(f \odot g_\sigma) \cup \mathbf{E}_{\min}(f \odot g_\sigma) \tag{82}$$

To illustrate these definitions we show in Figure 8 fingerprints for the same random 1D signal from the zero-crossings of second derivative in Gaussian scale-space, local extrema in multiscale dilation-erosion scale-space, and local extrema in multiscale closing-opening scale-space.

### A. Equivalence of Fingerprints

It is clear that the smoothed signal formed by the dilation-erosion is in general different from the smoothed signal formed by the closing-opening. However, the fingerprints are concerned only with the local extrema of these signals and it is not immediately apparent what the relationship is (if any) between fingerprints of the two scale-spaces.

Let us compare the dilation and the closing. From Proposition III.4 we see that (for $\sigma > 0$) a local maximum on the dilated signal corresponds directly in height and position to a local maximum on the underlying signal. Then from the geometrical interpretation of the closing we can also see that the same should apply for the closing operation. Furthermore, an identical condition of the structuring function "touching" the signal at a local maximum is responsible for the local maximum in both the dilated and the closed signals. A minima in the smoothed signal can be formed in slightly different ways between the dilation and the closing but Proposition V.2 will show that these too are equivalent.

To formalize these ideas we have the following propositions. Note that we have split the results between two propositions as the method of proof differs considerably between them.

**Proposition V.1** *Let the structuring function have a single local maximum at the origin. The following two statements are equivalent:*

1. $(f \oplus g_\sigma)(\mathbf{x})$ *has a local maximum at* $\mathbf{x} = \mathbf{x}_{\max}$ $\Longleftrightarrow$
2. $(f \bullet g_\sigma)(\mathbf{x})$ *has a local maximum at* $\mathbf{x} = \mathbf{x}_{\max}$

*Likewise, for the erosion and opening, the following two statements are equivalent:*

1. $(f \ominus g_\sigma)(\mathbf{x})$ *has a local minimum at* $\mathbf{x} = \mathbf{x}_{\min}$ $\Longleftrightarrow$
2. $(f \circ g_\sigma)(\mathbf{x})$ *has a local minimum at* $\mathbf{x} = \mathbf{x}_{\min}$

*Proof.* A proof of this proposition is given in the appendix. ∎

**Proposition V.2** *Let the structuring function have a single local maximum at the origin. The following two statements are equivalent.*

1. $(f \oplus g_\sigma)(\mathbf{x})$ *has a local minimum at* $\mathbf{x} = \mathbf{x}_{\min}$ $\Longleftrightarrow$
2. $(f \bullet g_\sigma)(\mathbf{x})$ *has a local minimum at* $\mathbf{x} = \mathbf{x}_{\min}$

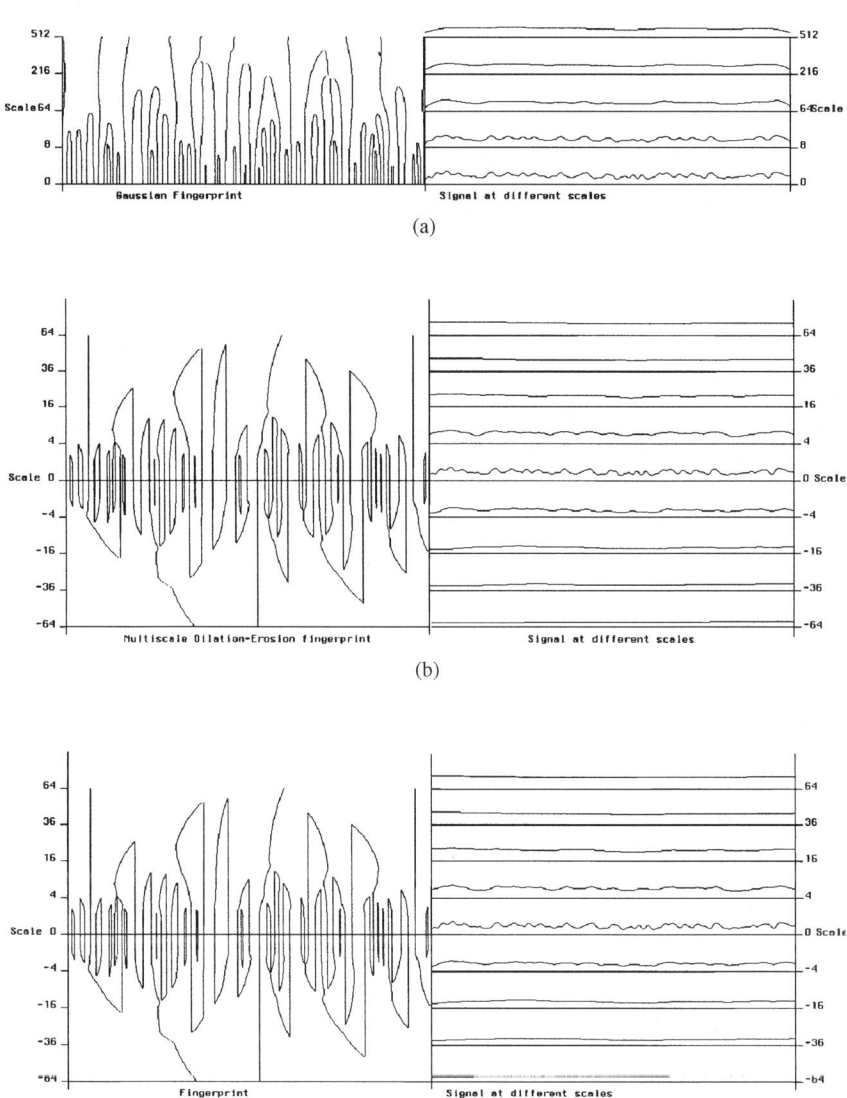

FIGURE 8. A comparison of fingerprints—(a) zero-crossings of second derivative in Gaussian scale-space; (b) local extrema in multiscale dilation-erosion scale-space; (c) local extrema in multiscale closing-opening scale-space. Figure 8(b) reprinted from (Jackway & Deriche 1996, © IEEE).

Likewise, for the erosion and opening, the following two statements are equivalent:

1. $(f \ominus g_\sigma)(\mathbf{x})$ has a local maximum at $\mathbf{x} = \mathbf{x}_{\max}$ $\iff$
2. $(f \circ g_\sigma)(\mathbf{x})$ has a local maximum at $\mathbf{x} = \mathbf{x}_{\max}$

*Proof.* A proof of this proposition is given in Section X. ∎

Together these propositions show that the full scale-space fingerprints for the dilation-erosion and the closing-opening are identical,

$$E^{\oplus}(\sigma) = E^{\circ}(\sigma) \qquad \text{for all } \sigma \in \mathbb{R} \qquad (83)$$

so we can drop the superscript and simply write $E(\sigma)$. Thus, the surfaces formed by the multiscale dilation-erosion and multiscale closing-opening of a multidimensional function, although different almost everywhere, have local extrema of the same height and at the same points.

This has computational implications, since the dilation is usually quicker to compute than the closing. Therefore, if it is desired to extract the morphological scale-space fingerprints from a signal, then the dilation-erosion scale-space is the most efficient way to do this.

### B. Reduced Fingerprints

Proposition III.4 shows that fingerprint paths corresponding to signal maxima (minima) do not change spatial position as scale is varied above (below) zero. Therefore, this subset of the full fingerprint consists of straight lines only, and so can be represented vary compactly. This leads us to define a subset (maxima only for positive scales, minima only for negative scales) of the full fingerprint called the *reduced* fingerprints and denoted by the subscript $r$:

**Definition V.3** *(Morphological Scale-Space Reduced Fingerprint):* The *morphological scale-space reduced fingerprint* is defined as:

$$E_r(\sigma) = \begin{cases} E_{\max}(f \oplus g_\sigma) & \text{if } \sigma > 0 \\ E_{\max}(f) \cup E_{\min}(f) & \text{if } \sigma = 0 \\ E_{\min}(f \ominus g_\sigma) & \text{if } \sigma < 0 \end{cases} \qquad (84)$$

Now, from Theorem III.1 (or, alternatively, Theorem IV.2), we find that in the reduced fingerprint,

$$\sigma_1 < \sigma_2 < 0 \Rightarrow E_r(\sigma_1) \subseteq E_r(\sigma_2) \subseteq E_r(0) \qquad (85)$$

$$\sigma_4 > \sigma_3 > 0 \Rightarrow E_r(\sigma_4) \subseteq E_r(\sigma_3) \subseteq E_r(0) \qquad (86)$$

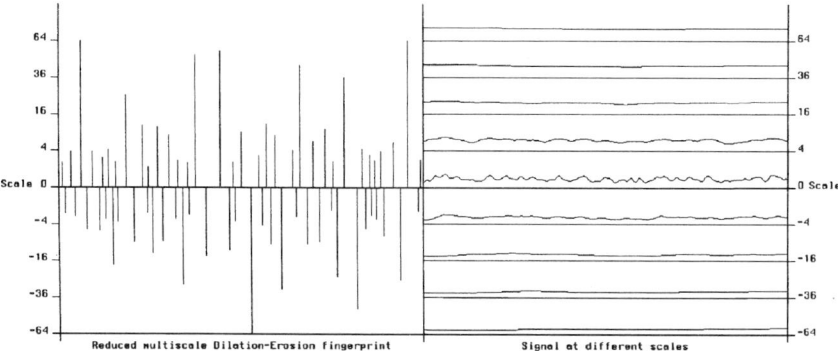

FIGURE 9. The reduced morphological scale-space fingerprint. Reprinted from (Jackway & Deriche 1996, © IEEE).

Because at any scale, $E_r(\sigma)$ is a point-set, the preceding set inclusions confirm that the reduced fingerprint consists only of vertical lines, beginning at zero-scale and extending into positive and negative scale-space until they end. This behavior can be seen in the example of a reduced morphological scale-space fingerprint shown in Figure 9.

This property makes the reduced fingerprint particularly easy to represent; all we need is to specify the position of each fingerprint line and the scale (positive or negative) at which it ends. This is easily stored as a list: Suppose we have a signal $f: R^n \to R$ with $k$ local extrema at $f(\mathbf{x}_i)$, $i = 1, 2, \ldots, k$, then, since each local extrema is the origin of a fingerprint line, we can represent the whole reduced fingerprint as a list of $k(n + 1)$-tuples, $(\mathbf{x}_i, \sigma_i)$, $i = 1, 2, \ldots, k$, where $\sigma_i$ is the scale associated with fingerprint line $i$.

## C. Computation of the Reduced Fingerprint

The practical importance of the reduced fingerprint is enhanced by the fact that we have developed an efficient method for computing the reduced morphological scale-space fingerprint, which does not involve actually computing the scale-space image or indeed any signal smoothing.

We can consider the $n$-tuple $(\mathbf{x}_i, \sigma_1)$ as assigning a scale $\sigma_i$ to the local extrema at position $(\mathbf{x}_i)$ in the original signal, where $\sigma_i$ represents the scale value at which the fingerprint line at $(\mathbf{x}_i)$ ends. From the geometrical interpretation of the morphological operations, we see that this $\sigma_i$ is the value such that structuring functions of $\sigma \leq \sigma_i$ "touch" the signal at $(\mathbf{x}_i)$, whereas those of scale $\sigma > \sigma_i$ do not. This is the property that is used in the algorithm which

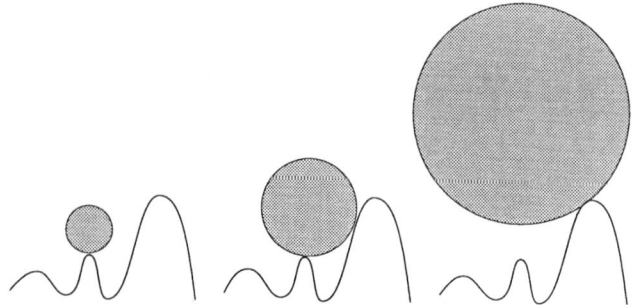

FIGURE 10. Finding the scale of a local maximum on a signal.

follows—for each extrema in the signal, we search for a structuring element satisfying this property and then assign its scale value to the fingerprint line.

It may help to visualize this operation as inflating a balloon resting on top of a local maxima (see Fig. 10). If the local maxima is lower than the global maximum, then eventually the balloon gets so big that it touches the surface elsewhere and, from then on, no longer rests on the local maximum. We equate the radius of the balloon (structuring function) at "liftoff" as the critical scale of the local maximum. Of course, if the signal local maximum is equal to the global maximum, we can immediately assign the scale to be $\infty$.

To formally present the algorithm, we consider a bounded discrete signal $f : Z^n \to [0, 1, \ldots, M]$ with $k$ local maxima at $f(\mathbf{x}_i)$ $i = 1, 2, \ldots, k$. To translate the description of the previous paragraph into a practical algorithm, we note that at the critical scale $\sigma_i$ for the local maximum at position $\mathbf{x}_i$, the structuring function with origin at $f(\mathbf{x}_i)$ and of scale $\sigma_i$ passes through a point in common with the surface at *at least one* other point. Let's denote any one of these other points by $f(\mathbf{x}_p)$. We need to find $f(\mathbf{x}_p)$, for by knowing this point we can immediately determine $\sigma_i$ by solving (fog $\sigma_i$) the equation,

$$f(\mathbf{x}_i) - g_{\sigma_i}(\mathbf{x}_p - \mathbf{x}_i) = f(\mathbf{x}_p) \tag{87}$$

Because $g_\sigma$ is convex, we can always find a function $g^{-1}(\mathbf{x})$ such that if $g_\sigma(\mathbf{x}) = z$ then, $\sigma = g^{-1}(\mathbf{x}, z)$. Using this function the solution to Eq. (87) is

$$\sigma_i = g^{-1}(\mathbf{x}_p - \mathbf{x}_i, f(\mathbf{x}_i) - f(\mathbf{x}_p)) \tag{88}$$

As an example, for spherical structuring functions given by:

$$g_\sigma(\mathbf{x}) = |\sigma|((1 - \|\mathbf{x}/\sigma\|^2)^{1/2} - 1) \qquad \|\mathbf{x}\| \le \sigma \tag{89}$$

we have

$$g^{-1}(\mathbf{x}, z) = -\frac{\|\mathbf{x}\|^2 + z^2}{2z^2} \tag{90}$$

giving

$$\sigma_i = -\frac{\|\mathbf{x}_p - \mathbf{x}_i\|^2 + (f(\mathbf{x}_i) - f(\mathbf{x}_p))^2}{2(f(\mathbf{x}_i) - f(\mathbf{x}_p))} \tag{91}$$

The remaining step is to find the point $\mathbf{x}_p$. We could conduct an exhaustive search by applying Eq. (88) to all the points $\mathbf{x}_j, j \neq i$ in the signal and taking the minimum of all the resulting scales,

$$\sigma_i = \min_{j \neq i}(g^{-1}(\mathbf{x}_i - \mathbf{x}_j, f(\mathbf{x}_i) - f(\mathbf{x}_j))) \tag{92}$$

Fortunately, there is generally no need to search all the points of the signal. For a start, if we are calculating the scale for the local maximum at $\mathbf{x}_i$, we can ignore all other points $\mathbf{x}_j$ in the signal for which $f(\mathbf{x}_j) < f(\mathbf{x}_i)$, as the structuring function cannot touch there. Second, as we progressively compute Eq. (92), if we denote the minimum scale found so far by $\hat{\sigma}_i$ then we only have to search the points in the region defined by,

$$\{\mathbf{x} : g_{\hat{\sigma}_i}(\mathbf{x} - \mathbf{x}_i) \geq f(\mathbf{x}_i) - M\} \tag{93}$$

where $M$ is the global maximum of the signal $f$. This is because, outside of this region, the magnitude of the structuring function is such that it cannot touch the signal. Note that as the search in Eq. (92) progresses and the minimum scale found so far decreases, the region defined by Eq. (93) also decreases, ensuring that the algorithm terminates. For this reason it is mot efficient to search over points $x_j$ in order of increasing radius from $x_i$. In practice, it easier to search along the sides of an expanding square around $x_i$. We can now write the algorithm:

**Algorithm 1** (To find the scale of the local maximum at $x_i$): Let $M$ denote the global maximum of signal $f(\mathbf{x})$.

**ENTRY POINT:** $\mathbf{x}_i$ is a local maximum of $f$

**Step 1**   $\sigma_i \leftarrow \infty$
**Step 2**   IF $(f(\mathbf{x}_i) = M)$, RETURN
**Step 3**   $R \leftarrow 0$
**Step 4**   $R \leftarrow R + 1$
**Step 5**   FOR all points $\mathbf{x}_j$ of radius $R$ from $\mathbf{x}_i$ DO
**Step 6**   $H \leftarrow f(\mathbf{x}_i) - f(\mathbf{x}_j)$
**Step 7**   IF $(H \geq 0)$ GOTO Step 10
**Step 8**   $\sigma_j = g^{-1}(\mathbf{x}_j - \mathbf{x}_i, H)$

**Step 9**   IF $(\sigma_j < \sigma_i), \sigma_i \leftarrow \sigma_j$
**Step 10**  ENDFOR
**Step 11**  IF $(g_{\sigma_1}(R) \geq f(\mathbf{x}_i) - M)$, GOTO Step 4
**RETURN:** value in $\sigma_i$

A similar algorithm finds the scale associated with the local minima of $f$.

An examination of Algorithm 1 shows that for each local extrema in the signal we search all the points in an $n$-dimensional volume given by

$$\{\mathbf{x} : g_{\sigma_i}(\mathbf{x} - \mathbf{x}_i) \geq f(\mathbf{x}_i) - M\} \tag{94}$$

Therefore, the algorithm is approximately of order $O(\sigma_i^n)$. If we let $\overline{\sigma^n} = \frac{1}{k}\sum_{i=1}^{k} \sigma_i^n$ then the computation for the extraction of the complete reduced morphological scale-space fingerprint is $O(k\overline{\sigma^n})$. An example code fragment in the computer language $C$ for the implementation of this algorithm for a 2D function (a range image) is presented in Section X.

Before we leave the computational details of fingerprint extraction, one further point deserves mention. The input to the fingerprint algorithm is a list of the positions of the local extrema in the signal. For computational purposes, where we deal with a discrete signal, two difficulties may occur.

First, because the range of the signal is discrete, it is possible (and common for some signals!) to have plateaus or areas of equal level. This presents difficulties as all the internal points of the plateau are *both* local maxima and local minima by definition—which is somewhat counterintuitive. We can avoid this problem by using only the *strict* local extrema in the fingerprint; that is, the central point must be higher (lower) than its neighborhood. Now, another problem occurs: For a flat-topped hill, there is no point higher than all its neighborhood, whereas, intuitively, a hill should possess a single local maximum. One solution, is to first find all the regional extrema, and then to follow this step by a procedure that represents each extrema by a single point near the center of its region. This center point can be found by a *shrinking algorithm* to shrink the connected regions down to single points (Rosenfeld, 1970). The outcome is that we have a list of isolated points describing the extrema in the signal, which seems an adequate solution to the problem.

Second, the definition of extrema is closely related to that of *neighborhood* of a point. For the square 2D lattice, there are two usual possibilities, the four horizontal and vertical neighbors, or the eight neighbors which include the diagonal points as well. The definition of neighborhoods in digital spaces is closely related to issues of connectivity (Rosenfeld, 1970; Lee and Rosenfeld, 1986); in particular it is well known that square lattices on 2D suffer from a number of deficiencies and that for issues such as neighborhoods the hexagonal lattice has a number of advantages (Serra and Lay, 1985; Bell et al., 1989).

To summarize the process of extracting the reduced fingerprint: We first find all the regional extrema in the signal, then we reduce connected regions in this extrema-map to isolated single points. This results in a list of coordinates $\mathbf{x}_i$, $i = 1, 2, \ldots, k$. This list is passed to the fingerprint extraction step, which adds the scale entries to the list and returns, $(\mathbf{x}_i, \sigma_i)$, $i = 1, 2, \ldots, k$. This list is the *morphological scale-space reduced fingerprint*.

A number of examples of the use of the reduced fingerprint for object recognition in range images have been presented in Jackway (1995a).

## VI. STRUCTURING FUNCTIONS FOR SCALE-SPACE

In Section II, we have treated the structuring function in a general way, introducing constraints only where mathematically needed to obtain desired theoretical results. From those results we may conclude so far that, depending on which of the properties of Sections III and IV we require, the structuring function should be continuous, convex, and have a single local maximum at the origin.

In this section we will collect together some further results that impact on the choice of structuring function.

### A. Semigroup Properties

We begin our discussion of structuring functions by looking at the *semigroup* property of dilation-erosion scale space.

We may view the multiscale dilation and erosion as *operators* on a signal, i.e.,

$$\mathfrak{D}_\sigma f = f \oplus g_\sigma \tag{95}$$

$$\mathfrak{E}_\sigma f = f \ominus g_\sigma \tag{96}$$

We have already seen in Section V.B and Proposition II.1 that the convex scaled structuring functions form a one-parameter continuous semigroup under the morphological dilation. Now using the chain rules for dilations and erosions Eqs. (15) and (16) lead directly to the semigroup property (Butzer and Berens, 1967) for the scale parameterized morphological operations.

$$\mathfrak{D}_{\sigma+\mu} f = \mathfrak{D}_\sigma \mathfrak{D}_\mu f \sigma, \mu \geq 0 \tag{97}$$

$$\mathfrak{E}_{\sigma+\mu} f = \mathfrak{E}_\sigma \mathfrak{E}_\mu f \sigma, \mu \geq 0 \tag{98}$$

$$\mathfrak{D}_0 f = \mathfrak{E}_0 f = f \tag{99}$$

A word of caution is needed here. We have dealt with the erosions and dilations separately, because we do not have the full *group* structure, which would require a negative scale operation to cancel out the effect of a positive scale operation. The erosion comes the closest to canceling the effect of a dilation, but erosion following dilation does not give the identity operator—it gives the closing!

$$\mathfrak{D}_\sigma \mathfrak{E}_\sigma f = f \bullet g_\sigma \neq f \qquad (100)$$

We should also note in passing that the closings and openings do not enjoy any such semigroup property; the sieving properties (26) and (27) ensure that instead of achieving an additive effect, successive openings or closings merely preserve the effect of the strongest structuring function.

Many authors consider the semigroup property of scale-space one of the fundamental foundations (Koenderink, 1984; Lindeberg, 1990). Lindeberg (1988) considers the semigroup property to be very important in his construction of a (linear) discrete scale-space. In fact he makes it one of his scale-space *axioms* from which the whole structure of the scale-space is developed.

The semigroup property of dilation-erosion scale space enables the signal at scale $\sigma + \mu$ to be obtained directly from the previous signal at scale $\sigma$ by repeated dilation (or erosion) and specifies how the global structure of the scale-space is related to the local structure.

In practical terms the semigroup property ensures that the result of smoothing a signal at some scale is independent of the path taken to arrive at that smoothing; that is, a signal smoothed at scale $\sigma_3$ may be obtained by smoothing the original signal $f_{\sigma_3} = f \circledast g_{\sigma_3}$ or by smoothing an already smoothed signal $f_{\sigma_3} = f_{\sigma_1} \circledast g_{\sigma_2}$ where $\sigma_3 = \sigma_1 + \sigma_2$. Subject to the propagation of numerical errors, the construction of a set of smoothed signals by incremental smoothing with a small scale structuring function may be computationally more efficient than using larger and larger-scaled structuring functions.

### B. A More General Umbra

In applications, due to the physical nature of the problem to be addressed, more constraints on the desired behavior of any scale-dependent operator may need to be imposed. The morphological operations depend on *shape* (from the Greek root *morphē* = shape). We continue our discussions of structuring functions by showing how we require the introduction of an additional parameter for the notion of the shape of a *function* to be properly defined. This leads to a generalization of the *umbra* concept of Sternberg. When working with *functions* (rather than *sets*) we need to be careful with the concept of shape, which is central to morphology.

Binary images are readily represented as sets (for example, the set of all white points) and mathematical morphology was originally set based (Matheron, 1975; Serra, 1982). However, gray-scale images are more naturally represented as functions where the value of the function at a point represents the image intensity. The umbra (Sternberg, 1986) provides a natural way to associate a set $U$ with a function $f(x, y)$ and was the first way in which the operations of mathematical morphology were extended to grayscale functions,

$$U[f] = \{(x, y, z) : z \leq f(x, y)\}. \tag{101}$$

Because morphological operations depend on shape, we can see a shortcoming of the umbra approach, as defined here, for physical signals. This is best illustrated with a more common function-to-set mapping—the drawing of a graph.

Consider the process of drawing the graph of a signal voltage $v(t)$ as a function of time. First, a scale is chosen for the $x$-axis to display the required time interval. Then a scale is chosen for the $y$-axis (relative to that of the $x$-axis) to cause the *shape* of the resulting graph to convey the required information to the viewer. Mathematically, we can view this operation as choosing appropriate scaling constants $\alpha$ and $\beta$ to relate a physical time-varying, signal voltage to a set of points $G$ in the plane (the graph). In symbols:

$$G = \{(x, y) : x = t/\alpha, y = v(t)/\beta\} \tag{102}$$

In the digitization of a physical signal (by sampling in time and quantizing the samples) the signal is mapped into the digital space $Z^2$. The scaling factors appear in the selection of sampling rate $(\alpha)$ and the gain of the A-to-D converter $(\beta)$.

We wish to emphasize here that *function shape* depends on the values of $\alpha$ and $\beta$ (see Fig. 11). More precisely, we can see that *scale* depends on $\alpha$ and *shape* on the ratio $\lambda = \alpha/\beta$. Because morphological operations are scale invariant, $\alpha(G \oplus B) = \alpha G \oplus \alpha B$, we can, without loss of generality, take the scale to be unity. This suggests that (103) be reparameterized as

$$G = \{(x, y) : x = t, y = \lambda v(t)\} \tag{103}$$

Thus, in applying mathematical morphology to grayscale images we consider Sternberg's (1986) umbra to be underparameterized. A more general formulation would be:

A gray-scale image is a gray-level function $f(x, y)$ on the points of Euclidean 2-space. A gray-level function can be thought of in Euclidean 3-space as a set of points $[x, y, \lambda f(x, y)]$, imagined as a thin, undulating, not necessarily connected sheet. A gray-scale image $f(x, y)$ is represented in the mathematical morphology by an umbra $U[f, \lambda]$ in Euclidean 3-space, where a point $p = (x, y, z)$ belongs to the umbra if and only if $z \leq \lambda f(x, y)$. Where $\lambda$ is a shape parameter (cf. Sternberg, 1986).

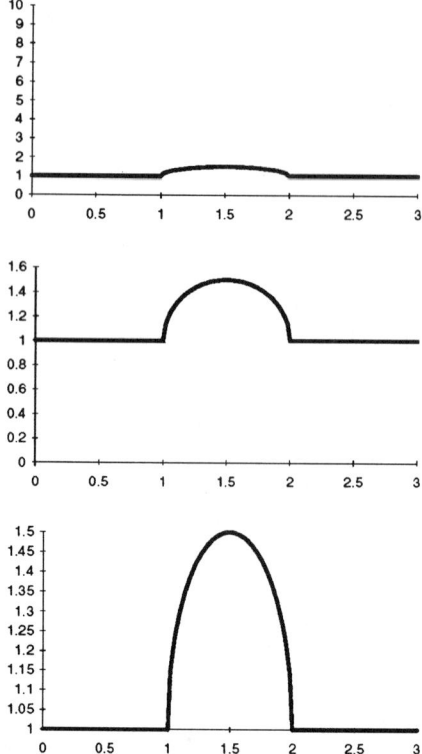

FIGURE 11. Shape depends on $X$, $Y$ scaling. Three graphs of the function $y = \sqrt{x^2 - 3x + 2}$ $1 \leq x \leq 2$.

This extra shape parameter, $\lambda$, presents a problem, however, due to the dimensional inhomogeneity between the spatial dimensions and the intensity dimension of the gray-scale image. The value of $\lambda$ is therefore undefined by the physical problem. It would be advantageous if the morphological operations were invariant with respect to the implicit value of $\lambda$. This type of shape invariance had received some attention in Verbeek and Verwer (1989) but the most thorough and general treatment is to be found in the dimensionality property of Rivest et al. (1992).

## C. Dimensionality

Rivest et al. (1992) have recently shown the importance in image processing and analysis of a concept they call dimensional consistency or *dimensionality*. Dimensional measurements on gray-scale images have a *physical significance*.

Suppose that an intensity image is mathematically modeled as a function $f(\mathbf{x})$, $\mathbf{x} \in \mathrm{R}^2$ into the closed segment $I = [0, 1]$. The intensity axis $I$ represents the irradiance (light intensity) at the image plane and is, therefore, not dimensionally homogeneous with the spatial dimensions. Any measurement with physical significance should not couple these physically different dimensions.

Scaling in the spatial dimensions is known as *homothety* and scaling in the intensity direction as *affinity*, indicating the fundamental physical difference between magnifying an image and brightening it (Rivest *et al.*, 1992). Making a measurement on an image consists in applying a functional on the image, where a *functional* is a global parameter associated with a function. The property of dimensionality applies to functionals; that is:

**Definition VI.1** *(Dimensional Functionals):* The functional $W$ is defined to be *dimensional* if there exists constants $k_1$, $k_2$ such that for all $\lambda_1 \lambda_2 > 0$:

$$W(\lambda_1 f(\lambda_2 \mathbf{x})) = \lambda_1^{k_1} \lambda_2^{k_2} W(f(\mathbf{x}))$$

where $\lambda_1$ is the affinity and $\lambda_2$ the homothety (Rivest *et al.*, 1992).

This relation restricts the way in which affinities and homotheties of an image affect dimensional measurements on it, and results in a decoupling between affinity and homothety measurements. For example, the *volume* of $f$, $V(f) = \int f(\mathbf{x}) \, d\mathbf{x}$, is dimensional because $V(f') = \lambda_1 \lambda_2^{-2} V(f)$ with $f' = \lambda_1 f(\lambda_2 \mathbf{x})$. If $f(\mathbf{x})$ is the irradiance at $\mathbf{x}$, then $V(f)$ has physical significance as the total radiant power.

Rivest *et al.* (1992) indicate that useful image processing operators should conserve dimensionality. It is commonly thought that *volumic* (nonflat) morphological structuring elements lead to the breakdown of dimensionality in morphological operations (Rivest *et al.*, 1992; Sternberg, 1986). This is true for fixed-scale structuring elements. However, we will show in the next section that if the morphological dilation-erosion scale-space is formed by using scaled *elliptic poweroid* structuring functions, which are in general volumic, any dimensional functional of the scale-space image is also a dimensional functional of the underlying image.

In scale-space filtering we work with the scale-space image, which is of higher dimensionality than the signal (Witkin, 1983). We can be careful to ensure that any operations used preserve dimensionality in the scale-space image. As an illustrative (somewhat nonpractical!) example, suppose we were to try matching 1D signals by counting the number of closed loops in their morphological scale-space fingerprints.

Because the existence and relative position of local extrema are invariant under homothety and affinity, the number of closed fingerprint loops is a *dimensional functional* on the scale-space image. Now, a moment's reflection shows that this is not the important point; the real question is whether or not our dimensional functional on the scale-space image $F$ (which after all is a construct) is a dimensional functional on the *original signal f* and thus has a *physical* significance.

So we need to find out how to construct dimensional functionals on $f$. First, we have extended the definition of dimensionality to scale-space images:

**Definition VI.2** *(Scale-Space Dimensionality):* $W$ is a dimensional functional on $F$ if given $\lambda_1, \lambda_2, \lambda_3 > 0$ there exists constants $k_1, k_2, k_3$ such that:

$$W(\lambda_1 F(\lambda_2 \mathbf{x}, \lambda_3 \sigma)) = \lambda_1^{k_1} \lambda_2^{k_2} \lambda_3^{k_3} (F(\mathbf{x}, \sigma))$$

We will show that if the dilation-erosion scale-space is constructed with the elliptic poweroid structuring functions, *all* dimensional functionals on $F$ are also dimensional functions on $f$.

Consider the scale-space formed by the multiscale dilation $F(\mathbf{x}, \sigma) = (f \oplus g_\sigma)(\mathbf{x})$, $\sigma > 0$.

**Proposition VI.1** *With elliptic poweroid structuring functions, $g_\sigma(\mathbf{x}) = -|\sigma| \times (\sqrt{\mathbf{x}^\top \mathbf{A} \mathbf{x}}/|\sigma|)^\alpha$, all dimensional functionals $W(F)$ are also dimensional functionals of $f$:*

*Proof.* A proof of this proposition can be found in Jackway (1995b). ∎

As a point of interest, as $\alpha \to \infty$ the circular poweroid structuring functions approach the flat (nonvolumic) cylindrical structuring element well known in gray-scale morphology and image processing (Nakagawa and Rosenfeld, 1978; Sternberg, 1986; Haralick *et al.*, 1987).

It is rather difficult to demonstrate the effect of nondimensionality in scale spaces as the effects are likely to be small. One difference, however, can be seen in the connectivity of fingerprints. As an example of the dimensionality property in 1D, we present Figure 12, which shows the differing effects of the use of dimensional (parabolic) and nondimensional (spherical) structuring functions in the computation of the multiscale dilation-erosion scale-space fingerprint of a certain signal and that same signal with an affinity. A *very* close examination of Figure 12 shows that with a nondimensional structuring function, the fingerprint of the stretched signal may be differently connected, whereas with the parabolic structuring function the fingerprint is merely compressed in the scale direction. We can obtain a functional by counting the closed loops of the fingerprints. In this case we have Figure 12c with 19 loops, and Figure 12d with 20 loops (the difference is in the 8th loop from the right) indicating the breakdown of dimensionality. In contrast (with a parabolic structuring function) both Figures 12e and 12f contain 19 loops.

If the signal was an intensity image and some image analysis operation, such as pattern recognition, was sensitive to the connectivity of the fingerprint,

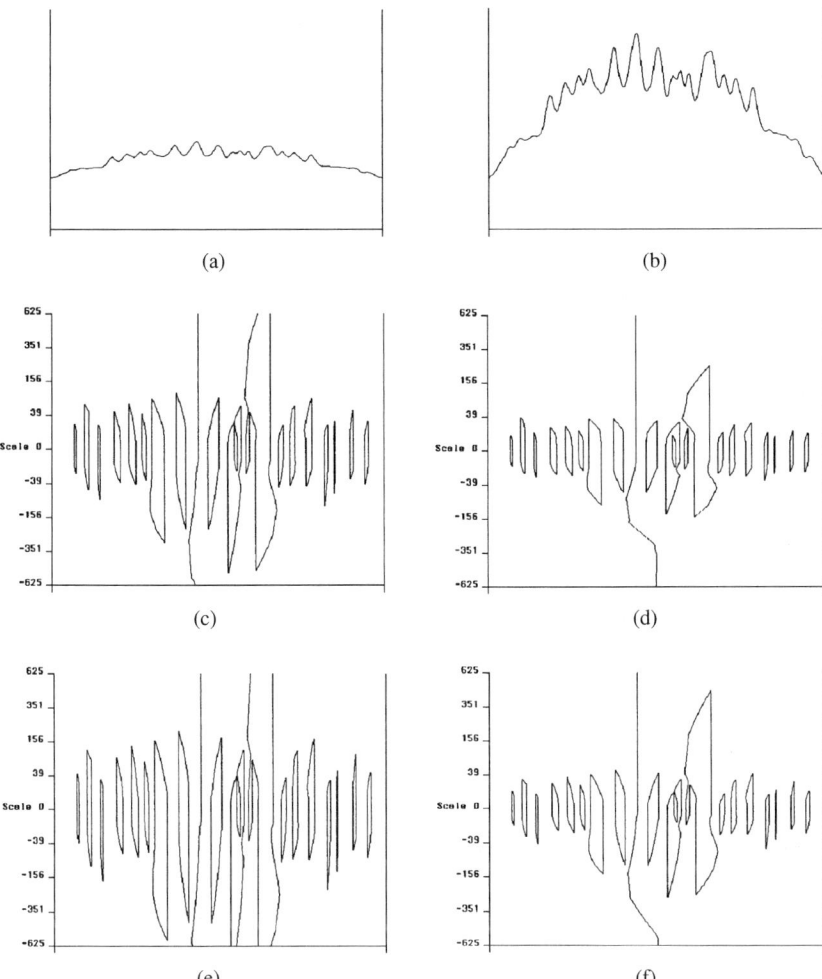

FIGURE 12. An example of dimensionality in scale-space. (a) A random signal. (b) This signal with an affinity of size 4.0. (c) and (d) The multiscale dilation-erosion fingerprints of the preceding signals with a nondimensional (spherical) structuring function. Note the connectivity and structure of the fingerprints differ because of the affinity: (c) has 19 closed loops and (d) 20 closed loops (the difference is in the 8th loop from the right) indicating the breakdown of dimensionality. (e) and (f) The multiscale dilation-erosion fingerprints of the preceding signals with a dimensional (parabolic) structuring function. Note the structure of the fingerprints remains similar (with 19 closed loops) indicating the conservation of dimensionality.

then with the nondimensional structuring function, the output would depend on the arbitrary scale chosen to represent the intensity dimension relative to the spatial dimensions of the image.

## D. The Poweroid Structuring Functions

We make the general observation that (for any fixed scale) the more *pointed* (low $\alpha$) structuring functions tend to give more emphasis to the *local* shape near a signal feature. Other constraints or requirements may dictate the choice of $\alpha$ and, hence, the structuring function.

The flat structuring function ($\alpha = \infty$) is commonly used because the morphological operations reduce to simply taking the maximum or minimum of the signal over some neighborhood (Nakagawa and Rosenfeld, 1978). That is, for flat structuring functions,

$$(f \oplus g_\sigma)(\mathbf{x}) = \bigvee_{\mathbf{t} \in G} \{f(\mathbf{x} - \mathbf{t})\}$$
$$(f \ominus g_\sigma)(\mathbf{x}) = \bigwedge_{\mathbf{t} \in G} \{f(\mathbf{x} + \mathbf{t})\} \quad (104)$$

The use of flat structuring functions is most appropriate for binary images where the morphological operations are indentical to those on point-sets. However, on gray-scale images, the use of flat structuring functions leads to flat regions in the output signal around the local extrema, and the local extrema are no longer exactly localized in position. This is certainly a disadvantage in multiscale dilation-erosion scale-space, because it is the local extrema of the output signal which are our scale-space features and exact localization is absolutely necessary.

There is a computational reason for the importance of the paraboloid structuring functions in particular. For a fixed scale, the 2D morphological dilation can be computed most directly by the *morphological convolution*

$$(f \oplus g_\sigma)(i, j) = \max_{(x,y) \in G} \{f(i - x, j - y) + g_\sigma(x, y)\} \quad (105)$$

where $G$ is a square neighborhood of $(i, j)$. If the size of this neighborhood is $r^2$, then the computational burden of the direct implementation of Eq. (105) is $B_d = O(r^2)$. The Landau symbol $O$ is often used to indicate computational complexity: $R(\phi) = O(\phi^x)$ means that $R(\phi)/\phi^x$ is bounded as $\phi \to \infty$ (Lipschutz, 1969).

However, for the 2D paraboloid structuring function, $g_\sigma(x, y) = -|\sigma|(x^2 + y^2)/\sigma^2$, we have a separability property:

$$g_\sigma(x, y) = g_\sigma^{(1)}(x) + g_\sigma^{(1)}(y) \quad (106)$$

where the 1D structuring function $g_\sigma^{(1)}(x) = -|\sigma|x^2/\sigma^2$. The max function has a similar property,

$$\max_{x,\,y \in G}\{f(x,y)\} = \max_{x \in G^{(x)}}\left\{\max_{y \in G^{(y)}}\{f(x,y)\}\right\} \quad (107)$$

where $G^{(x)}$ and $G^{(y)}$ are the projections of $G$ on the $x$- and $y$-axes. Combining these properties we get the desired result

$$(f \oplus g_\sigma)(i,j) = \max_{x \in G^{(x)}}\left\{\gamma(i-x,j) + g_\sigma^{(1)}(x)\right\} \quad (108)$$

where

$$\gamma(i,j) = \max_{y \in G^{(y)}}\left\{\gamma(i,j-y) + g_\sigma^{(1)}(y)\right\}. \quad (109)$$

The computation has been reduced to a sequence of two 1D morphological convolutions with a computational burden, $B_1 = O(r)$. The cost is that additional storage is required for the intermediate result $\lambda$.

This result has recently become known in the literature as the separable decomposition of structuring elements (Shih and Mitchell, 1991; Gader, 1991; van den Boomgaard, 1992; Yang and Chen, 1993). There are actually two kinds of separability involved here. First, that is additive separability where $g_\sigma(x,y) = g_\sigma^{(1)}(x) + g_\sigma^{(1)}(y)$; second, that is morphological separability, $g_\sigma(x,y) = g_\sigma^{(1)}(x) \oplus g_\sigma^{(1)}(y)$. Note in the result (108)–(109) we have used additive separability to obtain a morphological separability result. Recent work has in fact shown that for square morphological templates (i.e., discrete structuring elements) the two kinds of separability are in fact equivalent (Yang and Chen, 1993).

Yang and Chen (1993) show in a theorem that if $g(x,y)$ is additively separable of size $(2r+1) \times (2r+1)$, and it is convex, then it can be expressed as $g(x,y) = (k_1^h \oplus k_2^h \oplus \cdots \oplus k_r^h) \oplus (k_1^v \oplus k_2^v \oplus \cdots \oplus k_r^v)$, where $k_i^h$ is a horizontal 1D structuring element of size 3, and $k_i^v$ is a vertical 1D structuring element of size 3. The importance of this result is that by the chain rules for dilations (end erosions), if $g = k_1 \oplus k_2 \oplus \cdots k_r$, then $f \oplus g = (((f \oplus k_2) \oplus k_2) \oplus \cdots) k_r$. So the whole operation can be performed as a sequence of 1D three-point operations. The point to stress here is that to obtain all these nice results we need additive separability of the structuring function. Writing the 2D elliptic poweroids as

$$g(x,y) = -\left(\sqrt{\mathbf{x}'\mathbf{A}\mathbf{x}}\right)^2 = -\left(\sqrt{a_{11}x^2 + 2a_{12}xy + a_{22}y^2}\right)^2 \quad (110)$$

where

$$A = \begin{pmatrix} a_{11} & a_{12} \\ a_{12} & a_{22} \end{pmatrix} \quad (111)$$

The conditions necessary are therefore that (a) $a_{12} = 0$ ($A$ is a diagonal matrix), and (b) $\alpha = 2$. Therefore $g(x, y)$ must be a circular paraboloid ($a_{11} = a_{22}$), or an elliptic paraboloid with the major and minor axes of the ellipse aligned with the coordinate system $x$, $y$-axes (for $a_{11} \neq a_{22}$). In practical terms this is a very favorable property of the paraboloids.

Van den Boomgaard (1992) has shown that the elliptic paraboloid structuring functions (called the *quadratic structuring functions,* QSF) are closed with respect to morphological dilation (and erosion). This result is an extension of the semigroup property of Section VI.B to arbitrary QSF kernel matrices. In fact, van den Boomgaard (1992) argues that the elliptic paraboloids can be considered to be the morphological equivalent of the Gaussian convolution kernels because this class is dimensionally separable and closed with respect to dilation (and erosion), thereby establishing an equivalence between the parabolic structuring function in mathematical morphology and the Gaussian kernel in convolution.

## VII. A Scale-Space for Regions

In this section we will show how we can change the signal feature involved and still maintain the monotone property of scale-space. We will first extend the morphological scale-spaces from signal regional extrema (Corollaries III.1.2 and IV.2.2) to signal watershed regions via the *watershed transform.* Then, via *homotopy modification* of the gradient, we will further extend the scale-space property to watershed regions of the gradient function where we will demonstrate its application to multiscale segmentation.

The idea in both cases is that if we can find some transform of a signal that gives a new feature that is 1:1 to the signal regional extrema, then Corollaries III.1.2 and IV.2.2 ensure that this new feature also possesses a scale-space monotone property. Let's make this a proposition:

**Proposition VII.1** *If $C[R(f)]$ denotes the number of connected components in the point-sets of the regional extrema of a signal $f$, and there exists some measure $\#$ on the transforms $\psi_1(f)$ and $\psi_2(f)$ such that*

$$\#[\psi_1(f)] = C[R_{\max}(f)]$$
$$\#[\psi_2(f)] = C[R_{\min}(f)]$$

*then for any scales $\sigma_1 < \sigma_2 < 0 < \sigma_3 < \sigma_4$,*

*(a) $\#[\psi_2(f \circledast g_{\sigma_1})] \leq \#[\psi_2(f \circledast g_{\sigma_2})] \leq \#[\psi_2(f)]$*
*(b) $\#[\psi_1(f \circledast g_{\sigma_4})] \leq \#[\psi_1(f \circledast g_{\sigma_3})] \leq \#[\psi_1(f)]$*

*and*

(c) $\#[\psi_2(f \odot g_{\sigma_1})] \leq \#[\psi_2(f \odot g_{\sigma_2})] \leq \#[\psi_2(f)]$
(d) $\#[\psi_1(f \odot g_{\sigma_4})] \leq \#[\psi_1(f \odot g_{\sigma_3})] \leq \#[\psi_1(f)]$

*Proof.* A proof of this proposition is given in Section X. ∎

We will discuss two such transforms, the *watershed transform* and a certain *homotopy modification*.

## A. The Watershed Transform

*Watershed transforms* are used primarily for image segmentation and are part of the tools of mathematical morphology (Lantuéjoul, 1978; Serra, 1982; Vincent and Beucher, 1989; Beucher, 1990). The recent development of powerful and fast algorithms (Vincent and Soille, 1991) has further served to popularize the method.

For segmentation (edge detection) the idea is that the watershed lines of a surface tend to follow the "high ground," so that if we find the watershed transform of the gradient image, the watershed lines will follow the edges (regions of high gradient) in the image, thereby performing a useful segmentation of the image into regions of low-intensity gradient, which are regions without edges.

However, although watershed image segmentation methods are very powerful and general, in many applications they tend to oversegment (Vincent and Beucher, 1989).

We can decompose the watershed transform into catchment basins (watershed regions) $\mathbf{W}_i$ $i = 1, 2, \ldots, q$ and the watershed lines themselves $\mathbf{L}$:

$$WS(f) = \mathbf{W}_1 \cup \mathbf{W}_2 \cup \cdots \cup \mathbf{W}_q \cup \mathbf{L} \tag{112}$$

Suppose a function $f : \mathbf{D} \subset \mathbf{R}^2 \to \mathbf{R}$ possesses $q$ regional minima; i.e., $C[\mathbf{R}_{\min}(f)] = q$. We can, therefore, write:

$$\mathbf{R}_{\min}(f) = \mathbf{N}_1 \cup \mathbf{N}_2 \cup \cdots \cup \mathbf{N}_q \tag{113}$$

where we have identified the individual regional minima $\mathbf{N}_i$ $i = 1, 2, \ldots, q$.

The watershed transform of a surface possesses the following properties (Beucher, 1990):

1. The watershed lines delineate open connected regions, $\mathbf{W}_i$.
2. All points of the surface either belong to a region or fall on a watershed line $\mathbf{x} \in \mathbf{D} \Rightarrow (\mathbf{x} \in \mathbf{W}_i$ for some $i = 1, 2, \ldots, q)$ or $(\mathbf{x} \in \mathbf{L})$.

3. Each watershed region contains a single regional minimum and each regional minimum belongs to a single watershed region (its catchment basin). So we can make the correspondences

$$N_i \subset W_i \qquad \text{for all } i = 1, 2, \ldots, q \tag{114}$$

From these watershed properties, we can write

$$C[WS(f)] = C[R_{\min}(f)] \tag{115}$$

where $C[WS(f)]$ counts the number of catchment basins in $WS(f)$. This equation is of the form required by Proposition VI.2, so it is possible to create a monotone scale-space property for watershed regions.

**Theorem VII.1** (Scale-Space Monotone Property for the Number of Watershed Regions): *Let $C[WS(f)]$ denote the number of watershed regions of an image f, then for any scales $\sigma_1 > \sigma_2 > 0$,*

*(a)* $C[WS(f \ominus g_{\sigma_1})] \leq C[WS(f \ominus g_{\sigma_2})] \leq C[WS(f)]$;

*and*

*(b)* $C[WS(f \circ g_{\sigma_1})] \leq C[WS(f \circ g_{\sigma_2})] \leq C[WS(f)]$.

*Proof.* This theorem follows directly from Eq. (115) and Proposition VII.1. ∎

Although theoretically correct, unfortunately, this theorem is less useful in practice than it might at first appear. We have glossed over the role played by the function $f$ in the foregoing treatment, but we must consider it now.

To correspond to something useful (such as edges in the image), the watershed should be applied to the gradient of the original image. This suggests that $f$ should be the gradient image. However, the gradient surface is quite unlike the original image and smoothing the gradient with the multiscale morphological operations is not equivalent to smoothing the original image. For instance, small-scale features in the original can have arbitrarily high gradients and thus dominate the gradient image. Smoothing the gradient image will leave these dominant features until last but they should be the first to disappear. Moreover, because the gradient image we use is actually the *magnitude* of the gradient function, and the magnitude operation removes any symmetry between the shapes of the maxima and minima, many minima occur as narrow cusps at zero, and applying an erosion or opening to this image will not really help to analyze the signal.

No! To make sense, it is necessary to perform the multiscale smoothing on the original image; however, the watershed operation must still be performed on the gradient image. We therefore need a link between the two, which maintains the scale-space monotone property.

We can construct the necessary link by using Vincent's (1993) gray-scale reconstruction to modify the homotopy of the gradient image.

## B. Homotopy Modification of Gradient Functions

Loosely speaking, two functions (i.e., surfaces) are said to be *homotopic* if their hills, channels, and divides have the same relationship to each other in both the functions; that is, their watershed transforms will make the same pattern (NB: for a more precise definition see Serra, 1982, Def. XII-3).

Vincent's (1993) gray-scale reconstruction provides a way to modify the homotopy of a function based on the values of another function called *the marker function*. Basically, the grayscale reconstructions can remove designated (i.e., marked) extrema from a function while leaving the remainder of the function unchanged. The use of these reconstructions is now the standard way to apply marker functions to watershed segmentation (Beucher, 1990). A full discussion of the gray-scale reconstruction is beyond the scope of the present article but interested readers are urged to consult Vincent (1993), which gives full details, examples, and fast algorithms.

We will employ the *dual gray-scale reconstruction* $\rho_f^\star(g)$ to remove from $f$ the regional minima "designated" by $g$.

To set this up, suppose we have a bounded function $f(\mathbf{x})$, which has $r$ regional minima which we label arbitrarily, $\boldsymbol{R}_{\min}(f) = \boldsymbol{N}_1 \cup \boldsymbol{N}_2 \cup \cdots \cup \boldsymbol{N}_r$. Suppose we then select (mark) $s < r$ of these regional minima by choosing $s$ sets $A_i$ so that $A_i \subset N_i$ $i = 1, 2, \ldots, s$. Let the union of these sets be denoted by $A = \bigcup_{i=1}^{s} A_i$. Now we form the marker function,

$$g(\mathbf{x}) = \begin{cases} 0, & \text{if } \mathbf{x} \in A \\ \bigvee f, & \text{otherwise} \end{cases} \tag{116}$$

We then have the following proposition:

**Proposition VII.2** *If $f$ is a continuous bounded function $f : D \subset \mathrm{R}^n \to [0, B]$, and $g : D \subset \mathrm{R}^n \to [0, B]$, is constructed as outlined earlier, then if we reconstruct the homotopy modified function $f^R$ using the dual gray-scale reconstruction,*

$$f^R = \rho_f^\star(g) \tag{117}$$

*then*

$$\boldsymbol{R}_{\min}(f^R) = \boldsymbol{N}_1 \cup \boldsymbol{N}_2 \cup \cdots \cup \boldsymbol{N}_s \tag{118}$$

*and, therefore,*

$$C[\boldsymbol{R}_{\min}(f^R)] = s \tag{119}$$

*Proof.* A proof of this result can be found in Jackway (1996). ∎

To proceed we must now consider the relationship between a function $f(x, y)$ and the magnitude of its gradient $|\nabla(f)|(x, y)$. Assume that $f$ is of class $C^1$ (that is, its first derivative exists and is continuous), then the gradient is zero on any regional extrema of $f$. Note that if $f$ is not of class $C^1$ (or in the discrete case) $|\nabla(f)|$ can simply be defined (i.e., forced) to equal zero on the regional extrema of $f$. We have

$$(x, y) \in (\boldsymbol{R}_{\max}(f) \cup \boldsymbol{R}_{\min}(f)) \Rightarrow |\nabla(f)|(x, y) = 0 \quad (120)$$

Because $|\nabla(f)|$ is a nonnegative function, all its zero points belong to its regional minima, so,

$$|\nabla(f)|(x, y) = 0 \Rightarrow (x, y) \in \boldsymbol{R}_{\min}(|\nabla(f)|) \quad (121)$$

Equations (120) and (121) imply

$$(\boldsymbol{R}_{\max}(f) \cup \boldsymbol{R}_{\min}(f)) \subset \boldsymbol{R}_{\min}(|\nabla(f)|) \quad (122)$$

which shows that the regional extrema of $f$ are subsets of the regional minima of $|\nabla(f)|$. Therefore, we can use Proposition VII.2. By choosing appropriate marker functions related to selected regional maxima or minima of $f$ and using the dual reconstruction on the gradient function, we can modify the homotopy of the gradient function to possess regional minima corresponding *only* to the selected regional extrema of $f$, thereby providing the necessary link between the homotopy of image and its gradient image. This idea is illustrated in Figure 13.

We formalize the above modification of gradient homotopy in a proposition.

**Proposition VII.3** *If $f(x, y)$ is a bounded function of class $C^1$ $f : \boldsymbol{D} \subset \mathbb{R}^2 \to [0, B]$ with $C[\boldsymbol{R}_{\max}(f)] = p$, with $C[\boldsymbol{R}_{\min}(f)] = r$, and with gradient $|\nabla(f)|$, and suppose we select $q \leq p$ regional maxima $M_i, i = 1, 2, \ldots, q$ and $s \leq r$ regional minima $N_j, j = 1, 2, \ldots, s$ and form the marker function*

$$g'(x, y) = \begin{cases} 0, & \text{if } (x, y) \in \left(\bigcup_{i=1}^{q} M_i \cup \bigcup_{j=1}^{r} N_j\right) \\ \max\{|\nabla(f)|\}, & \text{otherwise} \end{cases} \quad (123)$$

*and modify the homotopy of the gradient function,*

$$|\nabla(f)|'(x, y) = \rho^{\star}_{|\nabla(f)|}(g')(x, y) \quad (124)$$

*then:*

$$\boldsymbol{R}_{\min}(|\nabla(f)|') \subseteq (\boldsymbol{R}_{\max}(f) \cup \boldsymbol{R}_{\min}(f)) \quad (125)$$

*and*

$$C[\boldsymbol{R}_{\min}(|\nabla(f)|')] = q + s \quad (126)$$

*Proof.* A proof of this result is found in Jackway (1996). ∎

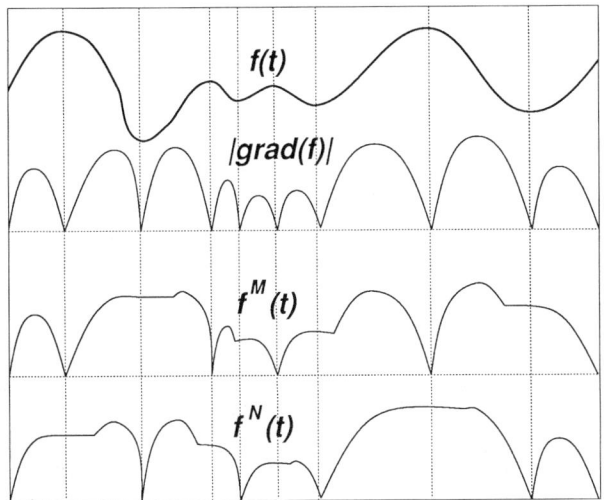

FIGURE 13. Modifying the homotopy of the gradient function. From top to bottom: a function, its gradient, the homotopy modified gradient function retaining only minima corresponding to maxima in the original function, and the homotopy modified gradient function retaining only minima corresponding to minima in the original function. Reprinted from (Jackway 1996 © IEEE).

Examples of similar homotopy modifications of gradient functions are presented in a more *ad hoc* manner in Beucher (1990), Vincent and Beucher (1989), and Vincent (1993).

If we set $q = p$ and $s = 0$ in Proposition VII.3, denoting the corresponding reconstructed magnitude-of-gradient by $|\nabla(f)|^+$, we obtain

$$C[R_{\min}(|\nabla(f)|^+)] = C[R_{\max}(f)] \quad (127)$$

and, because the regional minima of $|\nabla(f)|'$ by construction coincide with the regional maxima of $f$, we get the stronger result,

$$R_{\min}(|\nabla(f)|^+) = R_{\max}(f) \quad (128)$$

and, conversely, if $q = 0$ and $s = r$, denoting the corresponding reconstructed magnitude-of-gradient by $|\nabla(f)|^-$,

$$C[R_{\min}(|\nabla(f)|^-)] = C[R_{\min}(f)] \quad (129)$$

and

$$R_{\min}(|\nabla(f)|^-) = R_{\min}(f) \quad (130)$$

The effects of this homotopy modifications of the gradient and the resulting watersheds are shown in Figure 14.

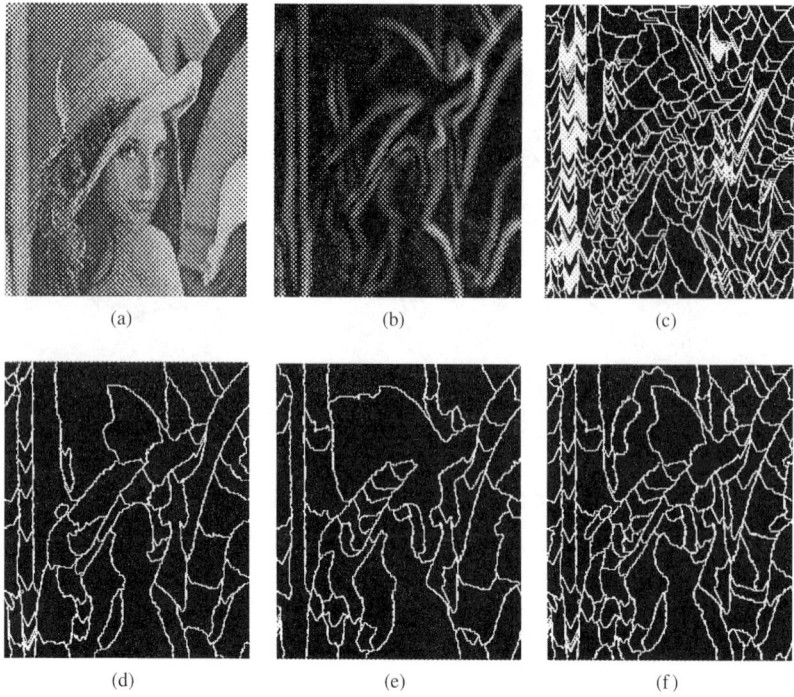

FIGURE 14. The watershed of the gradient of the homotopy modified "Lena" image. The images, from left to right, from top to bottom, show: (a) the original image, (b) its gradient, (c) the watershed of the unmodified gradient, (d) the watershed of the homotopy modified gradient retaining only minima corresponding to maxima in the original image, (e) the watershed of the homotopy modified gradient retaining only minima corresponding to minima in the original image, and (f) the watershed of the homotopy modified gradient retaining only minima corresponding to extrema in the original image.

## C. A Scale-Space Gradient Watershed Region

In the sequel we will use only the values $q = p$, $s = 0$ to make the minima of the gradient correspond to all the maxima of the image (for positive scales), or $q = 0$, $s = r$ to make the minima of the gradient correspond to all the minima of the image (for negative scales).

Instead of the original image $f$ we can use the preceding results on the smoothed image $f \circledast g_\sigma$. In particular, for $\sigma \geq 0$, we can find the watersheds of $|\nabla(f)|^+$, and for negative scales, the watersheds of $|\nabla(f)|^-$.

Replacing $f$ by $|\nabla(f)|^+$ and then $|\nabla(f)|^-$ in Eq. (115) we get

$$C[WS(|\nabla(f)|^+)] = C[R_{\min}(|\nabla(f)|^+)] \tag{131}$$

$$C[WS(|\nabla(f)|^-)] = C[R_{\min}(|\nabla(f)|^-)] \tag{132}$$

But using Eqs. (127) and (129) to substitute for $C[R_{\min}(|\nabla(f)|^+)]$ and $C[R_{\min} \times (|\nabla(f)|^-)]$ gives

$$C[WS(|\nabla(f)|^+)] = C[R_{\max}(f)] \qquad (133)$$

$$C[WS(|\nabla(f)|^-)] = C[R_{\min}(f)] \qquad (134)$$

which are transforms of the form required by Proposition XII.1. Therefore, we have a scale-space monotone property for gradient watershed regions.

**Theorem VII.2** (Scale-Space Monotonicity Property for the Number of Gradient Watershed Regions): *Let $C[WS(f)]$ denote the number of watershed regions of an image $f$, $|\nabla(f)|^+$ and $|\nabla(f)|^-$ the homotopy modified gradient images as defined earlier. Then for any scales $\sigma_1 < \sigma_2 < 0 < \sigma_3 < \sigma_4$,*

(a) $C[WS(|\nabla(f \oplus g_{\sigma_1})|^+)] \leq C[WS(|\nabla(f \otimes g_{\sigma_2})|^+)] \leq C[WS(|\nabla(f)|^+)]$;
(b) $C[WS(|\nabla(f \oplus g_{\sigma_4})|^-)] \leq C[WS(|\nabla(f \otimes g_{\sigma_3})|^-)] \leq C[WS(|\nabla(f)|^-)]$,

*and*

(c) $C[WS(|\nabla(f \odot g_{\sigma_1})|^+)] \leq C[WS(|\nabla(f \odot g_{\sigma_2})|^+)] \leq C[WS(|\nabla(f)|^+)]$;
(d) $C[WS(|\nabla(f \odot g_{\sigma_4})|^-)] \leq C[WS(|\nabla(f \odot g_{\sigma_3})|^-)] \leq C[WS(|\nabla(f)|^-)]$,

*Proof.* This theorem follows directly from Eqs. (133) and (134) and Proposition XII.1. ∎

With this theorem we have completed the formal development of a scale-space monotonicity theory for gradient watershed regions.

The full algorithm for obtaining a multiscale set of gradient watersheds is:

**Algorithm 2** (Gradient Watersheds):

```
   1. Select a set of scales of interest* {σ_k};
For each scale σ_k DO:
   2. smooth f to obtain f ⊕ g_{σ_k} using Eq.(52);
   3. find the regional minima, N_i, (for σ_k ≥ 0), or maxima,
      M_i, (for σ_k ≤ 0), of f ⊕ g_{σ_k}, and compute a suitable marker
      function g(x,y) (Eq. (123));
   4. compute the magnitude of gradient image |∇(f ⊕ g_{σ_k})|;
   5. modify the homotopy of this image (Eq. (125));
   6. find the watershed regions WS(|∇(f ⊕ g_{σ_k})|^±).
ENDDO:
```

---

*Note: The interesting scales may be prescribed by the application, or perhaps sampled (linearly or logarithmically) over some range. A set of scales at which regions vanish can be found without smoothing by Algorithm 1.

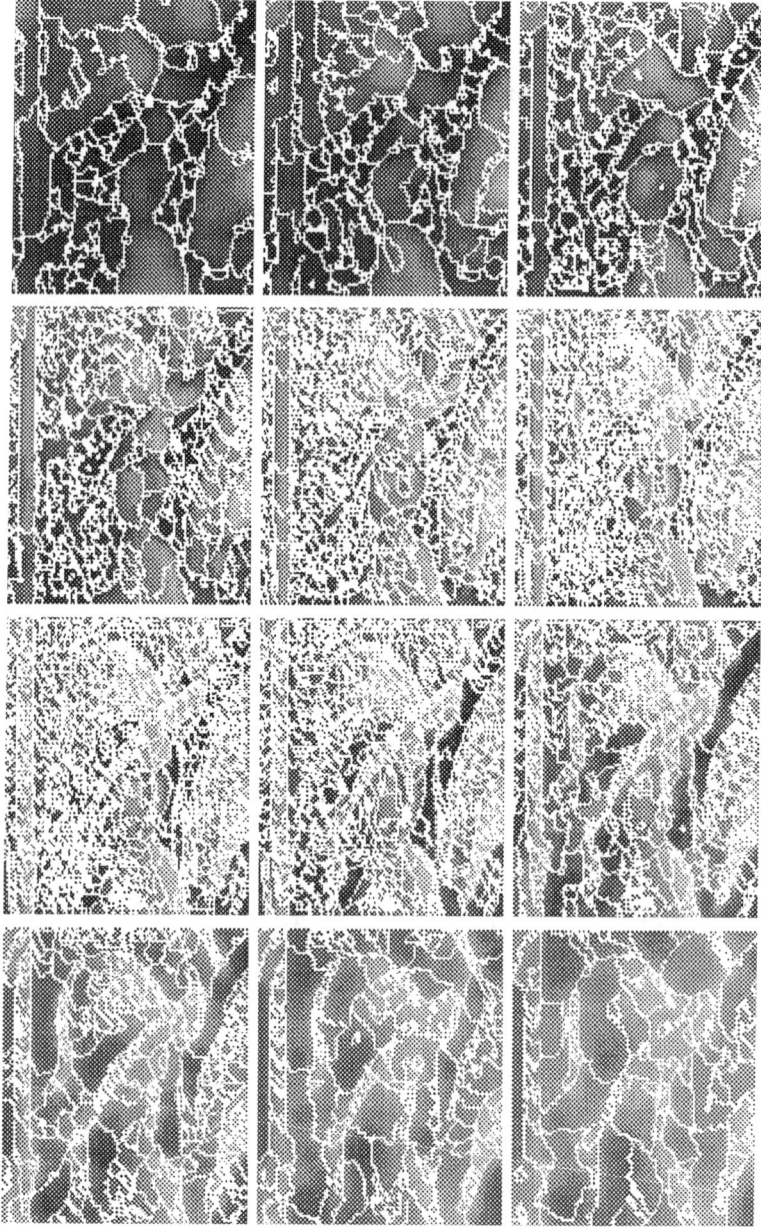

FIGURE 15. The closing-opening scale-space for gradient watershed regions. Homotopy modified gradient watersheds for the "Lena" image smoothed by multiscale closing-opening. From left to right, top to bottom, the scales are: $-2.5, -1.6, -0.9, -0.4, -0.1, 0.0-, 0.0+, 0.1,$ $0.4, 0.9, 1.6, 2.5$. A circular paraboloid structuring function was used.

As an example of the scale-space properties of the gradient watershed, we present Figure. 15.

### VIII. Summary, Limitations, and Future Work

#### A. Summary

Two scaled morphological operations, the multiscale dilation-erosion and the multiscale closing-opening, have been introduced for the scale-space smoothing of signals. These multiscale operations are translation invariant, nonlinear, increasing, and dependent on a real-scale parameter, which can be negative. The smoothed signals across all scales can be considered as a function on the so-called scale-space. This scale-space image exists for negative as well as positive scale and, thus, the information in the signal is more expanded than in the linear (Gaussian) scale-space image, which only exists for nonnegative scale. The scale-space image has good continuity and order properties. The position and height of extrema in the signal are preserved with increasing scale (maxima for positive scales and minima for negative scales), until they vanish at their characteristic scale. A monotone property for these signal features has been demonstrated. Fingerprint diagrams from these scale-spaces are identical and may be used to represent signals.

To summarize, the morphological scale-space differs from the Gaussian scale-space in that it:

- Possesses a monotone property in two and higher dimensions;
- Represents local extrema instead of zero-crossings;
- Exists for negative as well as positive scale.

With the morphological scale-spaces subsets of the full fingerprints, which are called reduced fingerprints, have been defined. The reduced fingerprint consists of local signal maxima for positive scale and local signal minima for negative scale. This reduced fingerprint:

- Consists of vertical lines only (since the position of signal features in not altered by the smoothing);
- Is equivalent to a set of $(n + 1)$-tuples, where $n$ is the dimension of the signal;
- Can be efficiently computed without signal smoothing by the algorithm presented.

The scale-spaces have been extended from point-set features—the extrema, to regions through the watershed transform. Through homotopy modification of the gradient function, the monotone has then been extended to gradient watershed regions.

## B. Limitations

It is important to note the limitations and restrictions on the proposed theory. In essence we return to the early days of scale-space theory by placing emphasis on the importance of signal features, and the tracking of these features through scale. Feature tracking is an idea which we believe is as yet underexplored.

Modern scale-space theory seems to concentrate on powerful mathematical results describing the axiomatic bases and the differential structure and invariants of the various scale-spaces. While important theoretically, causality in the form of the maximum principle on partial differential equations has taken precedence over a monotone principle for signal features. We have emphasized such a monotonic principle in this article.

Because we do not use an averaging filter for signal smoothing, the question of the sensitivity of this method to signal noise naturally arises. As the dilation and erosion depend on the extreme values of the signal in the neighborhood of a point, impulse noise in particular will upset the method. A high-amplitude impulse will be seen as a large-scale feature, especially if it is on a relatively flat region on the signal. In some applications this may be reasonable behavior; in others this may be unacceptable. In the end it is the application which determines if the definition of scale embodied in the proposed method is useful or otherwise.

Additionally, almost any high-frequency noise will introduce many new local extrema into the signal, causing many spurious features in the analysis. The glib answer is to say that if noise is present in the signal it should be filtered out *before* the signal is analyzed. This may indeed be appropriate in many cases, but realistically this sensitivity to noise may be one of the limitations of the approach.

## C. Future Work

The practical utility of our approach has yet to be demonstrated; in particular, work is needed on the computation, stability, inversion, and application of the full fingerprint and the watershed transform. In particular, which classes of signals are well represented by these quantities, and are these methods of use for signal compression?

In 1D, the noncreation of maxima (minima) implies the noncreation of minima (maxima) due to the interleaving of maxima and minima. However, in higher dimensions this does not necessarily hold. We have as yet no results on these other possible monotonic properties in higher dimensions.

In common with Gaussian scale-space, the theory of morphological scale-space has been developed in the continuous domain. Digital signal and image processing, however, involves discrete signals. The various results need to be formally obtained for digital signals.

Thinking out loud, we can make the following suggestions for future directions:

1. The formal extension of the theory to real and discrete functions on discrete spaces.
2. The inversion of the fingerprint and watersheds to reconstruct the original signal.
3. The use of multiscale morphology on the derivatives or integrals of multidimensional signals.
4. The use of transforms on the signal, before or after (or both!) scale-space analysis to alter the quantity represented by the scale dimension in the analysis.
5. A soft morphological scale-space based on the $k$th- and $(n - k)$th-order statistics.
6. ...!

### ACKNOWLEDGMENTS

Part of the material contained in this article was Ph.D. work under the supervision of V. V Anh, W. Boles, and M. Deriche from the Queensland University of Technology. The work on watersheds was supported by a grant from the University of Queensland.

## IX. APPENDIX

### A. Proof of Mathematical Results

#### 1. Proof of Proposition II.1

Because we consider only $\sigma, \mu \geq 0$, for clarity, we immediately drop the $|.|$ signs in (38). We start from the definition of convexity:

$$\lambda g(\mathbf{a}) + (1 - \lambda) g(\mathbf{b}) \leq g(\lambda \mathbf{a} + (1 - \lambda) \mathbf{b}), \qquad (0 \leq \lambda \leq 1) \quad (135)$$

Now make the following substitutions: $\lambda = \frac{\sigma}{\sigma+\mu}$; $(1 - \lambda) = \frac{\mu}{\sigma+\mu}$; $\mathbf{a} = \frac{\mathbf{x}-\mathbf{t}}{\sigma}$; $\mathbf{b} = \frac{\mathbf{t}}{\mu}$, so that

$$\frac{\sigma}{\sigma+\mu} g\left(\frac{\mathbf{x}-\mathbf{t}}{\sigma}\right) + \frac{\mu}{\sigma+\mu} g\left(\frac{\mathbf{t}}{\mu}\right) \leq g\left(\frac{(\mathbf{x}-\mathbf{t})+\mathbf{t}}{\sigma+\mu}\right) \quad (136)$$

This implies

$$\sigma g\left(\frac{\mathbf{x}-\mathbf{t}}{\sigma}\right) + \mu g\left(\frac{\mathbf{t}}{\mu}\right) \leq (\sigma+\mu)g\left(\frac{\mathbf{x}}{\sigma+\mu}\right) \qquad (137)$$

And we also note that,

$$\left\{\sigma g\left(\frac{\mathbf{x}-\mathbf{t}}{\sigma}\right) + \mu g\left(\frac{\mathbf{t}}{\mu}\right)\right\}_{\mathbf{t}=\frac{\mu\mathbf{x}}{\sigma+\mu}} = (\sigma+\mu)g\left(\frac{\mathbf{x}}{\sigma+\mu}\right) \qquad (138)$$

So,

$$g_\sigma \circledast g_\mu = \max_{\mathbf{t}}\left\{\sigma g\left(\frac{\mathbf{x}-\mathbf{t}}{\sigma}\right) + \mu g\left(\frac{\mathbf{t}}{\mu}\right)\right\}$$
$$= (\sigma+\mu)g\left(\frac{\mathbf{x}}{\sigma+\mu}\right)$$
$$= g_{\sigma+\mu} \qquad (139)$$

Thus, the semigroup property (Hille and Phillips, 1957) of convex structuring functions is proved.

## 2. Proof of Proposition II.2

From the property of the opening of the dilation, Eq. (17), we have

$$(f \oplus g) \circ g = f \oplus g \qquad (140)$$

and from the semigroup property for scaled structuring functions, Eq. (51), we have

$$g_{\sigma_2} = g_{|\sigma_2|-|\sigma_1|} \oplus g_1 \qquad (141)$$

Combining these equations we get the required result:

$$g_{\sigma_2} \circ g_{\sigma_1} = \left(g_{|\sigma_2|-|\sigma_1|} \oplus g_{\sigma_1}\right) \circ g_{\sigma_1}$$
$$= g_{|\sigma_2|-|\sigma_1|} \oplus g_{\sigma_1}$$
$$= g_{\sigma_2} \qquad (142)$$

## 3. Proof of Corollary III.1.2

Consider the case for positive scales, $0 < \sigma_3 < \sigma_4$. From the order properties for gray-scale dilation, Eq. (20), we have

$$(f \circledast g_{\sigma_4})(\mathbf{x}) \geq (f \circledast g_{\sigma_3})(\mathbf{x}) \geq f(\mathbf{x}) \qquad \text{for all } (\mathbf{x}) \qquad (143)$$

so with increasing scale the value of any fixed point $(f \circledast g_{\sigma_4})(\mathbf{x}_0)$ can never decrease.

A regional maximum is a connected component of the point-set of local maxima. A necessary condition for a regional maximum to exist is that *all* its points are local maxima. We may associate with each point of this set a scale $\sigma_i$ being the scale at which this point ceases to be a local maximum because one of its neighbors has exceeded its value. Then the whole regional maximum ceases to be a regional minimum at a scale of $\min\{\sigma_i\}$. Thus, each regional maximum exists for a range of scales of $0 \leq \sigma \leq \min\{\sigma_i\}$. Part (a) of the corollary follows. Part (b) follows from duality.

## 4. Proof of Proposition IV.1

From Proposition II.2, we find $g_{\sigma_1}$ is open with respect to $g_{\sigma_2}$, and $g_{\sigma_4}$ is open with respect to $g_{\sigma_3}$. Then, from Relation (28), the result follows.

## 5. Proof of Proposition IV.3

Equations (71) and (72) follow directly from Eq. (65). Relation (73) follows from Proposition IV.1, on the order (antiextensive) properties of the opening.

## 6. Proof of Proposition V.1

From Proposition III.4: $(f \oplus g_\sigma)(x_{\max})$ is a local maximum $\Rightarrow f(\mathbf{x}_{\max})$ is a local maximum, and $(f \oplus g_\sigma)(\mathbf{x}_{\max}) = f(\mathbf{x}_{\max})$. However, from Property (24) we have a sandwich result:

$$f(\mathbf{x}) \leq (f \bullet g_\sigma)(\mathbf{x}) \leq (f \oplus g_\sigma)(\mathbf{x}), \qquad \forall \mathbf{x} \in D \qquad (144)$$

therefore, $(f \bullet g_\sigma)(\mathbf{x}_{\max})$ is also a local maximum and $(f \bullet g_\sigma)(\mathbf{x}_{\max}) = f(\mathbf{x}_{\max})$.

To show the reverse relation, we appeal to the geometric interpretation of the closing. If $(f \bullet g_\sigma)(\mathbf{x}_{\max})$ is a local maximum, then the origin of the translated (reflected) structuring element at $\mathbf{x}_{\max}$ must be greater than the origin for the structuring element at all $\mathbf{x}$ in some $\epsilon$-neighborhood of $\mathbf{x}_{\max}$. Because the locii of this origin form the surface of the dilation operation, we have,

$$(f \oplus g_{\sigma_4})(\mathbf{x}) \leq (f \circledast g_{\sigma_4})(\mathbf{x}_{\max}) \quad \text{for all} \quad \mathbf{x} \in N(\mathbf{x}_{\max}, \epsilon) \qquad (145)$$

which shows that $(f \circledast g_\sigma)(\mathbf{x}_{\max})$ is a local maximum. This completes the proof of the first part of the proposition. Again, the second part follows from the morphological duality properties.

## 7. Proof of Proposition V.1

If $(f \oplus g_\sigma)(\mathbf{x}_{\min})$ is a local minimum, then the origin of the translated (negated) structuring function is lower than in the surrounding neighborhood. Then, because the negated structuring function has a local minimum at the origin and is convex, the union of the structuring functions in the neighborhood of $\mathbf{x}_{\min}$ has a minimum at $\mathbf{x}_{\min}$ and, because this union is the closing, $(f \bullet g_\sigma)(\mathbf{x}_{\min})$, is also a local minimum.

To show the reverse relation, we note that if $(f \bullet g_\sigma)(\mathbf{x}_{\min})$ is a local minimum; then, because the negated structuring function has a local minimum at the origin, and is convex, we have

$$(f \bullet g_\sigma)(\mathbf{x}_{\min}) = (f \oplus g_\sigma)(\mathbf{x}_{\min}) \tag{146}$$

Appeal to Property (24):

$$(f \bullet g_\sigma)(\mathbf{x}) \leq (f \oplus g_\sigma)(\mathbf{x}), \qquad \forall \mathbf{x} \in D \tag{147}$$

we see that $(f \oplus g_\sigma)(\mathbf{x}_{\min})$ must also be a local minimum. This completes the proof of the first part of the proposition. Once again, the second part follows from the morphological duality properties.

## 8. Proof of Proposition XII.1

Because $\#[\psi_1(f)] = C[R_{\max}(f)]$, and, $\#[\psi_2(f)] = C[R_{\min}(f)]$, then

$$C[R_{\max}(f_3)] \leq C[R_{\max}(f_4)] \Rightarrow \#[\psi_1(f_3)] \leq \#[\psi_1(f_4)] \tag{148}$$

and

$$C[R_{\min}(f_1)] \leq C[R_{\min}(f_2)] \Rightarrow \#[\psi_2(f_1)] \leq \#[\psi_2(f_2)] \tag{149}$$

Then the proposition parts (a) and (b) follow from Corollary III.1.2, parts (a) and (b), and the proposition parts (c) and (d) follow from Corollary IV.2.2, parts (a) and (b).

### B. Computer Code

A fragment of C code to extract the reduced morphological scale-space fingerprint from local maxima of a 2D function:

```
/*typedef struct
      {int x; int y; float scale;}scaleitemtype;        */
/*float f[N,N] holds the signal.                        */
/*scaleitemtype FP[K] holds the reduced fingerprint,    */
/*on input FP[] contains the co-ordinates of the        */
/*local maxima; on output also contains the             */
/*associated scales.                                    */
/*M is the global maximum of f[]                        */

for (i=1; i<=K; i++){
  xi = FP[i].x; yi = FP[i].y;
  sigi = MAXFLOAT;
  if (f[xi][yi] < M){
   R = 0;
   do{
     R++;
     xj = xi + R; yj = yi + R;
     for (j=1; j<=2*R; j++) {
      H = f[xi][yi] - f[xj][yj];
      if (H < 0.0){
          sigj = -(hypot(xi-xj, yi-yj) + H*H) / (2.0 * H);
          if (sigj < sigi) sigi = sigj;
       }/*if*/
      xj--;
     }/*for*/
     for (j=1; j<=2*R; j++){
      H = f[xi][yi] - f[xj][yj];
      if (H < 0.0){
          sigj = -(hypot(xi-xj, yi-yj) + H*H) / (2.0 * H);
          if (sigj < sigi) sigi = sigj;
     }/*if*/
        yj--;
     }/*for*/
     for (j=1; j<=2*R; j++){
      H = f[xi][yi] - f[xj][yj];
      if (H < 0.0){
          sigj = -(hypot(xi-xj, yi-yj) + H*H) / (2.0 * H);
          if (sigj < sigi) sigi = sigj;
     }/*if*/
     xj++;
    }/*for*/
    for (j=1; j<=2*R; j++){
      H = f[xi][yi] - f[xj][yj];
```

```
        if (H < 0.0){
          sigj = -(hypot(xi-xj, yi-yj) + H*H) / (2.0*H);
          if (sigj < sigi) sigi = sigj;
        } /*for*/
        xj++;
      } /*for*/
    } while ((R <= sigi)&&(R*R <= -H*(2.0*sigi + H)));
  } /*if*/
  FP[i]. scale = sigi;
}/*for i*/
```

## References

Alvarez, L., and Mazorra, L. (1994). Signal and image restoration using shock filters and anisotropic diffusion, *SIAM J. Numerical Anal.* **31**(2), 590–605.

Alvarez, L., and Morel, J. M. (1994). Formalization and computational aspects of image analysis, *Acta Numerica* **(1994),** 1–59.

Asada, H., and Brady, M. (1986). The curvature primal sketch, *IEEE T. Patt. Anal. Mach. Intell.* **PAMI-8**(1), 2–14.

Babaud, J., Witkin, A. P., Baudin, M., and Duda, R. O. (1986). Uniqueness of the Gaussian kernel for scale-space filtering, *IEEE T. Patt. Anal. Mach. Intell.* **PAMI-8**(1), 26–33.

Ballard, D., and Brown, C. M. (1982). *Computer Vision,* Upper Saddle River, N.J.: Prentice Hall,

Bartle, R. G. (1964). *The Elements of Real Analysis,* New York: Wiley.

Bell, S. B., Holroyd, F. C., and Mason, D. C (1989). A digital geometry for hexagonal pixels, *Image Vis. Comput.* **7**(3), 194–204.

Beucher, S. (1990). Segmentation tools in mathematical morphology, *in* P. D. Gader (ed.), *Image Algebra and Morphological Image Processing,* Vol. Proceedings SPIE Vol. 1350, pp. 70–84.

Bischof, W. F., and Caelli, T. (1988). Parsing scale-space and spatial stability analysis, *Comput. Vis. Graph. Image Process.* **42,** 192–205.

Butzer, P. L., and Berens, H. (1967). *Semi-Groups of Operators and Approximation,* Die Grundlehren der Mathematischen Wissenschaften in Einzeldarstellungen, Band 145, Berlin: Springer-Verlag.

Campos, J. C., Linney, A. D., and Moss, J. P. (1993). The analysis of facial profiles using scale space techniques, *Patt. Recog.* **26**(6), 819–824.

Carlotto, M. J. (1987). Histogram analysis using a scale-space approach, *IEEE T. Patt. Anal. Mach. intell.* **PAMI-9**(1), 121–129.

Chen, M.-H., and Yan, P.-F. (1989). A multiscaling approach based on morphological filtering, *IEEE T. Patt. Anal. Mach. Intell.* **11**(7), 694–700.

DePree, J. D., and Swartz, C. W. (1988). *Introduction to Real Analysis,* New York: Wiley.

DuChateau, P., and Zachmann, D. W. (1986). *Theory and Problems of Partial Differential Equations,* Schaum's Outline Series, New York: McGraw-Hill.

Florack, L. (1997). *Image Structure,* Dordrech: Kluwer.

Gader, P. D. (1991). Separable decompositions and approximations of greyscale morphological templates, *CVGIP: Image Understanding* **53**(3), 288–296.

Haar Romeny, B. T., Florack, L., Koenderink, J., and Viergever, M., Eds. (1997). *Scale-Space Theory in Computer Vision, First International Conference, Scale-Space'97, Utrecht, the Netherlands, July 1997. Proceedings,* Lecture Notes in Computer Science, LNCS 1252, Heidelberg: Springer-Verlag.

Haralick, R. M., Sternberg, S. R., and Zhuang, X. (1987). Image analysis using mathematical morphology, *IEEE Trans. Patt. Anal. Mach. Intell.* **PAMI-9**(4), 532–550.

Heijmans, H. J. A. M., and Ronse, C. (1990). The algebraic basis of mathematical morphology: I. Dilations and erosions, *Comput. Vis. Graph. Image Process.* **50,** 245–295.

Hille, E., and Phillips, R. S. (1957). *Functional Analysis and Semi-Groups,* American Mathematical Society Colloquium Publications Volume XXXI, Providence, R. Id.: American Mathematical Society.

Hiriart-Urruty, J.-B., and Lemarechal, C. (1993). *Convex analysis and minimization algorithms,* Berlin: Springer-Verlag.

Hummel, R. A. (1986). Representations based on zero-crossings in scale-space, *CVPR'86: Proc. IEEE Computer Society Conf. Comput. Vis. Patt. Recog.,* pp. 204–209. Miami, Fla.: IEEE Computer Society Press.

Hummel, R., and Moniot, R. (1989). Reconstructions from zero crossings in scale space, *IEEE T. Acoust. Speech Sig. Process.* **37**(12), 245–295.

Iijima, T. (1959). Basic theory of pattern observation, *Technical report,* Papers of Technical Group on Automa and Automatic Control, IECE, Japan. (in Japanese).

Iijima, T. (1962). Basic theory on normalization of pattern (in case of typical one-dimensional pattern), *Bulletin of the Electrotechnical Laboratory* **26,** 368–388.

Iijima, T. (1971). Basic equation of figure and observational transformation, *Systems, Computers, Controls* **2,** 70–77.

Jackway, P. T. (1992). Morphological scale-space, *Proceedings 11th IAPR Int. Conf. Patt. Recog.,* IEEE Computer society Press, Los Alamitos, CA, The Hague, The Netherlands, pp. C252–255.

Jackway, P. T. (1995a). *Morphological Scale-Space With Application to Three-Dimensional Object Recognition,* Ph.D. thesis, Queensland University of Technology.

Jackway, P. T. (1995b). On dimensionality in multiscale morphological scale-space with elliptic poweroid structuring functions, *J. Visual Comm. Image Represent.* **6**(2), 189–195.

Jackway, P. T. (1996). Gradient watersheds in morphological scale-space, *IEEE T. Image Process.* **5**(6), 913–921.

Jackway, P. T., Boles, W. W., and Deriche, M. (1994). Morphological scale-space fingerprints and their use in object recognition in range images, *Proc. 1994 IEEE Int. Conf. Acoust. Speech Sig. Process. (ICASSP'94),* Adelaide, pp. V5–8.

Jackway, P. (1998). *Morphological Scale-Spaces,* Vol. 99, pp. 1–64, San Diego: Academic.

Jackway, P. T., and Deriche, M. (1996). Scale-Space properties of the multiscale morphological dilation-erosion, *IEEE T. Patt. Anal. and Mach. Intell.* **18**(1), 38–51.

Jang, B. K., and Chin, R. T. (1991). Shape analysis using morphological scale space, *Proc. the 25th Annual Conf. Inf. Sc. Syst.,* pp. 1–4.

Jang, B. K, and Chin, R. T. (1992). Gaussian and morphological scale-space for shape analysis, *Asia Pacific Eng. J. (Part A)* **2**(2), 165–202.

Koenderink, J. J. (1984). The structure of images, *Biol. Cybernetics* **50**(5), 363–370.

Lantuéjoul, C. (1978). *La Squelettisation et son Application aux Mesures Topologiques des Mosaïques Polycristallines,* Ph.D. thesis, School of Mines, Paris.

Lee, C. N., and Rosenfeld, A. (1986). Connectivity issues in 2D and 3D images, *CVPR'86: Proc. IEEE Computer Society Conf. Comput. Vis. Patt. Recog.,* pp. 278–285, Miami, Fla.: IEEE Computer Society Press.

Lifshitz, L. M., and Pizer, S. M. (1990). A multiresolution hierarchical approach to image segmentation based on intensity extrema, *IEEE T. Patt. Anal. Mach. Intell.* **12**(6), 529–540.

Lindeberg, T. (1988). On the construction of a scale-space for discrete images, *Technical Report TRITA-NA-P8808,* Royal Institute of Technology, Sweden.

Lindeberg, T. (1990). Scale-space for discrete signals, *IEEE T. Patt. Anal. Mach. Intell.* **12**(3), 234–254.
Lindeberg, T. (1994). *Scale-Space Theory in Computer Vision,* Boston: Kluwer.
Lipschutz, M. M. (1969). *Theory and Problems of Differential Geometry,* Schaum's Outline Series, New York: McGraw-Hill.
Maragos, P., and Schafer, R. W. (1987a). Morphological filters—part I: Their set-theoretic analysis and relations to linear shift-invariant filters, *IEEE T. Acoust. Speech Sig. Process.* **ASSP-35**(8), 1153–1169.
Maragos, P., and Schafer, R. W. (1987b). Morphological filters—part II: Their relations to median, order statistic, and stack filters, *IEEE T. Acoust. Speech Sig. Process.* **ASSP-35** (8), 1170–1184.
Marr, D. (1982). *Vision,* Freeman, San Francisco.
Marr, D., and Hildreth, E. (1980). Theory of edge detection, *Proc. Royal Soc. Lond. B* **207**, 187–217.
Marr, D., and Poggio, T. (1979). A computational theory of human stereo vision, *Proc. Royal Soc. Lond. B* **204**, 301–328.
Matheron, G. (1975). *Random sets and Integral Geometry,* New York: Wiley.
Mokhtarian, F., and Mackworth, A. (1986). Scale-based description of planar curves and two-dimensional shapes, *IEEE T. Patt. Anal. Mach. Intell.* **PAMI-8**(1), 34–43.
Morita, S., Kawashima, T., and Aoki, Y. (1991). Patt. matching of 2-D shape using hierarchical descriptions, *Syst. Comput. Japan* **22**(10), 40–49.
Nacken, P. F. M. (1994). Openings can introduce zero crossings in boundary curvature, *IEEE T. Patt. Anal. Mach. Intell.* **16**(6), 656–658.
Nakagawa, Y., and Rosenfeld, A. (1978). A note on the use of local min and max operations in digital picture processing, *IEEE T. Syst. Man Cyber.* **SMC-8**(8), 632–635.
Nielsen, M., Johansen, P., Olsen, O., and Weickert, J., Eds. (1999). *Scale-Space Theories in Computer Vision, Second International Conference, Scale-Space'99, Corfu, Greece, September 1999. Proceedings,* Lecture Notes in Computer Science, LNCS 1682, Heidelberg: Springer-Verlag.
Perona, P., and Malik, J. (1990). Scale-space and edge detection using anisotropic diffusion, *IEEE T. Patt. Anal. March. Intell.* **12**(7), 629–639.
Protter, M., and Weinberger, H. (1967). *Maximum Principles in Differential Equations,* Upper Saddle Rever, N.J.: Prentice Hall.
Raman, S. V., Sarkar, S., and Boyer, K. L. (1991). Tissue boundary refinement in magnetic resonance images using contour-based scale space matching, *IEEE T. Med. Imag.* **10**(2), 109–121.
Rangarajan, K., Allen, W., and Shah, M. (1993). Matching motion trajectories using scale-space, *Patt. Recog.* **26**(4), 595–610.
Rivest, J.-F., Serra, J., and Soille, P. (1992). Dimensionality in image analysis, *J. Visual Comm. Image Represent.* **3**(2), 137–146.
Rosenfeld, A. (1970). Connectivity in digital pictures, *J. Assoc. Comput. Mach.* **17** (1), 146–160.
Rosenfeld, A., and Thurston, M. (1971). Edge and curve detection for visual scene analysis, *IEEE T. Comput.* **C-20**(5), 562–569.
Serra, J. (1982). *Image Analysis and Mathematical Morphology,* London: Academic Press.
Serra, J. (1988). *Image Analysis and Mathematical Morphology. Volume 2: Theoretical Advances,* Academic Press, London.
Serra, J., and Lay, B. (1985). Square to hexagonal lattices conversion, *Sig. Process.* **9**, 1–13.
Shih, F. Y., and Mitchell, O. R. (1991). Decomposition of gray-scale morphological structuring elements, *Patt. Recog.* **24**(3), 195–203.
Sporring, J., Nielsen, M., Florack, L., and Johansen, P. (1997). *Gaussian Scale-Space Theory,* Dordrecht: Kluwer.

Stansfield, J. L. (1980). Conclusions from the commodity expert project, *A.I. Lab Memo No. 601,* Massachusetts Institute of Technology.

Sternberg, S. R. (1986). Grayscale morphology, *Comput. Vis. Graph. Image Process.* **35**(3), 333–355. Scale space:

ter Haar Romeny, B. M., Florack, L. M. J., Koenderink, J. J., and Viergever, M. A. (1991). Its natural operators and differential invariants, *Proc. 12th Int. Conf. Image Process. Med. Imag. '91,* Lecture Notes in Computer Science V.511, Wye, UK, pp. 239–255.

Tsui, H.-T., Choy, T. T.-C., and Ho, C.-W. (1988). Biomedical signal analysis by scale space technique, *Proceedings of the 5th Int. Conf. Biomed. Eng.*

van den Boomgaard, R. (1992). *Mathematical Morphology: Extensions Towards Computer Vision,* Ph.D. thesis, University of Amsterdam.

van den Boomgaard, R., and Heijmans, H. J. (2000). Morphological scale-space operators: An algebraic framework, in *Mathematical Morphology and Its Applications to Image and Sigal Processing: Proceedings 5th International Symposium on Mathematical Morphology and its Applications to Image and Signal Processing, Palo Alto, USA, June 2000,* J. Goutsias, L. Vincent and D. S. Bloomberg, Eds., Kluwer Series in Computational Imaging and Vision, CIVI 18, pp. 283–290, Boston: Kluwer.

van den Boomgaard, R., and Smeulders, A. (1994). The morphological structure of images: The differential equations of morphological scale-space, *IEEE T. Patt. Anal. Mach. Intell.* **16**(11), 1101–1113.

Verbeek, P. W., and Verwer, B. J. H. (1989). 2-D adaptive smoothing by 3-D distance transformation, *Patt. Recog. Lett.* **9**, 53–65.

Vincent, L. (1993). Morphological grayscale reconstruction in image analysis: Applications and efficient algorithms, *IEEE T. Image Process.* **2**(2), 176–201.

Vincent, L., and Beucher, S. (1989). The morphological approach to segmentation: An introduction, *Internal Report C-08/89/MM,* School of Mines, Paris.

Vincent, L., and Soille, P. (1991). Watersheds in digital spaces: an efficient algorithm based on immersion simulations, *IEEE T. Patt. Anal. Mach. Intell.* **13**(6), 583–598.

Weickert, J. (1998). *Anisotropic Diffusion in Image Processing,* Stuttgart: Teubner-Verlag.

Weickert, J., Ishikawa, S., and Imiya, A. (1999). Linear scale-space has first been proposed in Japan, *J. Math. Imaging and Vision* **10**, 237–252.

Wilson, R., and Granlund, G. H. (1984). The uncertainty principle in image processing, *IEEE Transactions on Patt. Anal. Mach. Intell.* **PAMI-6**(6), 758–767.

Witkin, A. P. (1983). Scale-space filtering, *Proc. Int. Joint Conf. Art. Intell.,* Kaufmann, Palo Alto, CA, pp. 1019–1022.

Witkin, A. P. (1984). Scale-space filtering: a new approach to multi-scale description, in *Image Understanding 1984,* S. Ullman and W. Richards, Eds. pp. 79–95, Norwood, N.J.: Ablex.

Witkin, A., Terzopoulos, D., and Kass, M. (1987). Signal matching through scale space, *Int. J. Comput. Vis.* pp. 133–144.

Wu, L., and Xie, Z. (1990). Scaling theorems for zero-crossings, *IEEE T. Patt. Anal. Mach. Intell.* **12**(11), 46–54.

Yang, J.-Y., and Chen, C.-C. (1993). Decomposition of additively separable structuring elements with applications, *Patt. Recog.* **26**(6), 867–875.

Yuille, A. L., and Poggio, T. (1985). Fingerprints theorems for zero crossings, *J. Opt. Soc. Am. A* **2**(5), 683–692.

Yuille, A. L., and Poggio, T. A. (1986). Scaling theorems for zero crossings, *IEEE T. Patt. Anal. Mach. Intell.* **PAMI-8**(1), 15–25.

# The Processing of Hexagonally Sampled Images

## RICHARD C. STAUNTON

*School of Engineering, University of Warwick*
*Coventry CV4 7AL, United Kingdom*

| | |
|---|---:|
| I. Introduction | 192 |
|    A. Sampling | 192 |
|    B. Processor Architectures | 194 |
|    C. Binary Image Processing | 195 |
|    D. Monochrome Image Processing | 195 |
| II. Image Sampling on a Hexagonal Grid | 196 |
|    A. The Hexagonal Packing of Sensory Elements in the Eye | 196 |
|    B. Hexagon-Shaped Sensor Elements | 198 |
|    C. Two-Dimensional Sampling Theory | 199 |
|    D. Noise and Quantization Error | 205 |
|    E. Practical Aspects of Digital Image Acquisition | 205 |
|       1. CCD TV Camera | 209 |
|    F. Measurement of 2D Modulation Transfer Function and Bandlimit Shape | 211 |
| III. Processor Architecture | 218 |
|    A. Single-Instruction, Single-Data Computers (SISD) | 218 |
|    B. Parallel Processors | 221 |
|    C. Two- and Multidimensional Processor Arrays | 222 |
|       1. Fine-Grain Arrays | 222 |
|       2. Coarse-Grain Arrays | 225 |
|    D. Pyramid Processors | 225 |
|       1. Hexagonal Pyramids | 226 |
|    E. Pipelined Processors | 229 |
|       1. Pipelined Systems | 231 |
|    F. Hexagonal Image-Processing Pipelines | 233 |
| IV. Binary Image Processing | 237 |
|    A. Connectivity | 237 |
|    B. Measurement of Distance | 238 |
|    C. Distance Functions | 238 |
|    D. Morphological Operators | 239 |
|    E. Line Thinning and the Skeleton of an Object | 239 |
|    F. Comparison between Hexagonal and Rectangular Skeletonization Programs | 240 |
|       1. Skeletal Quality | 243 |
|       2. Program Efficiency | 245 |
|    G. Tomography | 246 |
| V. Monochrome Image Processing | 247 |
|    A. The Hexagonal Fourier Transform | 247 |
|    B. Geometric Transformations | 247 |
|    C. Point Source Location | 248 |
|    D. Image-Processing Filters | 248 |

|     |     |
| --- | --- |
| 1. Linear Filters | 248 |
| 2. Nonlinear Filters | 250 |
| E. Edge Detectors | 250 |
| F. Hexagonal Edge-Detection Operators | 251 |
| G. The Visual Appearance of Edges and Features | 252 |
| 1. Human Interpretation | 256 |
| VI. Conclusions | 256 |
| References | 260 |

## I. Introduction

This chapter argues the case for the hexagonal sampling of images. Historically, square sampling has always predominated even though at each stage in the development of digital image processing over the last 30 years good hexagonal alternatives have been advanced. Advantages have been shown for the hexagonal scheme in sampling efficiency, processing algorithms and parallel processors; these will be discussed in the following sections.

### A. Sampling

Classically, the brightness at a point in a continuous two-dimensional (2D) field, $b = f(x, y)$ where $x$ and $y$ are the horizontal and vertical distances of the point from the origin. This field can be considered to be sampled by a grid of delta functions to produce a spatially discrete set of brightness values, and these brightness values themselves can be discretized to form what is usually considered to be a digital image (Gonzalez and Woods, 1992).

Figure 1 shows two of many possible regular grids of delta functions that can be used for spatial sampling. The vertical spacing of each has been chosen to be identical. If the sampled brightness is included as an orthogonal vector at

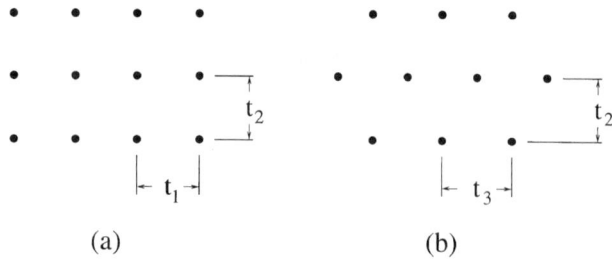

FIGURE 1. Sampling grids: (a) square; (b) hexagonal.

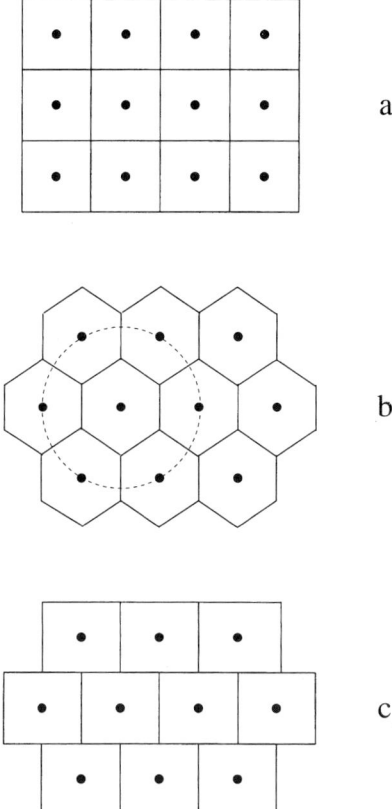

FIGURE 2. Image tilings: (a) square grid with squares tiles; (b) hexagonal grid with hexagonal tiles; (c) hexagonal grid with rectangular tiles.

each point then a digital image is formed. This image may be viewed on a TV monitor if the 2D space is tiled by picture elements (pixels), where each pixel is associated with one sampled value and is filled with the sampled brightness value. Referring to Figure 1a, if $t_1 = t_2$, then the image may be completely tiled by square pixels; in Figure 1b, if $t_3 = 2/\sqrt{3}t_2$, then the image may be completely tiled by regular hexagons or rectangles. The hexagonal tiling shown in Figure 2 has resulted in the term *hexagonal sampled image*.

Various aspects of hexagonal sampling are investigated in Section II. These include the evolution of hexagonal arrays of biological sensors, the effect of the choice of sampling grid and tiling on the appearance of the image, and the light-gathering properties of the sensor. To show an advantage of one sampling scheme over another in terms of the information content, the bandlimiting of the image by the sensor system must be investigated. An example of the measurement of the bandlimiting characteristics of some TV camera-frame grabber systems is given at the end of the section.

## B. Processor Architectures

Much image processing is accomplished using single-instruction, single-datum (SISD) computers (Flynn, 1966). A single-processor PC is an example of such a computer. Here, images are stored in semiconductor memory, disk files, or during computation in an array. For square sampled images the data map directly into a 2D array, each element of which can be accessed by a pair of row and column pointers that directly relate to the original position of the pixel. For hexagonally sampled images the data can again be readily stored, but the mapping of the data onto a square array within a program requires some care as described in Section III.

With multiprocessor systems, the processing task is divided and distributed among the processor elements (PE). This can simply be accomplished by organizing the PEs in a pipeline and assigning each a different task such as smoothing, edge detection, etc. Other architectures that readily allow such task divisions include hypercubes and shared memory machines. Another way to divide the task is to assign each PE to a local area of the image and then to allow it to sequentially apply separate tasks to that area. With this arrangement, the sampling grid shape can affect the way the PEs are interconnected for communication.

Image-processing tasks can be categorized at low, middle or high level (Luck, 1987). Low-level processes have input data that are associated with the original sampling grid and their output is also associated with it. The Sobel edge detector (Gonzalez and Woods, 1992) is an example of such a process. Middle-level processes again take data that are associated with the grid, but the output is often symbolic and not locked to the grid. A Hough transform (Illingworth and Kittler, 1988) that determines the angle of a straight line is an example of such a process. High-level processes have both input and output data that are not locked to the sampling grid. For example, in optical character recognition a process may take as an input a set of features including stroke end points and junctions, and output the ASCII code of the character.

The effect of the sampling grid on the structure of multiprocessor systems in discussed in Section III. A comparison between the processing of rectangularly and hexagonally sampled images by a pipeline processor has been presented.

## C. Binary Image Processing

With a binary image one brightness level is often used to distinguish foreground objects and the other is used to distinguish the background. However, in realistic images containing noise, some pixels invariably are incorrectly classified. With binary images many processing algorithms are concerned with how pixels are connected and hence a tile or pixel model of the image is used. Connectivity is easily defined for the hexagonal scheme (Rosenfeld, 1970) and holds for either tiling shown in Figure 2.

Hexagonal connectivity between the set of pixels in the object and the set in the background can both be defined as six-way connected. If the cluster of hexagons in Figure 2(b) are considered as an object, then connectivity between a central pixel and each of the six surrounding neighbors is identical apart from the orientation of the border between the pixels. For the square scheme, pixels can be considered to be part of an object if they are either four-way connected, that is, along a vertical or horizontal border, or eight-way connected where corner-to-corner connectivity is allowed. Background pixels can also be either four- or eight-way connected, but if the foreground is eight-way connected, the background must be four-way connected or visa versa; otherwise foreground and background features may cross over one another.

Many hexagonal processing algorithms for binary image processing have been researched and published. Section IV discusses some of the advantages and disadvantages of hexagonal operators and their square counterparts. A comparison between hexagonal and rectangular skeletonization programs is presented.

## D. Monochrome Image Processing

With gray-scale images, pixel, sampling point, and other models of the image structure have been used in the development of processes. Hexagonal counterparts of well-known square-grid process have been designed, and accuracy and computation speed comparisons have been made between the two schemes.

Many of the hexagonal algorithms designs have exploited the equidistance between neighboring sampling points rather than any notional pixel shape. Figure 2b shows a regular hexagonal grid with a circle imposed on the six nearest-neighbor sampling points that surround a central point. Algorithms often utilize masks of coefficients that are convolved with local areas of the image. Coefficient weights are often a function of distance from the center, and thus the symmetry of the hexagonal scheme can result in simplified processing. Some square scheme masks used with such convolution operators are separable but this does not apply to hexagonal scheme operators. This can partly remove the advantage of a hexagonal operator in some cases. Small area masks can be efficiently convolved with the image in the spatial plane, but greater efficiency can be achieved with large masks by initially transforming the mask and data to the Fourier plane. Efficient hexagonal scheme transforms (Rivard, 1977) have been developed that compare favorably with the square-system fast Fourier transform (FFT).

In Section V some simple hexagonal processing algorithms for gray-scale image processing are presented and their advantages and disadvantages with their square counterparts are discussed. The design of a simple hexagonal grid edge detector is discussed and its operation compared with that of the square-grid Sobel operator. Finally, comments on the visual appearances of hexagonally and rectangularly sampled images are made.

## II. Image Sampling on a Hexagonal Grid

### A. *The Hexagonal Packing of Sensory Elements in the Eye*

Biological and opthalmic observations on the human eye indicate that a hexagonal packing of retinal sensory elements has evolved in nature. This was a motivation for the study of hexagonal sampling schemes for computer vision covered in this chapter. Behind the eye, ganglion cells and neurons connect to the retinal sensory elements and to each other to provide processing of the image focused on the retina. Further image processing occurs in the visual cortex and the brain. Models of biological image processing have lead to the development of computer architectures such as artificial neural networks and pyramid processors for computer vision. However, in this section, the discussion is limited to the sensor element structure.

Helmholtz includes an anatomical description of the eye in his *Treatise on Physiological Optics* (Helmholtz, 1911, 1962). The higher orders of life have eyes capable of distinguishing both light and darkness and also form, hence, the eyes can have one of two forms. The first, common among insects, is a composite eye, in which sensory elements separated by opaque septa cover the surface of the eye. The elements at the surface of the eye are usually of a hexagonal and sometimes of a square shape. The second form of eye, as with the eyes of many vertebrates, has a lens that focuses light onto a retina. A section of a retina is shown in Figure 3. The retina is comprised of rod and cone sensory elements. In the human eye there are approximately 100 million elements of the smaller rod type, and 5 million cones that are distributed among the rods in varying densities depending on the particular part of the retina (Wandell, 1995). In the so-called yellow spot only cones are found, whereas towards the

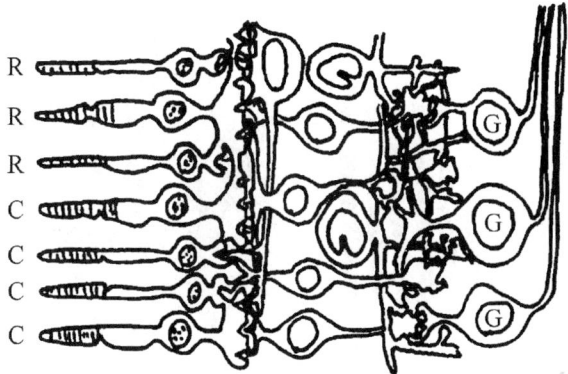

FIGURE 3. The human retina: R—rods; C—cones; G—ganglion cells.

periphery of the retina there are only rods. The rods primarily initiate low-level light vision and the cones initiate high-level light vision. Behind the surface layer of rods and cones are layers of fine fibers connecting these elements to a layer of ganglion cells. These cells perform many processes, one of which is to pass information to the optic nerve. Thus the retina is a complicated array of different types of sensory element and has a number of layers associated with detection and interconnection; possibly some image processing is also performed (Watson and Ahumada, 1989).

From the anatomic drawings in the Helmholtz treatise, it can be observed that the roughly circular sensory elements tend to pack together efficiently, which leads to a closely packed hexagonal lattice. Opthalmic experiments reported in Helmholtz's second volume prove this to be the case. In one experiment, Helmholtz set up a grating of light and dark lines of equal thickness viewed at various distances and under differing lighting conditions to measure the spatial resolution of the eye. His results indicated that two bright lines could only be distinguished if an unstimulated retinal element existed between the elements on which the images of the lines fell. This is in accordance with Nyquist's sampling theorem (Nyquist, 1928). He also noted that for grid spacings close to the resolution limit of the eye, the lines appeared wavelike or modulated with repeated thick and thin sections as shown in Figure 4. From this effect he inferred that the cone sensors, the only type of sensor in the high-resolution part of the retina, were packed in a hexagonal pattern.

Images of sections through cone sensors in the yellow spot of a human retina (Curcio *et al.,* 1987) show a roughly hexagonal shape for each cone,

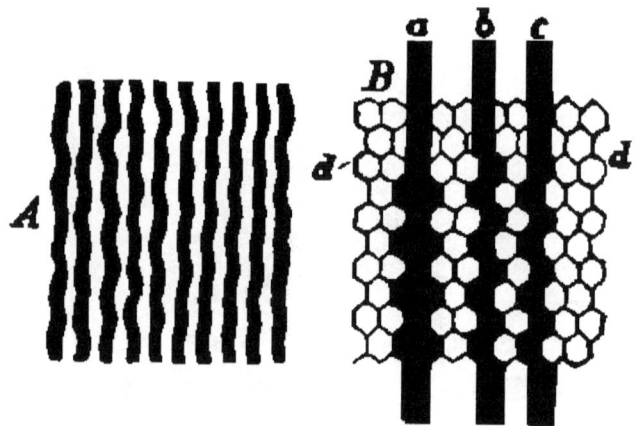

FIGURE 4. The wavelike appearance of parallel lines when viewed close to the eyes resolution limit and the hexagonal sensor pattern that produces this effect. (*Source: Helmholtz Treatise.* Helmholtz, 1911).

and sections through regions containing only rods show a hexagonal shape for each rod (Wandell, 1995). However, where cones exist in mixed regions their shape becomes more circular (Wandell, 1995).

## B. Hexagon-Shaped Sensor Elements

For any vision system, be it electronic, photochemical (Mitchell, 1993), biological or other, there are certain design parameters that can be optimized to increase its usefulness for a particular purpose or in a particular environment. Scenes with low light level can be best imaged using sensors with active areas that completely tile the image plane and that have long integration times or low shutter speeds. These techniques together with the use of large area sensors will increase the brightness signal-to-noise ratio (SNR). However, a smaller number of larger sensors will result in a lower spatial accuracy and a longer integration time in motion artifacts or missed events.

The 2D shape of the sensor elements and the geometry of the sampling grid will have an effect on the efficient acquisition of the image. The sensor element shape and any analogue signal processing by, for example, the lens, will 2D bandlimit the signal before digitization. In the general case, the oversampling of a spatial frequency bandlimited signal will not provide any increase in information, just more data.

## C. Two-Dimensional Sampling Theory

Before a computer can process an image, the image must be sampled and then the quantity sampled digitized. Real-world scenes can be considered as continuous 2D brightness fields. These brightness fields can be transformed to the Fourier plane and their spatial-frequency components analyzed. The magnitude of these transformed images can be considered as 2D signals and plotted against vertical spatial frequency and horizontal spatial frequency (Gonzalez and Woods, 1992), and their spectra can be analyzed. The phase information can be analyzed in a similar way. The image of the scene is focused onto a detector and then sampled. The sampling process can be considered as a 2D convolution between the continuous image and a grid of delta functions. For 1D signals, the sampling theorem (Nyquist, 1928) states that if a signal is to be perfectly recovered from its sampled version, then there must be no frequency component in the presampled signal that is greater than one-half the sampling frequency. A more recent theorem (Petersen and Middleton, 1962) allows consideration of multidimensional signals, and for a 2D image can be stated as follows: A brightness function whose Fourier transform is zero outside all but a finite area of the Fourier plane can be everywhere reconstructed from its sampled values, provided that this finite area and its periodic extensions in the Fourier plane are nonoverlapping. Any real-life continuous scene will contain spatial-frequency components throughout the spectrum and the direct sampling of such a brightness field would result in frequency aliasing where frequencies above half the sampling frequency will be folded about the half-sampling frequency and superimposed on the lower-frequency components. This aliasing results in corruption of the discrete image and makes it impossible to perfectly reconstruct the continuous image.

It is important to bandlimit the 2-D signal before sampling so that in the sampled signal the magnitude of its spectrum tends towards zero before components from periodic extensions of the spectrum interfere with the signal and cause aliasing. Figure 5 shows an example of the spectrum of a discrete 2D signal. The central hill at the origin is identical to the spectrum of the continuous signal, and the other hills are some of the closer periodic extensions of this. Here, the hills do not overlap so there will not be any aliasing. However, the gaps between the hills are indicative of inefficient sampling in that the vertical and horizontal sampling frequencies could be reduced by a factor of approximately two before aliasing would occur. The spectrum shown in Figure 5 results from an image sampled on a square grid. The periodically extended hills are located on a square grid in the Fourier plane with each centered on integer multiples of the horizontal and vertical sampling frequencies. Other sampling grids will lead to other extension patterns in the Fourier plane.

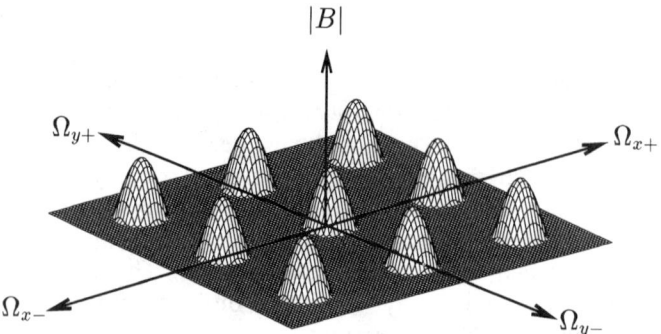

FIGURE 5. Example of the spectrum of a 2D discrete signal.

The conical shape of each hill has a circular cross section and is known as a circularly bandlimited signal (Mersereau, 1979). If the cross section is taken at the base of the hill, then all the signal information will be contained within this 2D bandlimit region. The efficient packing of these all-inclusive band regions has been studied (Petersen and Middleton, 1962).

Figure 6 shows the 2D spectrum periodicity for an octagonal bandlimited signal on a skewed grid. The regions are quite separate and the sampling

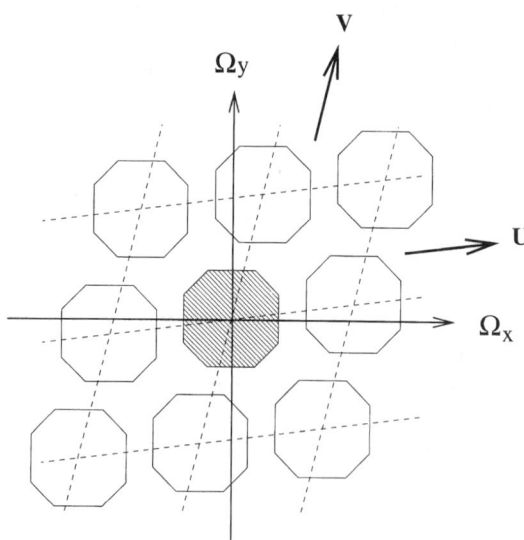

FIGURE 6. An octagonal band region, shown hashed, and some of its periodic extensions on a skewed grid.

efficiency could be increased by reducing the sampling frequency in the **U** and **V** directions. However, such octagonal regions will not pack together to completely tile the plane as do some other shapes located at certain positions.

Image sampling is usually achieved with a periodic sampling grid, and that grid is often square, but it is sometimes rectangular and occasionally hexagonal. The skewed sampling grid has been shown to be the general periodic grid (Petersen and Middleton, 1962), of which the foregoing are only special cases. We can now determine which is the most efficient grid. The minimum number of sampling points required to completely cover the image so that no information is lost must be found. This number will be a function of the grid geometry and the bandlimit of the image signal. If the bandlimit shape can be found and it completely tiles the Fourier plane, then we will have 100% efficiency. In theory there are many shapes that will completely tile a plane, including a square, a rectangle, a hexagon, an octagon with a small square extension in one corner, and a triangle that is alternately inverted. In practice, the bandlimit shape will be determined by the shape and characteristics of the sensor and any optical preprocessing by, for example, the lens. Theorists often choose a circular bandlimit shape to work with because then the spatial frequency is limited equally in each direction throughout the image plane. This means that a feature presented to the imaging system and detected at one angle would be equally well detected if presented at any other angle.

Early work (Petersen and Middleton, 1962) showed that circularly bandlimited images can be most efficiently sampled on a regular hexagonal grid as the bandlimit regions pack optimally in the Fourier plane. Such a packing is shown in Figure 7a. Petersen and Middleton (1962) quote and efficiency of 90.8% for the regular hexagonal grid compared with a maximum efficiency of 78.5% for the square grid.

Mersereau (1979) calculated that 13.4% fewer samples are required when a circular bandlimited signal is sampled on a hexagonal grid than when sampled on a rectangular grid. He continued to investigate bandlimit shapes in the Fourier plane. If a 2D continuous image is given by $f_c(x, y)$, where $x$ is the horizontal and $y$ is the vertical distance from the origin, then a discrete rectangularly sampled image can be described by

$$f_d(n_1, n_2) = f_c(n_1 t_1, n_2 t_2) \qquad (1)$$

where $t_1$ and $t_2$ are the horizontal and vertical sampling intervals as shown in Figure 1a, and $n_1$ and $n_2$ are integer indexes to the image array. If $F_c(\Omega_x, \Omega_y)$ is the continuous image $f_c(x, y)$ transformed to the Fourier plane, then the image is bandlimited within a shape $S$ if

$$F_c(\Omega_x, \Omega_y) = 0, \qquad (\Omega_x, \Omega_y) \ni S \qquad (2)$$

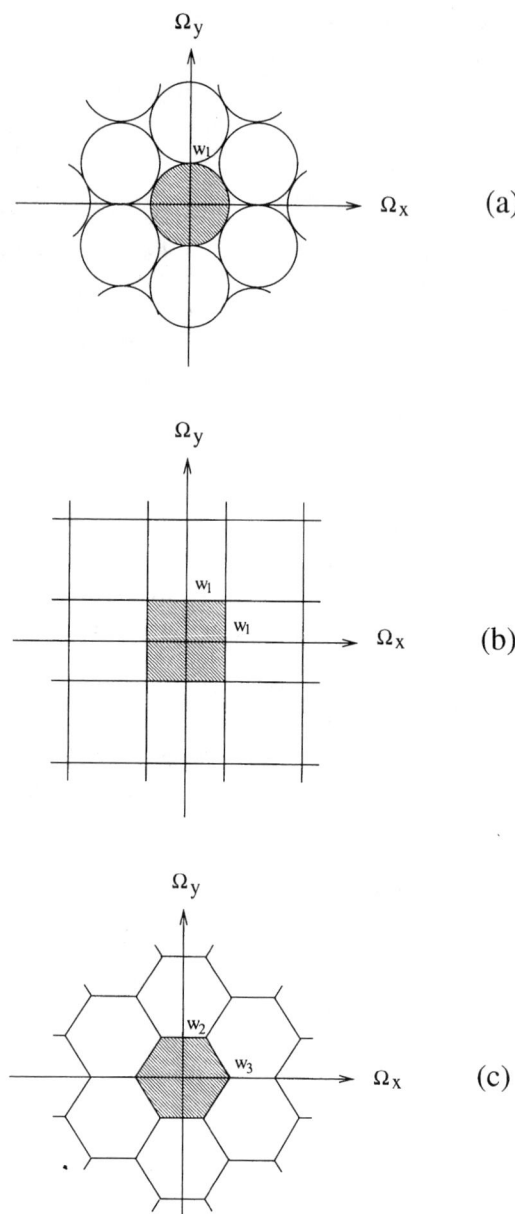

FIGURE 7. Tilings of the Fourier plane: (a) circular, (b) square; (c) regular hexagonal.

For the continuous image to be completely recoverable from a rectangularly sampled image, it must be bandlimited within the rectangular region defined by

$$\omega_1 < \pi/t_1 \qquad \omega_2 < \pi/t_2 \tag{3}$$

where $\omega_1$ is the horizontal and $\omega_2$ is the vertical bandwidth in rad · m$^{-1}$. If square sampling has been employed, then $\omega_1 = \omega_2$ and the band region will be square as shown in the crosshatched region in Figure 7(b). For the discrete image the Fourier plane will be tiled with periodic extensions of this base region with each square centered on coordinates that are $2n$ multiples of $\omega_1$, where $n$ is an integer. With a square band region, it is interesting to note that the image will have a frequency response at $\pm 45°$ that is $\sqrt{2}$ times that for the horizontal direction.

A hexagonal sampling theorem has been developed (Mersereau, 1979). A hexagonally sampled image can be described by

$$f_d(n_1, n_2) = f_c((n_1 - n_2/2)t_3, n_2 t_2) \tag{4}$$

where $t_2$ and $t_3$ are defined in Figure 1b, $n_1$ is an integer index along the horizontal axis, and $n_2$ is an integer index along an oblique axis at $120°$ to the horizontal. The vertical spacing of this grid ($t_2$) has been chosen to be the same as the vertical spacing of the previous rectangular grid. If the hexagonal grid is regular, then $t_3 = 2/\sqrt{3}t_2$. For the continuous image to be completely recoverable from a regular hexagonally sampled image, it must be bandlimited within the hexagonal region defined by

$$\omega_2 < \pi/t_2 \qquad \omega_3 < 4\pi/3t_3 \tag{5}$$

where, as shown in Figure 7c, $\omega_3$ is the horizontal and $\omega_2$ is the vertical bandwidth in rad · m$^{-1}$. Substituting for $t_3$ in Eq. (5), $\omega_3 < 2\pi/\sqrt{3}t_2$, and the maximum values of $\omega_2$ and $\omega_3$ are related by

$$\omega_{3,\,\text{max}} = \frac{2}{\sqrt{3}} \omega_{2,\,\text{max}} \tag{6}$$

The horizontal extent of the band region is larger than the vertical extent.

The hashed regions shown in Figures 7b and c represent the largest band regions for images that can be sampled on square and hexagonal grids. In practice, the image may be bandlimited to any arbitrary shape, but if this fits within the appropriate hashed region, then the bandlimiting will be sufficient to enable the image to be perfectly reconstructed. Considering a circular bandlimited image, then as shown in Figures 8a and b, the band region can be made to fit exactly within the square or hexagonal region by adjusting the common grid parameter $t_2$. The circle more completely covers the hexagonal region than the

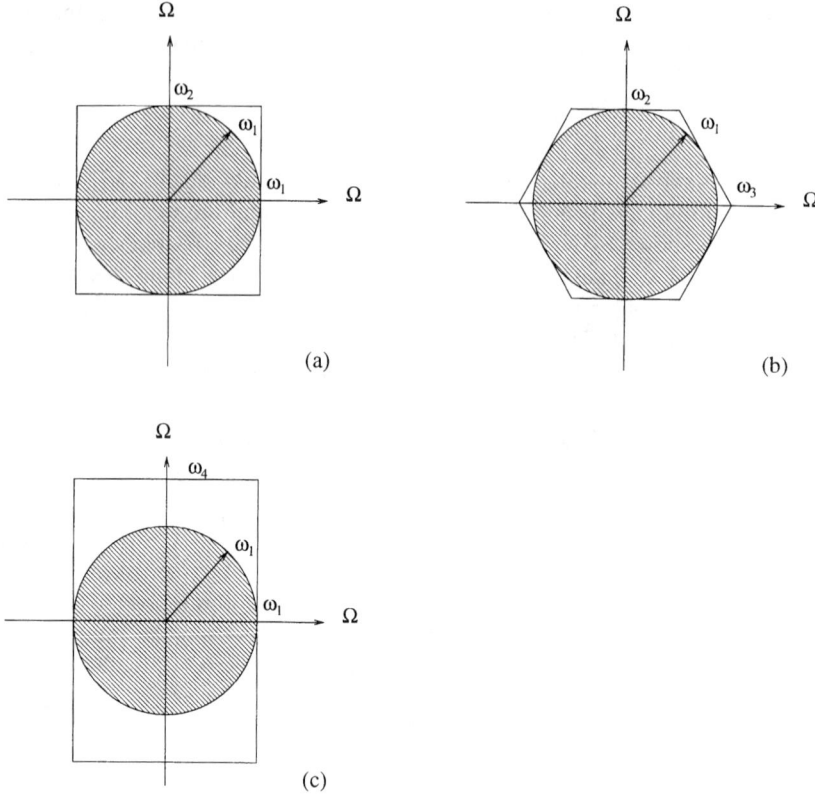

FIGURE 8. Utilization of available bandwidth by a circular bandlimited image sampled on various grids: (a) square; (b) regular hexagonal; (c) rectangular.

square and there is less wasted bandwidth. The circularly bandlimited image of radius $\omega_1$. can be sampled by either grid, the maximum spatial frequency will be equal in any direction within the sampled image, and frequency aliasing will not occur. The vertical spacing of each grid is identical $t_2$, but the horizontal spacing on the hexagonal grid is larger, resulting in a 13.4% saving in sampling points and an advantage for the hexagonal grid. Figure 8c shows a rectangular band region containing the circular region of radius $\omega_1$. The rectangular case has been fully analyzed elsewhere (Mersereau, 1979), but graphical observation indicates poor utilization of the available bandwidth. On the other hand, if the image was square bandlimited, the square grid would have an advantage. The bandlimiting of the image must be investigated before an advantage for one grid can be identified.

## D. Noise and Quantization Error

Noise from a number of sources can corrupt an image. Before sensing, low- and high-frequency lighting can modulate the image. Atmospheric distortion, rain, and vibration of the sensor can also add noise. Electronic noise can be additive or multiplicative, and introduced at the sensor or by the electronics. Quantization error will be introduced in both the spatial and brightness digitizations of the image. The average quantization error can be estimated (Kamgar-Parsi, 1989), and its effect on various image-processing operations can be evaluated.

Quantization error can be estimated for hexagonal grids of sensors (Kamgar-Parsi, 1992). The average error and the distribution of a function on an arbitrary number of independently quantized variables can be estimated and used to compare the relative noise sensitivity of hexagonal and square sampling grids. It has been shown (Kamgar-Parsi, 1992) that depending on the image-processing operation, the effects of hexagonal quantization error can be between 10% below to 5% above that for a square sampling grid quantization error. Finally, it is concluded here that there is little difference between the effects of the quantization error for the two systems.

## E. Practical Aspects of Digital Image Acquisition

Digital image acquisition systems generally provide several serially organized functions, including (1) continuous (analogue) image forming, (2) antialias filtering, (3) spatial discretization, (4) analogue-to-digital conversion (ADC), and (5) signal processing. In addition, reconstruction, the forming of a continuous image at the output of a digital system, is also considered by some researchers (Burton *et al.,* 1991) when estimating the quality of a system. Functions (1) and (2) can readily be considered together and are sometimes referred to as the image-gathering section (Burton *et al.,* 1991). Functions (3) and (4) cover spatial and brightness discretization and are often referred to jointly as digitization. Function (5) refers here to processes such as amplification or impedance matching within the electronics of the system.

The transfer functions of the image-gathering, sampling and reconstruction sections can be analyzed separately and then cascaded to determine the total effect on the reconstructed image as a part of the design. Sometimes it is possible to make a total system measurement (Staunton, 1998).

By making this separation between the analogue and digital sections of the system, we can consider that the image-gathering components bandlimit the analogue image before digitization (Staunton, 1996b). The shape of the bandlimit region can be determined, and, as discussed in Section II.C must be known before the most efficient sampling grid can be chosen.

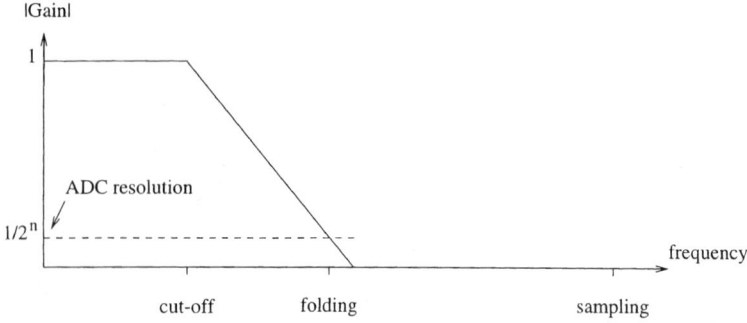

FIGURE 9. One-dimensional antialiasing filter.

Antialias filtering is important if the image is to be perfectly reconstructed. Such filters of various orders can readily be designed for time-varying voltage signals, and Figure 9 shows the modulus of the gain against frequency for such a 1D filter together with the sampling and folding frequencies. The slope in the cutoff region is determined by the order, and the design will typically require aliased components to be reduced to less than the resolution ($1/2^n$) of the ADC above the folding frequency, where $n$ is the number of bits in the ADC output.

For an imaging system the nature of the antialias filter will be determined by the physics of the imaging being undertaken. It will be 2D and the magnitude of its gain can be plotted as a series of 2D contours in the Fourier plane. Ideally, to avoid aliasing, the magnitude of the gain contour that indicates that the filter output is below the resolution of the ADC should coincide with the baseband spatial-frequency limit imposed by the sampling grid. The baseband region is shown crosshatched in Figures 7b and 7c for square and hexagonal grids. In practice, a circular bandlimit region that lies within the ideal band region will be used to ensure equal resolution in each direction. Often the antialiasing filter cutoff frequency is determined only by the focusing and limitations of the lens and the receptive area of the sensor. An example of an optical design is given in Section II.F.

The sensor array is discrete. It samples the image, but the finite receptive area of each sensor also smooths it. The sampling function can be considered as the convolution of the continuous image with a grid of Dirac delta functions. This can be expressed mathematically be Eqs. (1) and (4), or in a vector form (Ulichney, 1987; Burton et al., 1991) by

$$s(x) = \sum_n \delta(x - Vn) \qquad (7)$$

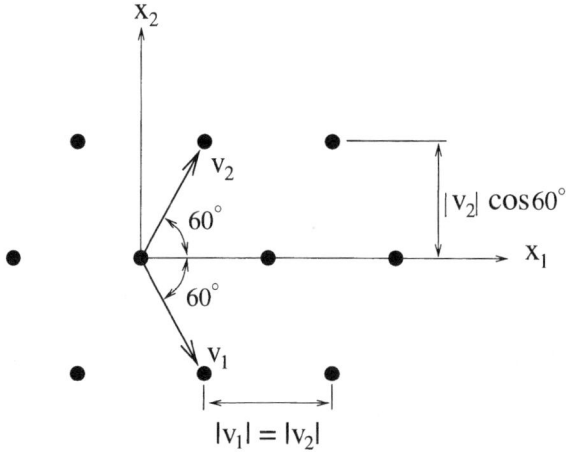

FIGURE 10. Regular hexagonal grid with sampling vectors.

where $\mathbf{n}$ is a 2D integer column vector, $\delta x$ is a delta function, and $\mathbf{V}$ is a 2D sampling matrix defined by $\mathbf{v}_1$ and $\mathbf{v}_2$ which are linearly independent column vectors, where

$$\mathbf{v}_1 = \begin{bmatrix} v_{11} \\ v_{21} \end{bmatrix}, \qquad \mathbf{v}_2 = \begin{bmatrix} v_{12} \\ v_{22} \end{bmatrix}, \qquad \mathbf{V} = \begin{bmatrix} v_{11} & v_{12} \\ v_{21} & v_{22} \end{bmatrix} \qquad (8)$$

The angle between $\mathbf{v}_1$ and $\mathbf{v}_2$ sets the geometry of the sampling grid, that is, 90° for rectangular and 120° for regular hexagonal, and their moduli set the distance between samples. Figure 10 illustrates the geometry of the regular hexagonal grid.

Images can be formed from scenes reflecting or emitting electromagnetic radiation from any or several parts of the spectrum. No use is made of the frequency information in monochromatic images, but for color images and other multidimensional images, brightness planes are stored for each of several frequency bands. Visible, infrared (IR), Xray, radio and ultraviolet images are commonly captured and processed. Other image sources employ ultrasound, seismic waves, surface-point contact measurements and atomic-particle emissions. In each case a large sensing area can increase the sensitivity of the detector and improve the singal-to-noise ratio (SNR), but it may reduce the maximum spatial frequency that can be captured. Focusing devices (lenses) can improve the situation. In many imaging cases the sensor transforms an energy signal into a voltage signal that can be further processed.

FIGURE 11. Hexagonal-faced photomultiplier tube.

The sensor designer may begin with the idea of completely tiling the image plane with sensors, as in the retina of the human eye, and then leading the electrical connections away from the rear. Hexagonal packing may be advantageous as has been found with radar systems (Sharp, 1961), point contact measurement (Whitehouse and Phillips, 1985), and for medical gamma cameras. Figure 11 shows a hexagonal-faced photomultiplier tube from such a camera.

A solid-state hexagonal sensor array has been researched for X-ray imaging (Neeser *et al.,* 2000). They have fabricated a 64 × 64 array based on an FET technology, with each element shaped as a nearly regular hexagon of size 50 $\mu$m × 42 $\mu$m.

A completely tiled sensor array can be analyzed using a pixel model. Each sensor element can be considered to provide the brightness information for one pixel. This implies that the sensor is a perfect integrator over its entire surface, that there is no signal leakage between sensor elements, and that there is no radiation scattering within the array that can result in more than one element responding to a single photon. In practice, these three conditions are seldom true. With integrated sensor technologies such as charge coupled devices (CCD) and CMOS, it is not easy to make large numbers of electrical connections to the rear of the array, and circuits are often laid out alongside the sensor elements of effect data transfer. The active areas of the sensor can be kept large compared to the communications and power circuits, but the pixel model is effectively further compromised. A wafer scale image-processing system has been proposed with connections made to the rear of the wafers, but this did not extend to the hundreds of thousands of connections required for pixel-to-pixel transfer (Nudd *et al.*, 1985).

*1. CCD TV Camera*

CCD image arrays can be 1D or 2D, with 2D image capture being achieved with a 1D array by scanning the object past it. Images are often large, with 2D arrays of $512 \times 512$ or $682 \times 512$ (4:3 aspect ratio) the most readily available. These sensors discretize the image spatially, but the brightness value remains an analogue value. There are various array architectures (Batchelor *et al.*, 1985), but the interline transfer (ILT) device is the most popular. The image is focused onto the sensor area and during the acquisition phase the elements store an electric charge that is inversely proportional to the intensity of the light falling on them multiplied by the exposure time. In the readout phase the electric charge is transferred to storage registers that run parallel to the columns of sensor elements. This arrangement is illustrated in Figure 12.

Once the charge is transferred, the sensor elements can begin to receive the next image and the storage registers can begin to communicate the current image to the camera electronics. The registers are analogue devices and rely on multiphase clocks to shift the charges and synchronize the process. The column storage registers shift the data a row at a time into an output register,

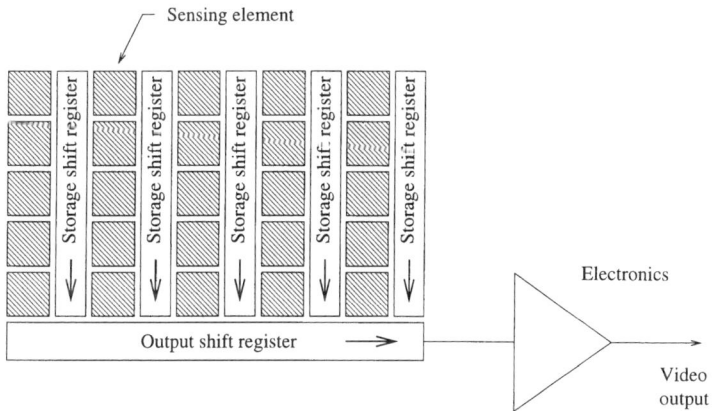

FIGURE 12. Interline transfer CCD sensor.

which in turn shifts the data to produce the raster scan output stream to the camera electronics. The electronics should include a reconstruction filter to correctly reproduce the image (Oakley and Cunningham, 1990), as well as amplification and impedance matching circuits. The camera output is therefore a time-varying continuous voltage signal. Considering a frame of this signal as a 2D image, then it is horizontally continuous and vertically discrete. There are many errors associated with CCD sensors (Schroder, 1980). In particular, there will be light reflection and charge leakage within the array, and frequency bandlimiting caused by the electronics. The area of the light-sensitive elements can be maximized with respect to the shifting elements, but complete tilings of the surface are not possible.

The tile and grid shape can be chosen by the designer. Square and rectangular shapes predominate for both, but a small ($8 \times 8$) hexagonal tiled, hexagonal grid sensor array has been fabricated (Hanzal *et al.*, 1985). Large RAM devices are often fabricated with cells on a hexagonal grid to save space. The technology to fabricate hexagonal grids exists.

The sensor array discretizes the image. At this stage the bandlimiting of the image can be analyzed so that the best shape can be chosen for the sensor element, and to ensure that there is no signal aliasing. The modulation transfer function (MTF) of the individual components, that is, the atmosphere, the lens and the CCD elements, can be estimated theoretically using simplistic models, and a composite figure is obtained. The MTF is analogous to the modulus of the frequency response of a system for processing time-varying signals. If the distance between the object and the lens is not great the MTF of the atmosphere can be neglected (Tzannes and Mooney, 1995).

The ideal sensor element integrates the light intensity over its active area and can thus be considered a low-pass spatial filter. If the element is rectangular, then its horizontal 1D MTF can be found by Fourier transforming the square profile 1D window of width $x$ m

$$MTF_{\text{element}}(f) = \left| \frac{\sin(\pi f x)}{\pi f x} \right| \quad \text{m}^{-1} \tag{9}$$

The spatial cutoff frequency is given by

$$fc_{\text{element}} = \frac{0.443}{x} \quad \text{m}^{-1} \tag{10}$$

The model is simplistic and provides only a 1D MTF. Techniques exist for measuring the MTF of individual sensor elements within an array (Sensiper *et al.*, 1993).

The lens is the final component to be analyzed. Its primary purpose is to focus the image, but in addition it acts as a low-pass spatial filter and can reduce aliased components. Both diffraction and aberration limiting occur within the lens (Ray, 1988). Diffraction limiting results in a high spatial-frequency cutoff

$$fc_{\text{diff}} = \frac{1}{1.22\lambda N} \quad \text{m}^{-1} \tag{11}$$

where $\lambda$ is the wavelength of the electromagnetic (EM) radiation and $N$ is the $f$-number of a circular aperture. A smaller aperture thus results in a lower cutoff frequency. If the aperture is circular, then the resulting 2D bandlimiting will be circular.

A 1D profile through a circular 2D MTF can be calculated (Gaskill, 1978):

$$MTF\left(\frac{f}{fc_{\text{diff}}}\right) = \frac{2}{\pi}\left\{\arccos\left(\frac{f}{fc_{\text{diff}}}\right) - \frac{f}{fc_{\text{diff}}}\left[1 - \left(\frac{f}{fc_{\text{diff}}}\right)^2\right]^{1/2}\right\} \quad \text{m}^{-1} \tag{12}$$

There are various aberrations that limit the frequency response of a lens. For monochromatic light these are spherical aberration, coma, astigmatism, curvature of field an distortion (Ray, 1988). The lens designer uses multiple elements to correct these aberrations, but the lenses that are often used in cost-effective TV systems still exhibit such defects. Aberration limiting results in a cutoff frequency that is proportional to the $f$-number of the aperture, with a wide aperture resulting in a low cutoff frequency. The cutoff frequency can be calculated for thin lenses (Black and Linfoot, 1957), but calculations are complicated by the choice of definition for "in focus" and by the compounding of thin elements. A computer-aided design (CAD) system or practical measurements should be employed.

Figure 13 shows MTFs for an ideal sensor element plotted using Eq. (9), a diffraction-limited lens ($f$ 8, visible wavelength) plotted using Eq. (12), and the product of these two that can be considered as the system MTF. The frequency axis has been normalized to the Nyquist frequency of the array. The frequency-limiting components of this system are not providing sufficient filtering to remove aliased components, and the response is still greater than 0.4 at the Nyquist frequency. These simple theoretical techniques are limited in that they do not include aberration-limiting or 2D information. Practical methods of measuring the 2D MTF exist, and the results can be compared with the theoretical calculations.

### F. Measurement of 2D Modulation Transfer Function and Bandlimit Shape

The MTF of continuous optical processing systems can be measured using traditional techniques (Ray, 1988), but these fail with digital acquisition systems due to signal aliasing. Various techniques have been researched to overcome

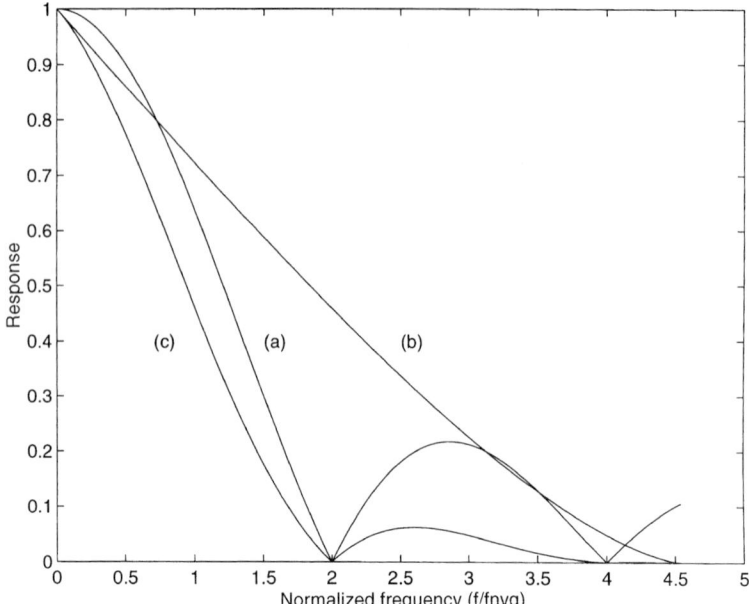

FIGURE 13. MTFs for: (a) ideal sensor element; (b) diffraction-limited lens (f-8, visible wavelength); (c) system MTF (product of a and b).

the problems introduced by discrete sensor arrays. The simplest method is the knife-edge technique, and it is suitable for use here to measure the bandlimit region and the filtering of aliased components. The method involves a shifting technique to produce a high-resolution profile across an image edge, and this renders the technique insensitive to geometric distortions. Geometric distortion information is important for applications such as image restoration, and for these, alternative techniques should be used (Zandhuis et al., 1997; Boudin et al., 1998).

The measurement can be extended to include the frame-grabber digitizer as indicated in Figure 14. The measurement now encompasses two discrete stages, the sensor array and the frame-grabber ADC. The array digitization is 2D, but the ADC is operating on a partly discrete raster-scanned image and only digitizes in the horizontal direction. The bandlimit region measurements can be analyzed to show the contributions from each system component.

The knife edge is provided by a long straightedge object, the image of which is dark on one side and bright on the other. It is focused onto the sensor array and a TV frame is grabbed. The MTF is calculated from the stored image that is a smoothed version of the input step. The technique works by aligning an edge slightly off vertical or horizontal. In this way, the straight edge cuts each

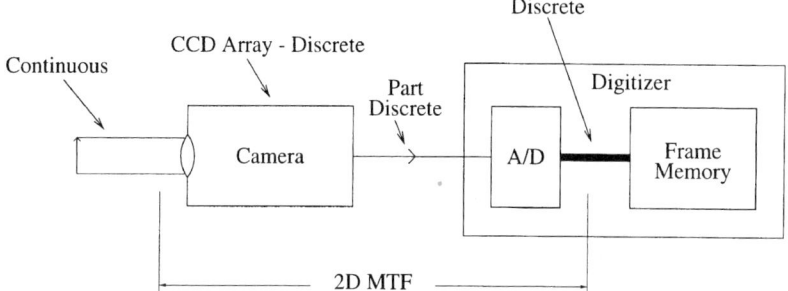

FIGURE 14. TV camera-digitizer system.

element along the line of the edge so that it records a slightly different brightness value than its neighbor. Assuming the edge to be straight, edge profiles along the edge can be aligned and a single high-resolution profile known as the edge spread function (ESF) is assembled from them. Because this is high resolution, this edge contains nonaliased information beyond the sampling frequencies of the array and the ADC.

Early implementations of the technique (Reichenbach *et al.*, 1991) were limited to MTF measurements in the vertical and horizontal directions and required several parallel edges. The use of spatial domain calculations (Tzannes and Mooney, 1995) enabled a single edge to be used, and the consideration of plane waves and interpolation has enabled 2D measurements to be made (Staunton, 1996b, 1997a, 1998).

It is important to set up the acquisition system to be as linear as possible for the technique to be effective. The automatic gain control of the camera must be disabled, and the gamma correction removed. The ESF can be differentiated and transformed to the Fourier plane to give the transfer function of the system, the modulus of which is the 1D MTF for the particular orientation of the edge profile; MTFs can be obtained for several edge-profile orientations and combined to form a 2D MTF (Staunton, 1997a, 1998).

A comparison of measured 2D MTFs has been made between six acquisition systems (Staunton, 1998). The systems were made from combinations of three cameras and two frame grabbers. The component specifications as obtained from the manufacturers data sheets are as follows:

Camera A:  2/3 in. CCD array. Square element shape. Sample spacing: 10 $\mu$m horizontal and vertical. Resolution: 756 × 581 elements. Lens: Fixed focal length, 16 mm.

Camera B:  1/2 in CCD array. Resolution: 752 × 582 elements. Lens: Fixed focal length, 16 mm.

Camera C:  1/3 in CCD array. Resolution: 750 elements horizontal, vertical not stated. Lens: Fixed focal length, 16 mm.
Frame Grabber X:  Frame store: 512 × 512. Aspect ratio: 1:1.
Frame Grabber Y:  Frame store: 512 × 512. Aspect ratio: 4:3.

Figure 15a shows measured and simulated 1D MTFs for one acquisition system. The measured MTF cuts off at a lower frequency than the simulated one. This is to be expected as the simulation of the MTF of the lens did not include aberration limiting, and only ideal CCD array characteristics were used. The measured response at the Nyquist frequency is still 0.2 and significant aliasing will occur. Figure 15b shows a typical high-resolution ESF from which the MTF would have been calculated.

Figure 16 shows 1D MTFs obtained for edge profiles oriented in 15° steps from 0° to 90° to the horizontal. The cutoff frequency increases with the angle of the profile, reaching a maximum for a vertical profile. The reduced cutoff frequency in the horizontal direction could be caused by filtering in the camera electronics or by an antialiasing filter in the frame grabber. These circuits operate only on the raster-scanned signal.

Figure 17 shows a quadrant of a 2D MTF where the results for edge profiles at angles other than those given in Figure 16 have been found by interpolation.

Figure 18 shows slices through the 2D MTFs for each camera-frame-grabber system. The slices are located at the −3dB modulation level and have been normalized to the vertical Nyquist frequency of the CCD array of camera A. The vertical cutoff frequency for each combination is between 0.37 and 0.69, whereas the horizontal cutoffs are between 0.27 and 0.48. The vertical cutoff is limited mainly by the lens and the CCD element area, whereas horizontally, the camera and frame-grabber electronics also provide limiting. The different horizontal and vertical charge-shifting registers in the CCD array (Fig. 12) may also lead to differences in the horizontal and vertical responses.

The system combinations—camera A, grabber X; camera A, grabber Y; camera C, grabber X; camera C, grabber Y—each show an increase in cutoff frequency with increasing edge-profile angle. The bandlimiting is not circular. The horizontal cutoff frequency of camera A is probably being limited by a reconstruction filter in the output electronics of the camera as the cutoffs are very similar for connections to grabber X and grabber Y. The horizontal cutoffs for camera B and camera C are nearly indentical when connected to the same frame grabber. The differences here are dependent on the frame grabber and could be caused by filtering in the input circuitry of the grabber. The traces for the systems, including camera B, are nearly circular and thus there is an advantage in using a hexagonal sampling grid. The grid pattern can be realized

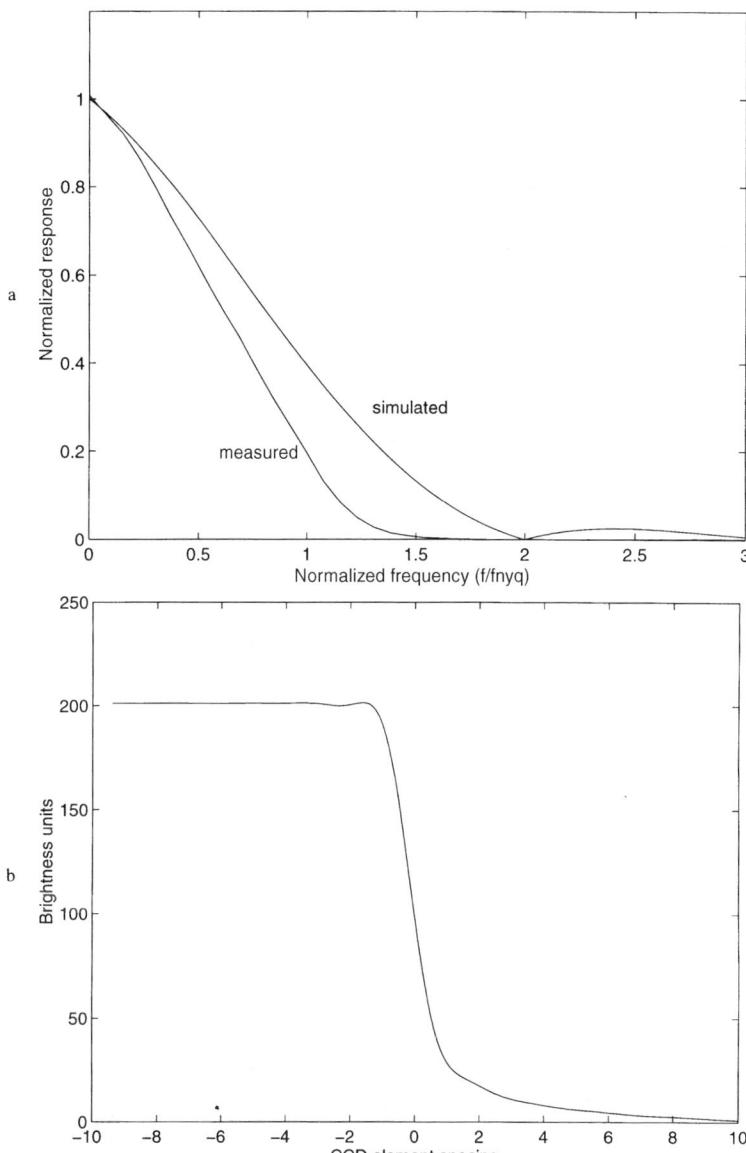

FIGURE 15. Camera A, frame grabber X: (a) a typical MTF; (b) a typical ESF. (*Reprinted from IEE Proc. Vision, Image and Signal Processing,* **145**(3): 229–235. Staunton, R. C. (1998). Edge operator error estimation incorporating measurements of CCD TV camera transfer function, with permission from the IEE Publishing Department.)

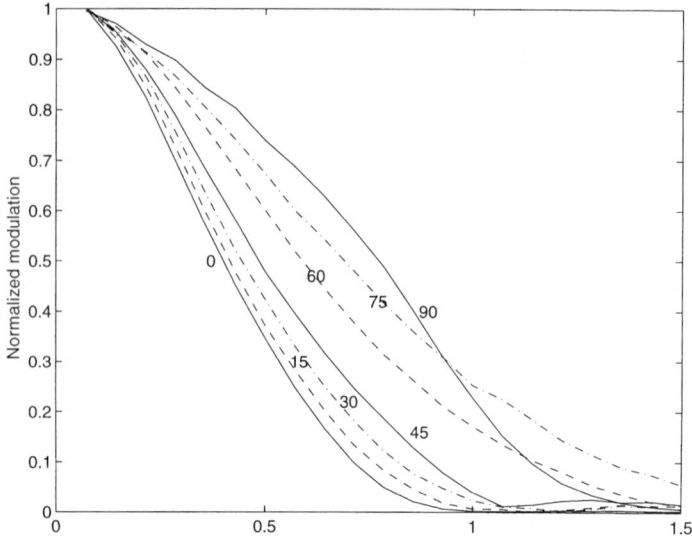

FIGURE 16. The 1-D MTFs obtained from edge normals at angles of 0° to 90° to the horizontal. Camera A, frame grabber X. (*Reprinted* from *IEE Proc. Vision, Image and Signal Processing,* **145**(3): 229–235. Staunton, R. C. (1998). Edge operator error estimation incorporating measurement of CCD TV camera transfer function, with permission from the IEE Publishing Department.)

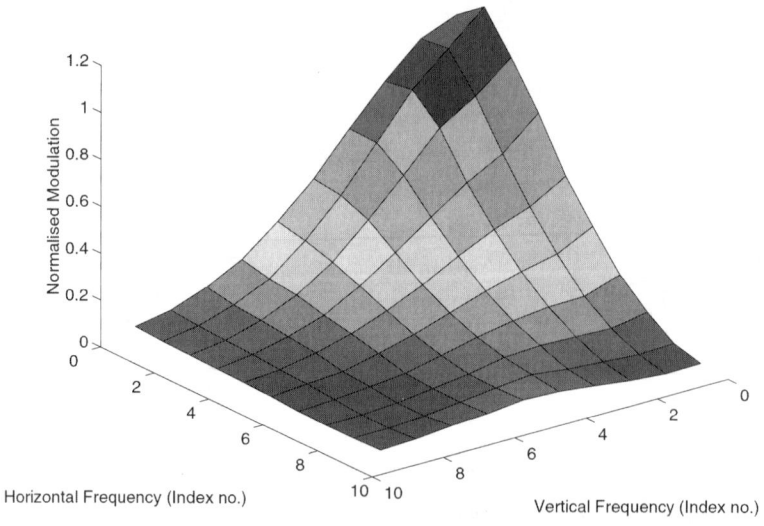

FIGURE 17. Quadrant of a 2D MTF interpolated from the data in Figure 16.

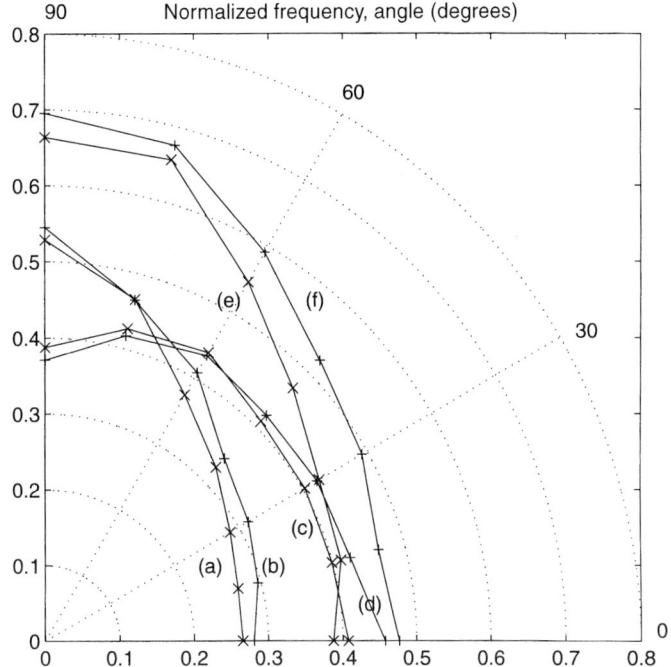

FIGURE 18. Polar plots of the −3 dB modulation points of the 2D MTFs obtained from edge normals at angles of 0° to 90° to the horizontal: (a) camera A, grabber Y; (b) camera A, grabber X; (c) camera B, grabber Y; (d) camera B, grabber X; (e) camera C, grabber Y; (f) camera C, grabber X. (*Reprinted* from *IEE Proc. Vision, Image and Signal Processing,* **145**(3): 229–235. Staunton, R. C. (1988). Edge operator error estimation incorporating measurements of CCD TV camera transfer function, with permission from the IEE Publishing Department.)

by the digitization circuits of the grabber by adding a half-sampling period at the beginning of each line in alternate TV fields.

If the square CCD element shape was adjusting the shape of the 2D MTF, then a deviation would be expected in the trace at 45°. No such deviations were observed, indicating that other limitations were dominant. This has shown that square CCD elements do not necessarily require a square sampling grid for optimum performance.

The systems containing camera A or camera C ideally require more antialias filtering in the vertical direction. This would also even up the horizontal and vertical responses. Such filtering is difficult to achieve physically without defocusing the lens or allowing vertical charge leakage between CCD elements. The images produced by the systems containing Camera B are nearly circularly bandlimited and can be sampled most efficiently on a hexagonal grid.

## III. Processor Architecture

The objective of digitizing the image is usually so that it can be processed using a digital computer. This section considers the storage of image data and the spatial relationship between the data. In particular, the square and hexagonal sampling schemes are compared and the advantages and disadvantages of processing them with computers of various architectures are discussed. A detailed comparison of some specific image-processing algorithms is given in Sections IV and V.

Parallel computer architecture is a large research area. The parallel processing of images is a smaller area, and the parallel processing of hexagonally sampled images is even smaller. However, most machines can process hexagonal images, but with varying degrees of efficiency. Surveys of parallel computer architectures include that of Fountain (1987), which provides an in-depth study of systems up to 1986. A special 1988 issue of the *Proceedings of the IEEE* on computer vision, edited by Li and Kender (1988), provides survey papers on architecture (Cantoni and Levialdi, 1988; Maresca *et al.*, 1988). There has been a special section in the *IEEE Transactions on Pattern Analysis and Machine Intelligence,* on computer architecture (Dyer, 1989). A more recent general survey that includes a new taxonomy of processors has been published (Ekmecic *et al.*, 1996).

### A. Single-Instruction, Single-Data Computers (SISD)

This is the conventional computer (Flynn, 1966). The program and data are stored in memory, and the memory is addressed in the correct order so that each particular datum is accessed and operated on as required. As reviewed in Section II.C, a circularly bandlimited image that has been hexagonally sampled will contain 13.4% less data for an equivalent information content than a square sampled image, but the addressing of the data will be less straightforward.

Programs running on such computers need to have the image data or their subsets stored in indexable arrays because the value of an output pixel is often a function of several pixels in the input image. A 2D square sampled image can be mapped one-to-one into an integer-indexed array where an increment of the index represents a step of one sampling distance in the image. The indexing of a hexagonal image stored in an array is less straightforward. A hexagonal pattern could be set up in a square array by filling only every other cell and shifting this pattern by one cell on alternate rows. Then the array would be twice as large, require double increments of the row address pointers, and the warp introduced would mean that either the horizontal or vertical increments would no longer be equivalent to the sample spacing.

$$x + y + z = 0. \tag{13}$$

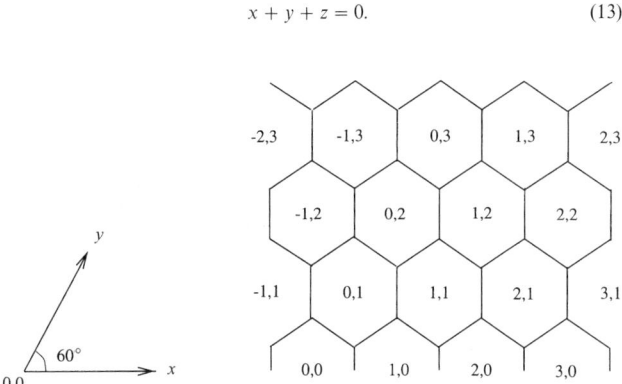

FIGURE 19. Hexagonal pixels and indexing scheme.

If the hexagonal image stored in the array is addressed with 60° or 120° axes, then indexing is possible (Mersereau, 1979). An example of such indexing is shown in Figure 19. For practical use within the computer program the pixel addresses can be mapped to complex numbers (Bell *et al.,* 1989). Alternatively, for small local area calculations, the hexagonal data can be mapped directly into a square array and different convolution masks used depending on whether the central pixel of the area is on an odd or even scan line (Staunton and Storey, 1990). Figure 20 shows a seven-neighbor hexagonal local area where six neighbors are equidistant from a central element, and the position shifting of the neighbors that occurs as the central element is located on either an odd or even scan line within a square 3 × 3 array. With such a scheme two sets of convolution masks are needed for each image-processing operation, although each is applied to only one-half of the image.

A simple three-integer coordinate scheme has been researched (Her, 1995) and is known as the symmetrical hexagonal coordinate frame. Three integers

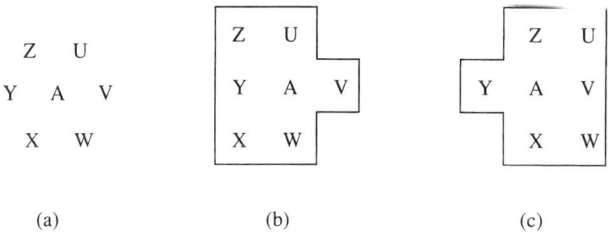

FIGURE 20. A seven-neighbor hexagonal local area: (a) the neighbor's positions within the image; (b) odd-row array positions; (c) even-row array positions.

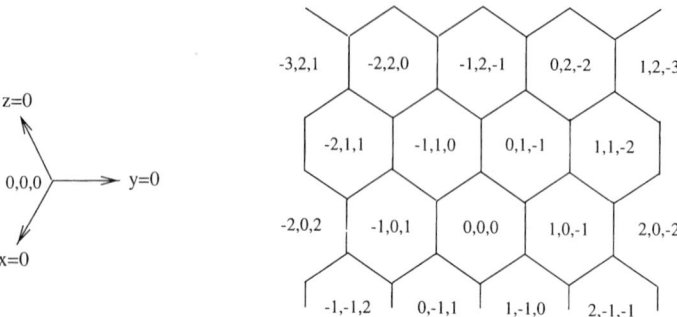

FIGURE 21. Three-integer index scheme.

point to each pixel. The frame overcomes difficulties experienced with other coordinate systems found when designing some types of processing operator. Figure 21 illustrates the coordinate indexing, with the center of the image shifted to the origin at $x = 0, y = 0, z = 0$. With this scheme $y$ points to the image scan line number, and $x$ to the individual pixels along the scan line. The image is planar, and if the origin is located centrally,

$$x + y + z = 0 \qquad (13)$$

The image plane cuts through a 3D Cartesian space, and each indexed image point coincides with an integer-indexed point in the 3D space, as illustrated in Figure 22. This can be useful when processing operators, and especially geometric transformation matrices are being designed. However, loading the image into a 3D array would lead to poor memory utilization. Thus for good

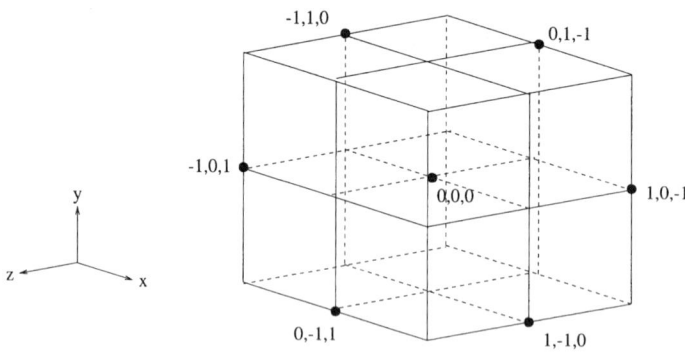

FIGURE 22. Three-integer index scheme sampling points embedded within a 3-D Cartesian coordinate grid.

memory utilization, the image should be loaded into a 2D array that is indexed by the oblique (60°) system. Now, allowing for the shift of the origin to the center of the image, $x$ and $y$ for the three-integer coordinate scheme are identical to the two coordinates in the oblique scheme as displayed in Figure 19.

The indexing of set members in a 3D color space has been researched (Deng et al., 2001). The 3D space is an extension to the usual 2D hexagonal space. The distance between clusters of colors is important, and so the close approximation of the "hexagonal" voxel to a sphere permits efficient indexing and processing of color information.

In conclusion, a hexagonally sampled image can be efficiently stored within the memory of a SISD computer. When stored in program arrays the image can be efficiently indexed and data pointers easily calculated. In the general case the three-integer index scheme is the most efficient, but the oblique axis and the method involving the direct mapping of data into a square array, and the shifting of convolution masks may also be used.

## B. Parallel Processors

This section contains a survey of some of the computer architectures that have been used for image processing that have either been designed to process hexagonally sampled images, or that have hexagonal interconnections between processors, but that have been designed to process high-level information that is not locked to a sampling grid. Two-dimensional arrays of fine- and coarse-grain processors are discussed, as are pipeline, vector, pyramid, hypercube, and shared-memory devices. A more general review of parallel architectures for image processing has been published (Downton and Crookes, 1998).

If the total image-processing task is considered, parallel processing can be applied in various ways: (1) The image can be divided into local areas, possibly overlapping, and a processor assigned to each area; (2) processes can be pipelined so that one processor completes the first task on the whole image and passes the resultant image on to the next processor for further processing, while waiting for the next image; (3) a pyramid of planes of 2D arrays of processors can be constructed in which partly processed images are passed up to the next level for further processing, with a reduction in the number of processors and interconnections at each level; and (4) a particular task may be readily performed on a general purpose array, hypercube, vector, or shared-memory processor.

Computational or execution bandwidth can be defined as the number of instructions processed per second, and the storage bandwidth as the retrieval rate of data from memory (Flynn, 1966). Latency can be defined as the total time associated with a process from excitation to response for a particular data. In practical terms, the latency is the number of processor clock cycles that elapse between the input of a datum and the output of the processed result.

## C. Two- and Multidimensional Processor Arrays

An array of processor elements (PEs) is a group of elements that operates in parallel to process a set of data. Consider intially a simple image-intensity transform where each member of a data set $b_{i,j}$ is multiplied by a scaling constant $K$. Each transformed pixel $a_{i,j} = Kb_{i,j}$. An array of PEs of size $i \times j$ could transform the entire array in one clock operation period. However, in many image-processing problems, $a_{i,j}$ would be a function of pixels within a local or global area. To facilitate these operations, interconnection is provided between PEs. The topology of the interconnections determines the dimensionality of the PE array. Figure 23 shows two examples of 2D interconnection topology. The eight-way interconnection of PEs has also been realized.

Three-dimensional interconnection involves the vertical stacking of such 2D planes, or the formation of a torus (Li and Maresca, 1989). Multidimensional interconnection topologies are also realizable. The PEs in an array can be single instruction, multiple data (SIMD), or multiple instruction, multiple data (MIMD); MIMD implies a high level of processor autonomy, but some autonomy is possible for PEs within the SIMD definition (Maresca et al., 1988).

### 1. Fine-Grain Arrays

Fine-grain arrays are more likely to be used for low-level image processing where there is an advantage in associating the array structure with the

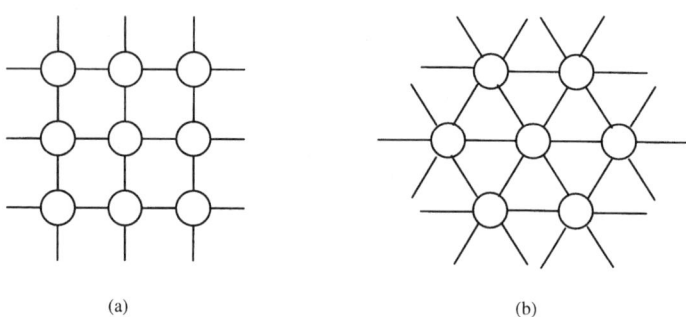

FIGURE 23. Two-dimensional processor array interconnection topology: (a) rectangular; (b) hexagonal.

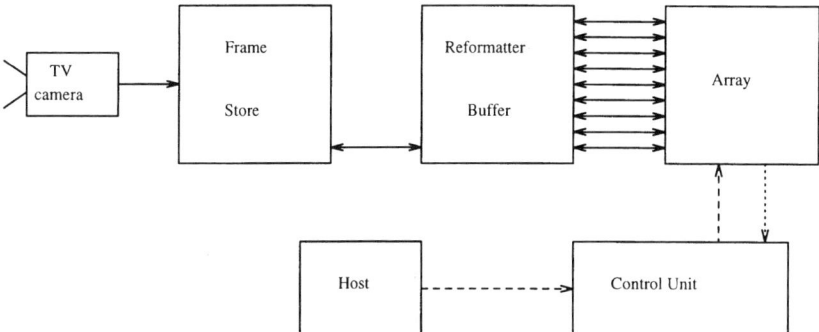

FIGURE 24. A fine-grain array system.

original data structure and applying connections between elements in local areas. Hexagonal interconnections where each processor connects to its six nearest neighbors have been realized. Some of these machines are listed in the following. Fine-grain array machines are especially useful for morphological image processing. Simple SIMD PEs are often used because arrays are large, many PEs can be integrated onto each VLSI device, and the single instruction processing reduces communication overhead. Figure 24 shows a fine-grain array together with some of the other units required to realize a system.

The control unit broadcasts image-processing instructions to the PEs in the array and receives back busy, finished signals from each. Many array processors are used with a raster-scanned input device such as a TV camera. The TV picture is captured by a frame grabber and then reformatted so that efficient array loading can be achieved. Loading such images requires the hardware overhead of the reformatting logic and a loading time overhead. Loading is often achieved by transferring a complete column of data from the reformatting logic to the array and rippling this and subsequent columns across the array until all columns are filled. In the Clip4 array (Fountain, 1987) the data and control paths are separated so that image loading can be performed concurrently with image processing. This requires an additional hardware overhead. Some examples of fine-grain arrays that incorporate hexagonal interconnections are as follows.

a. *Clip 4*

The Clip4 system (Fountain, 1987) was a fine-grain SIMD processor that embodied the features of Figure 24. It was developed from Clip2 and Clip3, which also allowed hexagonal interconnections between PEs. The Clip4 chip, which is used to assemble the array, was designed in 1974, and was limited by the fabrication technology available. Limitations, included the number of transistors per device (5000), the packaging (40 pin), and the clock speed (5 MHz). The resulting device contained eight PEs and has been used to build arrays of from $32 \times 32$ to $128 \times 128$ elements. The arrays can be connected in square or hexagonal 2D meshes. The processor data width was 1 bit, and as an example of the processing speed, an 8-bit addition could be performed in 80 $\mu$S. The processor could perform Boolean operations; a 32-bit RAM was provided in each PE, and input gating was used on the near-neighbor input connections to facilitate efficient morphological operator implementation. Individual PEs could be switched off by certain processes, and a global propagation function allowed data to be passed through the array 50 times more efficiently than if propagation was limited to near-neighbor-only communication. Clip4 arrays have been used for many applications (Duff, 1985). A process involving the measurement of the rate of growth of biological cell cultures was possible for a large number of samples, as computation could be performed in less time than it took physically to change the sample (Fountain, 1987).

b. *Illiac III*

The Illinois pattern recognition computer, Illiac III (McCormick, 1963), allowed four-, six-, and eight-way interconnections between PEs. It was used for analyzing bubble chamber traces.

c. *The PSC Circuit*

This was a programmable systolic processor that had three 8-bit input channels and three 8-bit output channels (Fisher *et al.,* 1983). Each PE contained an arithmetic and logic unit (ALU), multiplier, microcode store and sequencer, and RAM. It could be connected two, four and six ways into arrays.

d. *Silicon Retina with Correlation-based, Velocity-tuned Pixels*

This is a hexagonal architecture implemented on a CMOS chip (Delbruck, 1993). Visual motion computation is implemented using an analogue space-time algorithm.

e. *Analog Neural Network*

This has been used for image processing (Kobayashi *et al.,* 1995). Various interconnection topologies were researched, including hexagonal.

f. *Kydon* (Bourbarkis and Mertoguno, 1996)

This is a multilayer image-understanding system. The processors in the lower-level arrays are connected in a hexagonal mesh.

## 2. Coarse-Grain Arrays

With coarse-grain arrays, one PE will be associated with many data, or large local areas of pixels. In some systems memory may be shared among PEs. The PEs are likely to be sophisticated microcomputers, and considerable processing and communication autonomy will be devolved to them. The array is likely to be a MIMD processor. Communication overhead limits the number of processors that can be inserted in an array to obtain faster processing. For some processes, such as low-level image processes, it may be advantageous to divide the image space and assign one PE to each local area. For higher-level processes, the computer programmer may perceive an advantage in redistributing the processing in a different way across the array. This is easier with a shared-memory system. Each of the relatively sophisticated communication channels supported by the PEs require significant chip area for their implementation. This results in early devices being limited to four-way interconnection, as, for example, in the transputer (Inmos Ltd., 1989). With the development of hypercubes, etc., connectivity has increased again. Six-way interconnectivity has been reported for a system referred to as HARTS (Dolter *et al.*, 1991).

The optimum interconnectivity of these arrays is not primarily a function of the sampling grid of an original image. Arrays of PEs are used to speed up processing, but as the number of PEs added increases, the communications bandwidth limits the increase in speed. The interconnection topology can be chosen to optimize processing speed.

### D. Pyramid Processors

A typical pyramid architecture is shown in Figure 25. At the base of the pyramid, level 1 is the input image. This is connected upward to level 2 so that four level-1 pixels connect to one level-2 pixel. This is known as a quadrature pyramid. Binary, hexagonal, 16-way, and other connection systems have also been realized. In its simplest from the structure may be a pyramid of memory elements so that reduced resolution images are stored at each level, as with the pipelined parallel machine (Burt *et al.*, 1986). Pyramids of PEs are also realizable with architectural variations in the types of PE at each level and in the autonomy of control. Processor elements communicate between neighbors within their level, and also pass data upward to their associated PEs at the higher level. Some processes require that data pass both up and down the pyramid (Watson and Ahumada, 1989). In addition, PEs can work autonomously within the pyramid, or control can be passed down layer to layer from the apex.

In one type of pyramid, the PEs are of the same type in each level and the arrays can be coarse grained (Handler, 1984) or, more usually, fine grained

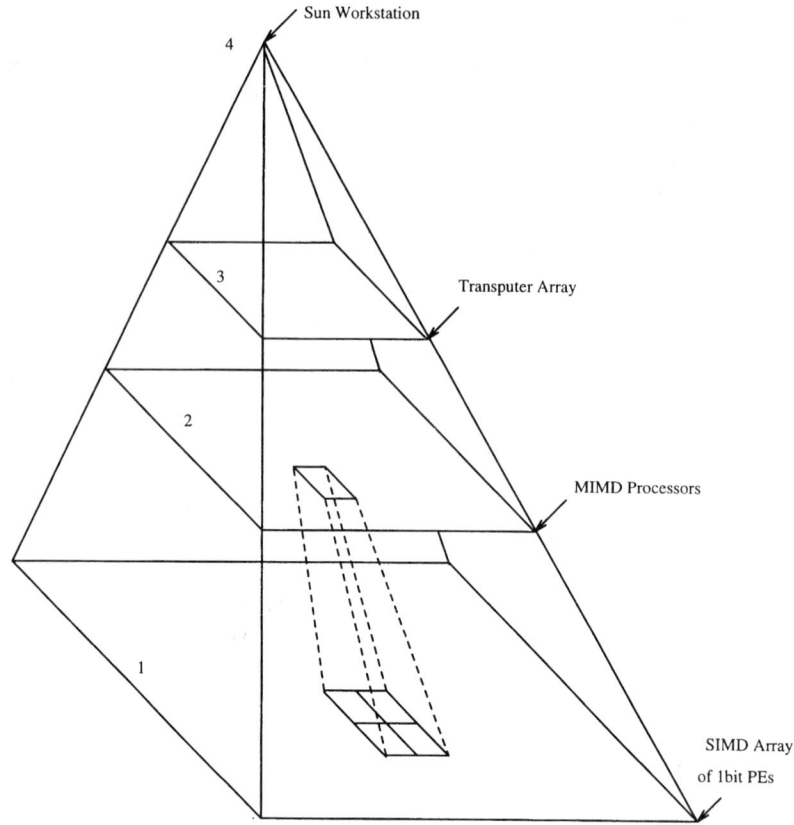

FIGURE 25. A typical pyramid architecture.

(Tanimoto *et al.*, 1987). In another type, different PEs will be incorporated at the different levels (Nudd *et al.*, 1989), where level 1 is populated by a $256 \times 256$ array of SIMD PEs, level 2 by a $16 \times 16$ MIMD array, level 3 by a $8 \times 8$ transputer array, and level 4 by a host Sun workstation. Pyramid processors are efficient architectures for image-under standing systems. The input is any general image at level 1 and the output would be a description of the scene in the form of, for example, a list of objects at the highest level.

1. *Hexagonal Pyramids*

Hartman and Tanimoto (1984) investigated a hexagonal pyramid data structure for image processing. Level 1 was tiled with hexagons, but each hexagon

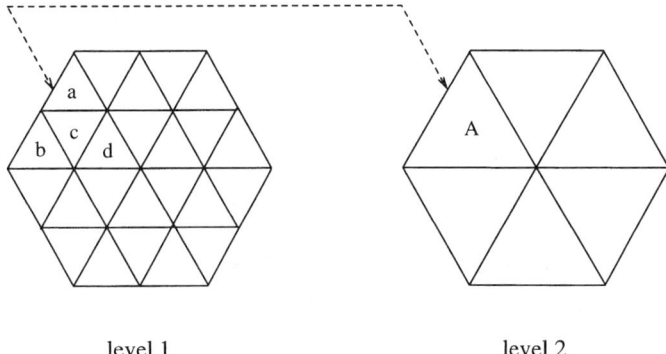

FIGURE 26. Hartman and Tanimoto's pyramid structure.

was subdivided into six equilateral triangular pixels. Four triangles were then combined to give a single equilateral triangle at the next level, as shown in Figure 26, where level-1 triangles [a, b, c, d] combine to give a level-2 triangle [A]; PEs could also be arranged in such a scheme, but the basic triangular pixel scheme is difficult to sample directly with a raster-scanned device as the image line spacings would be uneven. Resampling hexagonally sampled data to the triangular scheme could be achieved relatively easily.

A hexagonal pyramid structure that models the processing structure of the human visual cortex has been researched (Watson and Ahumada, 1989). Anatomically, behind the hexagonally packed retinal sensors are a layer of retinal ganglion cells, which, in the center of the retina, connect one-to-one with the sensors (Perry and Coney, 1985). The ganglion cells can also be considered to be connected on a hexagonal grid. The $2.10^6$ ganglion cells connect to the visual cortex that contains approximately $10^9$ neurons. Physiological experiments have shown that between the retina and the visual brain the image undergoes a sequence of transformations, and sets of cells in the cortex can be identified with these various transforms. The research considers a transform performed by the ganglion cells and a subsequent one performed within the cortex. The ganglion cells transfer spatial and brightness information. Their transfer function is broadband and they provide local adaptive gain control. The transform within the cortex is different. The cells are narrowband and employ a so-called hybrid space-frequency code to convey the position, spatial variation, and orientation of a region. The process in this group of cells has been modeled by a hexagonal orthogonal-oriented quadrature pyramid.

The image transform performed in the cortex can be considered as image coding and the aim of the research was to model the transform with a pyramid constructed from elements that were themselves modeled on known

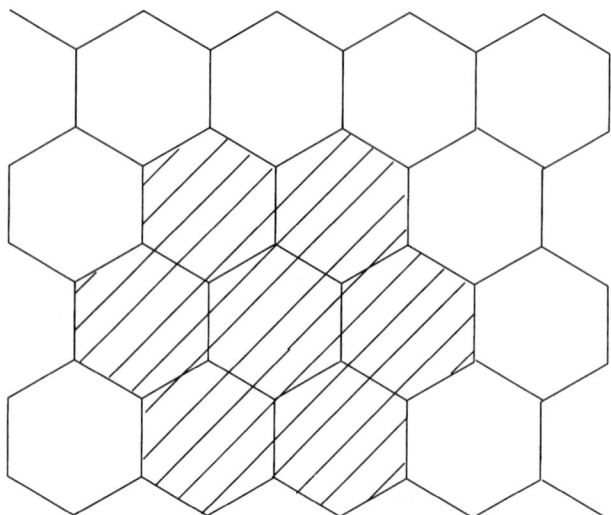

FIGURE 27. A seven-element local area that produces one value in a reduced resolution image.

physiological components. The pyramid had a hexagonal lattice input layer, the transform was invertible, and the overall process was found to be efficient.

The input image is passed on by the retinal ganglion cells to the lowest level of the pyramid. This level can be considered to be tiled with hexagonally shaped pixels. The transformation to the next highest level in the pyramid involves taking a group of seven of these pixels (the shaded area in Fig. 27) and producing one output pixel that contains a vector of values from a set of seven kernels, one of which produces the average brightness value of the local area, and the other six of which are bandpass and localized in space, spatial frequency, orientation, and phase. Each low-level pixel only contributes to one next-level pixel, so the next level contains only one-seventh the number of pixels, and so on, until the apex of the pyramid is reached. The resulting hexagonal pyramid structure is shown in Figure 28. In this figure, the input image lattice is represented by the vertices and centers of the smallest hexagons and the highest level, which is also the lowest resolution image, is represented by the largest, thickest line hexagon.

At the highest level there may only be one pixel, but the vector associated with it encodes all the image information and can be decoded back down the pyramid to reconstruct the original image. The model produces results that agree reasonable closely with physiological measurements, but some modifications, such as using larger kernels, are needed to produce a better match.

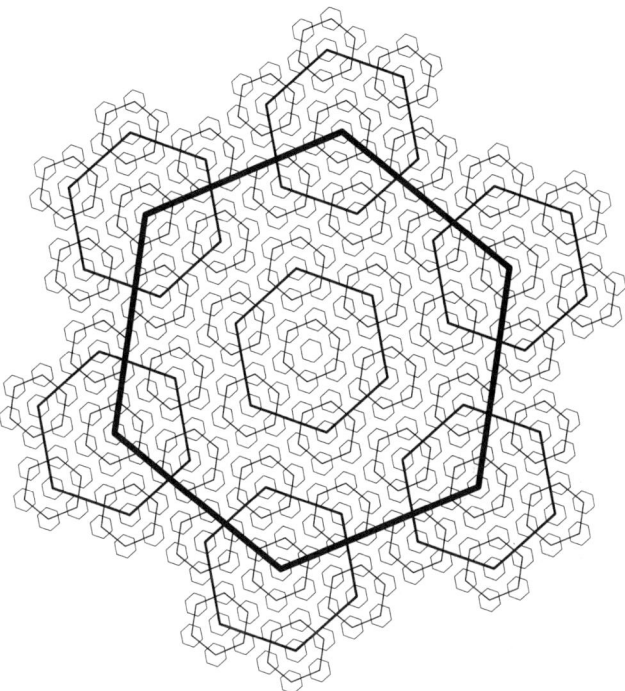

FIGURE 28. The hexagonal pyramid structure. (Generated using the program listed in the Appendix of Watson and Ahumada's 1989 paper.)

The hierarchical subsampling of images that give an image on a hexagonal grid that can then be used to develop fractal regions has been researched (Lundmark *et al.*, 2001). They chose a support-area shape around each sampling point that enables fractal patterns to be developed. When the original image is also hexagonally sampled, a hexagon-shaped region can be grown with fractal borders.

### E. Pipelined Processors

With pipelined processors there is a single stream of data from the memory or input device, and the stream passes serially through several PEs, each of which performs a different operation on the data before the data are finally sent to their destination. This is shown diagrammatically in Figure 29. At a particular time PE3 is operating on data (0), PE2 is operating on data (1), and PE1 is operating on data (2).

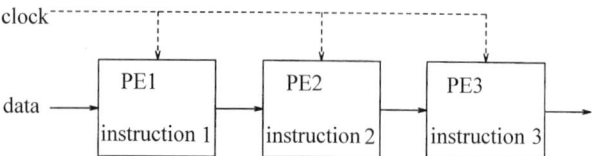

FIGURE 29. A pipeline processor.

Each PE is operating on a different data set and computing a different process. The pipeline can be termed a parallel processor. Programming flexibility can be compromised by such an arrangement. Events are separated in time, as with the SISD processor, but the sequence of instructions performed in a single pipe does not allow branching or easy rescheduling of instruction order. Efficient computation is achieved by applications where the same set of instructions must be applied to large sets of data.

With images captured under controlled lighting conditions, local image-processing operations can be sufficient. Local image-processing operations do not require a knowledge of the complete frame of an image, but only of a group of adjacent pixels. If these operations are performed on a pipeline processor there is no need to store a complete image. Such a processor may operate directly on the serial data stream from the digitized output of a raster-scanned device. The PEs must operate in real time, but as the operations are very simple, only a few lines of the image must be stored in line-length digital-shift registers within each PE. Figure 30 shows a pipeline processing element that stores two lines of the video image.

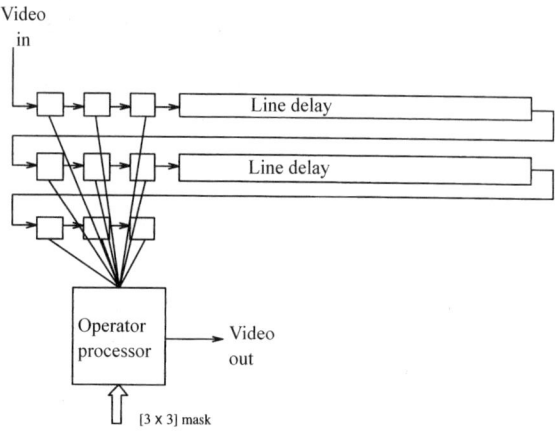

FIGURE 30. A pipeline processor element.

In this example, 3 bytes of each of the two line-storage registers, together with 3 bytes from the previous line, from a 3 × 3 local image area on which processing is performed. Real-time processing operations are performed on this array of elements and a new value for the center pixel is calculated and output to from the output video stream. Storing more lines of an image enables larger local areas to be used. For example, a 5 × 5 pixel area could be used by storing four video lines. Processes of increased complexity can be achieved by cascading a number of PEs in a string. The pipelined processor operates in real time, although an increasing latency is introduced by the successive line delays.

Many pipelined image-processing machines containing PE architectures based on that of Figure 30 have been reported in the literature. Few are limited to contain only this simple PE design but also have general-purpose ALU and lookup table elements. If the "Warp" system (Annaratone *et al.*, 1986; Crisman and Webb, 1991) is considered as a pipeline, then each PE can also contain local memory to aid multipass algorithm calculation. However, this feature causes Wrap to be classified as a 1-D MIMD array. Some pipelined systems are listed in the following, together with notes on those designed to process hexagonally sampled images.

## 1. *Pipelined Systems*

Early pipelined image processors have been reviewed (Preston *et al.*, 1979). The basic PE architecture of Figure 30 is evident in these systems, but of those early pipelined image processors, only the Cytocomputer was capable of real-time processing. Some more recent systems are reviewed in what follows.

Recirculating pipeline systems are characterized by having only short pipelines of PEs, and by individual PEs in the line being of a different hardware construction. Frame stores are used to enable data to be recirculated through the pipeline as shown in Figure 31, and many video and system buses are employed for data communication and system control. These systems are capable of performing a wide variety of complicated image processes, some of which can be classified as being in the midlevel vision range. For example, the convex hull process has been realized (Bowman, 1988). First, data are scanned horizontally out of one frame store, processed, and restored in the second frame store. The data are then scanned vertically out of the second frame store, processed, and restored in the original frame store.

*Cytocomputer* (Lougheed and McCubbrey, 1980). The PE design conforms closely to Figure 30, but with a programmable operator function. The hardware of the PEs in the pipeline is identical. The operator processor is limited to

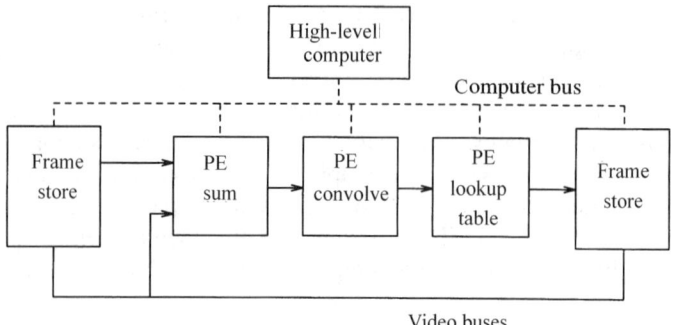

FIGURE 31. A recirculating pipeline system.

morphological and logical operations on a 3 × 3 local area. Scan line lengths up to 2048 pixels can be accommodated, and PEs are constructed on individual circuit boards from large scale integration (LSI) and very large-scale integration (VLSI) components.

*Cyto-HSS* (Lougheed, 1985). This is a recirculating system based on a pipeline of Cytocomputer PEs.

*PIPE* (Luck, 1986; 1987). Each PE contains three lookup table operators, image-combining units, a 3 × 3 arithmetic or Boolean local operator, and output crossbar switching logic. Images with resolutions from 256 × 256 to 1024 × 1024 can be processed. The crossbar switching enables data to process normally along the pipeline, to be switched in reverse direction along the pipe, or for the PEs to operate independently. Morphological and filtering operations are possible.

*University of California machine* (Ruetz and Broderson, 1986). This system provides a custom chip set for the designer. Each PE function is realized by a different VLSI chip. Advantages include real-time operation and potential cost reduction through the use of VLSI, but the nonprogrammability of PEs has led to a dynamic inflexibility and a requirement to design a different chip for each PE function.

*University of Strathclyde* (McCafferty *et al.*, 1987). The system uses LSI components, operates in real time and employs sophisticated image-processing algorithms for edge detection.

*University of Belfast* (McIlroy *et al.*, 1984). This system contains a real-time PE incorporating LSI logic devices that perform the Roberts edge-detection algorithm.

*TITAN* (Lenoir *et al.*, 1989). In this design the PE has been implemented on a gate array. It is capable of several binary and gray-level morphological operations. The local operator size is 4 × 3 pixels.

*Elor Optronics Ltd.* (Goldstein and Nagler, 1987). This is a pipeline processor system for detecting surface defects in metal parts. Each PE is a single-board SIMD computer.

*Kiwivision* (Bowman and Batchelor, 1987). There are three PEs in a recirculating pipeline. Each PE performs a different set of operations. The first PE is a 16-bit ALU, the second, a general-purpose local filter, operating on a $3 \times 3$ local area, and the third, a lookup table processor. In *Kiwivision II* (Valkenburg and Bowman, 1988), a pipeline of Datacube PEs feed an Inmos transputer array.

*Datacube* (Datacube Inc., 1989). A series of single board PEs have been produced that can be configured as a recirculating pipeline.

*PREP* (Wehner, 1989). Here, several parallel recirculating pipelines are used to speed processing by operating on distinct areas of the image.

*IDSP* (Minami *et al.*, 1991). This is a four-pipeline system implemented on a single VLSI chip. Additions and subtractions are allowed between data in each pipe. Applications: Video codec.

*Cheng Kung University* (Sheu *et al.*, 1992). This system uses a pipeline architecture to perform gray-scale morphological operations. It is suitable for VLSI implementation.

*Pipeline Processor Farm* (Downton *et al.*, 1994) *System.* This contains several pipes. Applications include general image processing and coding.

*Chung Cheng Institute* (Lin and Hseih, 1994) *Modular System.* This contains three pipelines. It works in real time on $512 \times 512$ images. It is suitable for VLSI implementation. Applications: Template matching.

*New Jersey Institute of Technology* (Shih *et al.*, 1995). This pipeline architecture has been implemented as a systolic system. Applications: Recursive morphological operations.

*Jaguar* (Kovac and Ranganathan, 1995). This is a fully pipelined single-chip VLSI device used for color JPEG compression of images of up to $1024 \times 1024$ pixels.

*Texas Instruments Pipelines* (Olson, 1996). These have been discussed and with particular emphasis on how to program them.

*Yonsei University* (Lee *et al.*, 1997) *Real-time System.* This is used for HDTV applications and can perform edge detection. Another paper (Lee *et al.*, 1998) includes a discussion on de-interlacing and color processing.

## F. Hexagonal Image-Processing Pipelines

*GLOPR* (Golay, 1969) was a pipeline for processing *hexagonally* sampled images. It operated on a seven-element local area that was passed to it from a host computer. It contained delay lines and could process images up to

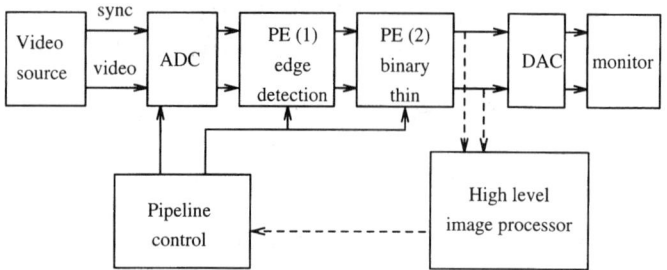

FIGURE 32. A simple pipeline processor consisting of two PEs.

128 × 128 in size at 3 $\mu$S per pixel operation (Preston et al., 1979). It was used extensively for processing medical images, and was produced commercially as the Perkins-Elmer, Diff3. It could also perform many image-processing tasks including the basic morphological operations (Preston, 1971).

A *University of Warwick* pipeline system that can process hexagonal or square sampled images has been designed (Storey and Staunton, 1989, 1990; Staunton and Storey, 1990). The specification required operation at the video rate, construction from reconfigurable hardware PEs, and a VLSI implementation. A lack of resources has allowed only a simulation of the pipeline to be completed.

The PE was designed at a functional level that could be configured to provide one of a number of image-processing operations. The initial device operated on a 3 × 3 local area. The processed images could be viewed directly on a TV monitor or transferred to a computer for high-level image processing. The PE has been designed to operate on sampled images up to 512 × 512 in size.

Figure 32 shows a simple pipeline comprising an analogue-to-digital converter, two PEs, a digital-to-analogue converter and a control unit; PE (1) is performing edge detection and PE (2) binary line thinning. A novel feature of the PE design was that an image-processing operation such as convolution, edge detection, median filtering, gray-level morphological, or a binary operation could be completely performed with a single PE in one pass of the image data.

Figure 33 shows the PE input and output signals. The clock is at the pixel rate. There are two 8-bit image data input channels to each PE. In the figure, one channel is connected to the output of the previous PE in the pipeline, and the other to a second source, which could possibly be the output from a second pipeline. The two channels are combined arithmetically inside the PE. Within the PE the image datum is clocked at the pixel rate through the various processing stages. A pair of unprocessed data is clocked in, and a processed datum clocked out from the PE with every clock pulse. The bandwidth of the

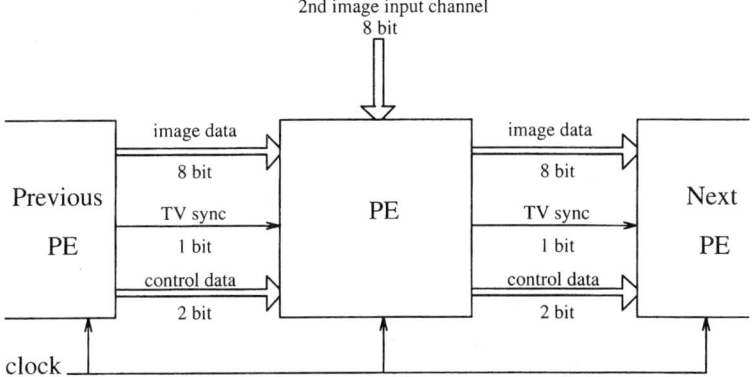

FIGURE 33. PE input and output signals.

PE is equal to the video rate, but the pipelining of the processes within the PE introduces a latency equal to an integer number of pixel clock periods.

The PE hardware is based on that shown in Figure 30. A block diagram of the basic PE is shown in Figure 34. The video image enters the PE as a raster-scanned 1D stream and the $3 \times 3$ local image area is assembled by employing two TV line length, 8-bit wide, digital-shift registers. The image adder allows two separate images to be combined at the PE input.

To enable hexagonally sampled images to be processed, the horizontal sampling spacing was increased by a factor of $2/\sqrt{3}$, and the first sampling point on alternate lines delayed by half-a-point spacing. By definition only even-numbered scan lines are delayed and the first image line is numbered one. The number of points per scan line is reduced by the $2/\sqrt{3}$ factor, giving typical image sizes of $721 \times 625$ or $443 \times 512$ pixels in comparison with the equal resolution square-sampled image sizes of $833 \times 625$ and $512 \times 512$. With hexagonal sampling the data rate is reduced by the $2/\sqrt{3}$ factor from 13.0 MBytes $\cdot$ s$^{-1}$ to 11.3 MBytes $\cdot$ s$^{-1}$.

For use in recirculating pipelines the only system modifications are to the initial image frame-grabbing module. Data can be processed by the pipeline at the designed 13.0 MBytes $\cdot$ s$^{-1}$ rate and thus-stored hexagonally sampled images can be processed in 13.4% less time than equivalent square images.

Changes to the PE architecture were minimal so as not to affect hexagonal processing, and extra taps were added to the line delays to reflect the reduced number of pixels per line. The operator processor was also modified. For square-sampled data, some operations performed by the processor require the convolution of the nine image pixels comprising the local area with one or

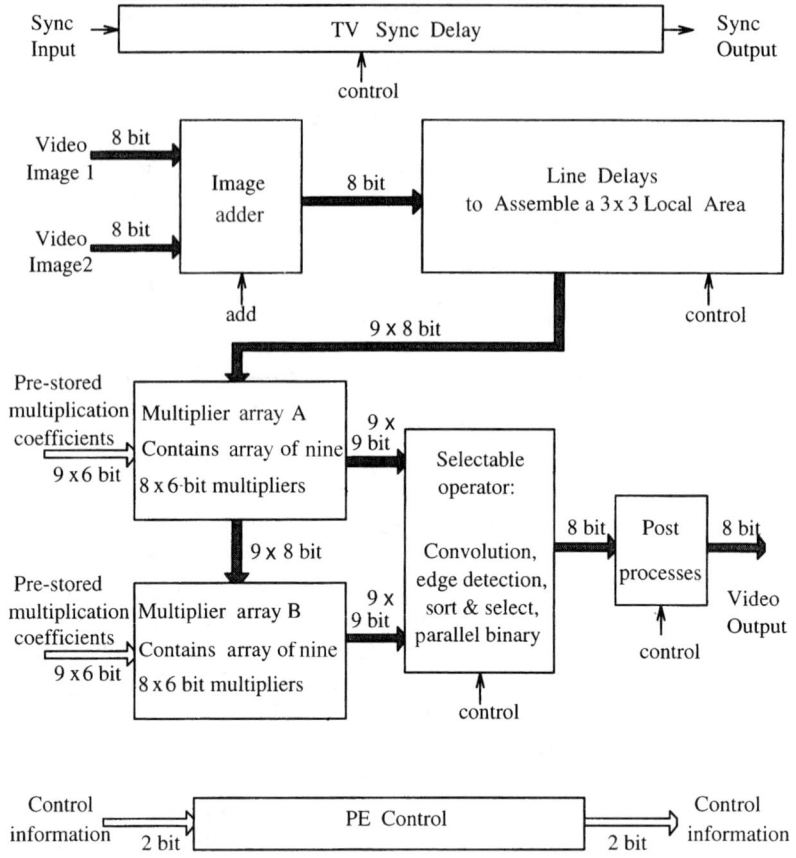

FIGURE 34. A block diagram of the pipeline processor element.

more 3 × 3 arrays of constants stored within the processor. The equivalent for hexagonally sampled data requires convolution with a seven-element array. With the foregoing system modifications for hexagonal data, the position of the central pixel with respect to the six neighbors changes within the grid of nine input pixels on alternate scan lines. This is illustrated in Figure 20.

It was necessary to store an extra set of convolution coefficient arrays within the PEs operator processor and to toggle between sets on alternate lines. This required the line-synchronization signals to be detected and the extra control signal to be processed by the control unit. The convolution coefficient magnitude range was identical to the square range as was the scaling capability provided. In practice the amount of scaling was less as fewer coefficients are employed.

For the processing of the hexagonal edge detector, changes were needed to the square-system edge-detection hardware module to reflect the different magnitude equations. For the other operators implemented within the PE the modifications were minimal.

In conclusion, with a pipeline processor, the processing time for a real-time video image will be unaltered for a particular operation regardless of whether square or hexagonal digitization is employed. One image is processed in one frame time, although a latency delay period is introduced by the string of PEs in the pipeline. Even so, there are still advantages for hexagonal digitization in pipelined systems despite the requirement of some extra control information. The line delays can be reduced in length by 13.4% and the PE master clock can be reduced by the same factor. The shorter line delays reduce process latency and the size of the circuit. In a recirculating pipeline system hexagonally sampled images will be able to be processed in 13.4% less time than square-sampled images. As the local image area contains only seven elements for a hexagonally sampled image, many of the processing modules would be simpler than for a square-system PE. For example, only seven multipliers would be required as opposed to nine in each multiplier array module.

## IV. Binary Image Processing

With hexagonal binary image-processing operator design, the simple six-way connectivity definition is exploited, and usually an equivalent hexagonal operator will be smaller and more easily computed than its square grid counterpart. Many hexagonal processing algorithms for binary image processing have been researched. As discussed in Section II.C there will be fewer samples (pixels) covering a given area, but if the hexagonal operators are simpler, or the processes are recursive, greater savings may be possible. The basic binary image-processing operations are described in textbooks (Davies, 1990). Some processes for the hexagonal grid are reviewed in what follows.

### A. Connectivity

In determining if a group of pixels is connected together to form an object, a definition of connectivity must first be stated. On a hexagonal grid, all neighboring sampling points, with associated pixels touching a central pixel, are equidistant from the central sampling point. If the pixel shape is hexagonal, then all the nearest neighbors touch the central pixel along equal length sides.

This scheme is known as six-connectedness. Hexagonal grids with rectangular pixels, as shown in Figure 2(c) can also be defined as six-connected.

On a rectangular grid, there are four nearest-neighboring pixels, but four additional pixels touch the central pixel at each corner. There are two definitions of connectivity:

- Four-connectedness, where only edge-adjacent pixels are neighbors; and
- Eight-connectedness, where corner adjacent pixels are also considered as neighbors.

A problem arises because the connectivity of the background pixels can also be considered. Now, if the four-connected definition is used on both foreground and background, some pixels will not appear in either set. A simple closed curve should be able to separate the background and object into distinct connected regions, but this is not the case. Again, if the 8-connected definition is used, some pixels will appear in both sets. One solution is to use four-connectedness for the object and eight-connectedness for the background. Another is to define a six-connectedness that involves only two corners.

The hexagonal systems unambiguous definition is more convenient. Connectivity is an important consideration in many image processes, especially where groups of pixels are being considered for membership of a particular feature, or the edges of a feature are being traced out and coded. The use of connectivity in shrinking and edge-following algorithms has been explored (Rosenfeld, 1970). Consideration has been given to the more general topological properties of digitized spaces, and in particular to connectivity and the order of connectivity (Mylopoulis and Pavlidis, 1971).

## B. Measurement of Distance

Useful measurements include the distance between points, the dimensions of a part, the area of an object, its perimeter, etc. (Rosenfeld and Kak, 1982). Connectivity evaluation, counting, and edge following also are important operations.

## C. Distance Functions

Distance functions are used in shape analysis. The distance of each pixel in an object from the boundary of the object is measured and overlaid on the binary image of the object, and this information is then analyzed (Rosenfeld and Pfaltz, 1968). Metrics using a four-, six-, or eight-way connectivity have been compared. The eight-way distance involves a $\sqrt{2}$ step for diagonals. The hexagonal six-way function was found to give a better approximation to Euclidean distance than the other functions (Luczak and Rosenfeld, 1976).

Recently, efficient distance transforms researched for the rectangular grid have been extended for use on the hexagonal grid (Mehnert and Jackway, 1999). The algorithm requires only integer arithmetic and can be implemented on both SISD and pipeline architectures.

### D. Morphological Operators

Mathematical morphology is an approach to computer vision based on searching for the shape of an object or the texture of a surface. Morphological operators are applied repeatedly to the image to remove irrelevant information and to enhance the essential shape of the objects within the scene. These methods are based on set theory. Operator design and application have been considered by various researchers (Matheron, 1975; Serra, 1982, 1986, 1988; Haralick et al., 1987).

Hexagonal sampling grids and morphological image processing have been strongly linked since they were first introduced. Hexagonal parallel pattern transformations involving morphological operations have also been reported (Golay, 1969; Preston, 1971). The main reason researchers chose hexagonal sampling was to avoid the ambiguous connectivity definitions between pixels on a square array. One of the most active researchers in this area, Serra (1982, 1986, 1988), makes extensive use of the hexagonal grid, preferring it to the square because of the connectivity definition, its large possible rotation group on the grid, and the simple processing algorithms that result.

### E. Line Thinning and the Skeleton of an Object

The skeleton or medial axis of a shape can be used as a basis for object recognition. In particular, it is often used in optical character-recognition systems. There are several steps involved in the process:

- Thresholding: The gray-level image is converted to a binary image in such a way as to maintain the shape.
- Thinning: The shape, which may have a width of several pixels, is analyzed, or eroded, to find a 1-pixel thick line that fits centrally within it.
- Line tracking: The thinned lines are chain coded.

- Line segmentation: The chain-coded information is converted to vector form. This point is the limit of the skeleton-forming process.
  Subsequent processes analyze the vectors to identify junctions and then the object.

Variations on this procedure exist and a large number of algorithms have been developed for the processes at each step. Surveys of these algorithms have also been reported (Smith, 1987; Lam *et al.*, 1992). There are two main classes of thinning algorithm, namely, iterative and methodical. In iterative methods a local area on the edge of the object is examined, and the central pixel of the area is removed if certain rules designed to preserve the connectivity of the final skeleton are obeyed. The process is repeated on the image in a way that removes pixels equally from both sides of the object until no further pixels are changed. The resulting skeleton is connected, its pixels are a subset of the original object's pixels, and it can be sufficient for many recognition tasks. Jang and Chin (1990) have used mathematical morphology to formally define thinning, and produced a set of operators that are proved to produce single-pixel thick connected skeletons. However, this resulting skeleton may lie only approximately in the correct place. The methodical algorithms aim to ensure a correctly positioned skeleton, but the iterative methods can produce sufficiently accurate skeletons for many applications; as they compute efficiently and can be easily realized using local operators, they have been applied to many problems.

Deutsch (1972) reports similar thinning algorithms developed for use with rectangular, hexagonal, and triangular arrays and has compared their operation. The triangular algorithm produced a skeleton with the least number of points, but it was sensitive to noise and image irregularities. The hexagonal algorithm was the most computationally efficient, produced a skeleton with fewer points than the rectangular algorithm, and was easily chain coded. Deutsch concluded that of the three algorithms, the hexagonal was optimal. Other hexagonal skeletonization algorithms have been reported (Meyer, 1988; Staunton, 1996a).

### F. Comparison between Hexagonal and Rectangular Skeletonization Programs

A comparison (Staunton, 1996a) has been made between an algorithm designed for the rectangular grid (Jang, 1990) and a similar one designed for the hexagonal grid.

There are many rectangular grid algorithms for the iterative removal of border pixels. Jang and Chin's (1990) was used for this comparison as it was designed using a mathematical framework based on morphological set transforms. Using these it can be proved that the final skeleton will conform to most of the following properties:

1. It will contain a number of single-pixel width lines.
2. Each skeletal element will be connected to at least one other. The skeleton will contain no gaps.
3. Skeletal legs will be preserved.
4. It will be accurately positioned.
5. Noise-induced pixels will be ignored, i.e., limbs will not be formed towards single-pixel edge protrusions.

It has been possible (Staunton, 1996a) to use similar mathematics to design a hexagonal algorithm that was close in operation to Jang and Chin's (1990). In each case the analysis led to the design of a set of thinning templates as shown in Figure 35 for the rectangular algorithm and Figure 36 for the hexagonal algorithm. Further analysis proved that the templates can be applied in parallel pairs to the image for the hexagonal case, and parallel triplets for the rectangular case. If the skeleton is to be positioned correctly the pairs must be applied in a particular order, as shown in Figure 37 for both the rectangular and hexagonal algorithms.

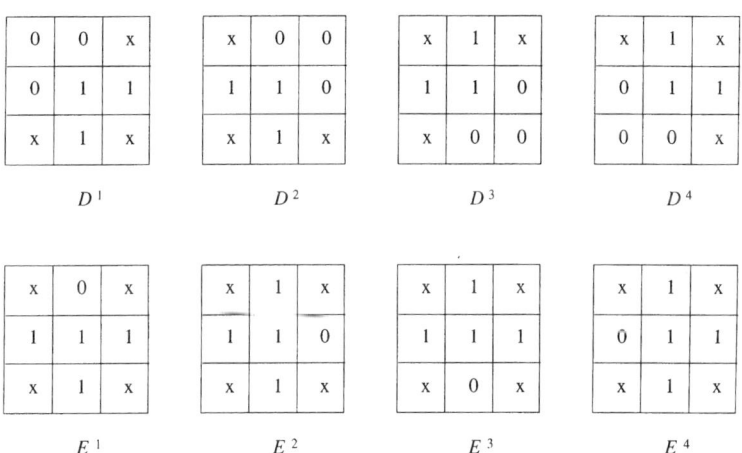

FIGURE 35. Rectangular scheme thinning templates $\mathbf{D} = \{D^1, D^2, D^3, D^4\}$, and $\mathbf{E} = \{E^1, E^2, E^3, E^4\}$. (Reprinted from R. C. Staunton (1996a). An analysis of hexagonal thinning algorithms and skeletal shape representation, *Pattern Recognition*, **29**(7): 1131–1146; Copyright (1996), with permission from Elsevier Science.)

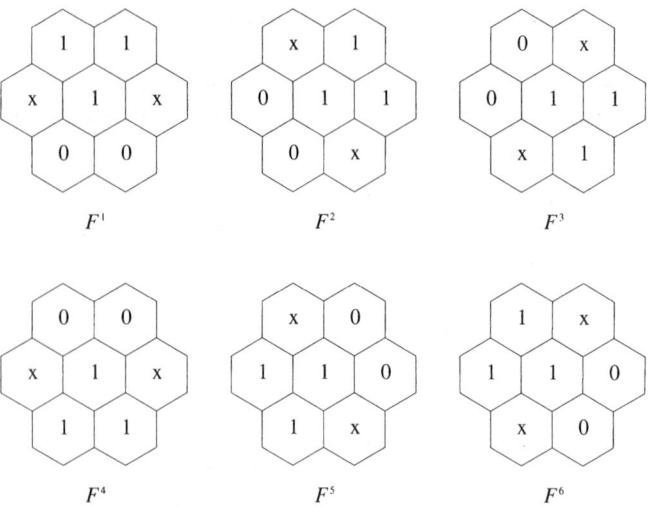

FIGURE 36. Hexagonal scheme thinning templates $\mathbf{F} = \{F^1, F^2, F^3, F^4, F^5, F^6\}$. (Reprinted from R. C. Staunton (1996a). An analysis of hexagonal thinning algorithms and skeletal shape representation, *Pattern Recognition,* **29**(7): 1131–1146; Copyright (1996) with permission from Elsevier Science.)

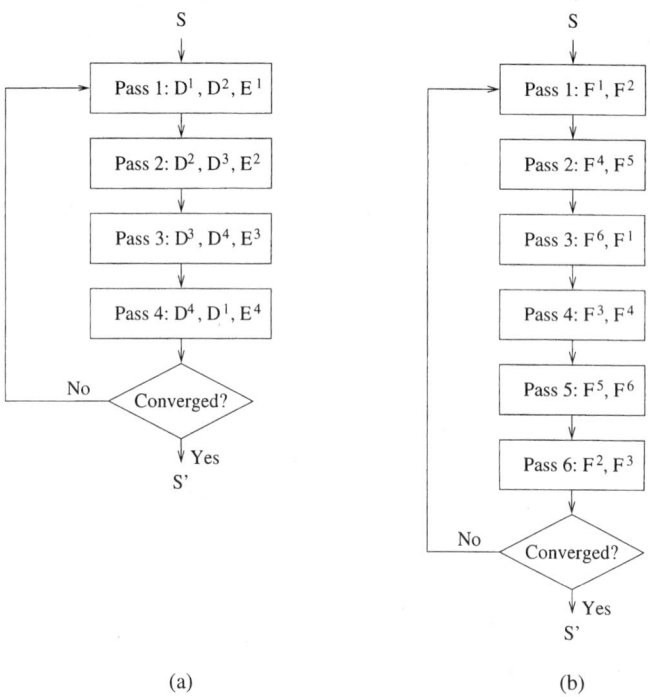

FIGURE 37. Template application order for thinning algorithms: (a) rectangular, (b) hexagonal.

The templates can alternatively be applied sequentially, and for the hexagonal case this produces a better preservation of skeletal legs and a slightly more accurate positioning of the skeleton. However, the parallel application of the templates resulted in a converged skeleton in approximately half the time required for the sequential application of the templates.

1. *Skeletal Quality*

Figure 38 shows some examples of the skeletons of four geometric shapes and a sample of text digitized on a rectangular grid. Figure 39 shows the same

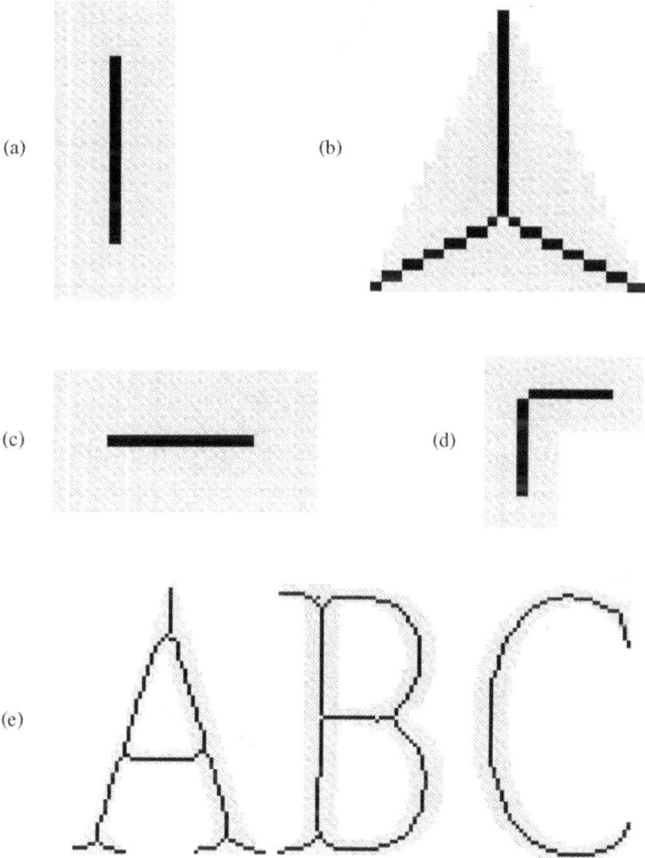

FIGURE 38. Skeletons produced by the new rectangular algorithm.

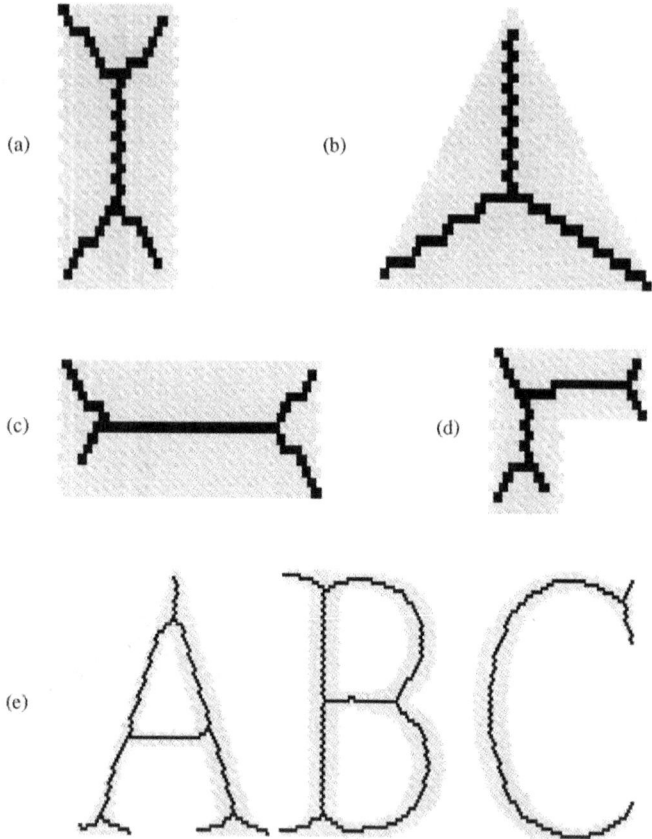

FIGURE 39. Skeletons produced by the hexagonal algorithm.

examples digitized on a hexagonal grid. These images can be compared and evaluated with respect to the five good skeletal qualities listed earlier in this section.

Properties 1 and 2 hold for each algorithm. Property 3 concerns the preservation of skeletal legs. These are not preserved to 90° corners by the rectangular algorithm, and there is some shrinkage to the corners introduced by the hexagonal algorithm. This shrinkage was less with the sequential version of the hexagonal algorithm (Staunton, 1996a).

Property 4 concerns the accurate positioning of the skeleton. Both algorithms have made good attempts at positioning the major axes of each shape, but few minor axes have been preserved by the rectangular algorithm. Considering the skeleton of Figure 38a, the single component could have been produced by any one of many rectangular shapes, but the information remaining can only indicate the position of the shape and the orientation of its major axis. The hexagonal skeletons contain more limbs and more information. The extra limbs within the rectangular shapes digitized on the hexagonal grid give an indication of their original size. Other rectangular algorithms have been researched that retain skeletal branches (Guo and Hall, 1989). The triangular shapes of Figures 38b and 39b have both been processed to produce good skeletons. The rectangular algorithm has shifted the center of gravity, or branch junction, towards the base of the triangle. The hexagonal algorithm has produced an unusual step pattern in the lower left limb, and some limb shortening.

Both algorithms have produced good skeletons of the text. The long thin strokes and the acute corners have not resulted in missing legs using the rectangular algorithm. The text images are from real scanned text, whereas the geometric shapes were computer generated. Property 5 states that noise-induced pixels should be ignored. The text images contain single pixel protrusions in the edges of each letter that can be defined as noise. The hexagonal algorithm was insensitive to these noise pixels, some of which can be observed in Figure 39e on the cross stroke of the "A" and the bottom of the "B." The rectangular algorithm was sensitive to the noise and limbs were formed to these pixels. A two-pass morphological filter was designed to remove the noise that resulted in the acceptable skeletons seen in Figure 38e, but introduced a processing overhead.

## 2. *Program Efficiency*

Both the rectangular and hexagonal algorithms can be computed on parallel machines, or alternatively, for a SISD machine the templates to be applied in parallel can be logically combined and then computed. The hexagonal templates compute more quickly as they have only seven elements as opposed to nine for the rectangular ones. The removal of a complete layer of pixels from the outside of an object can be referred to as an iteration of the algorithm. For each application of the rectangular four-pass scheme illustrated in Figure 37, the "corner" templates are applied twice and the "edge" templates once, whereas each application of the six-pass hexagonal scheme applies each template twice.

Table 1 compares the number of applications of the algorithm required to produce a converged skeleton from the images presented in Figures 38 and 39. In each case the hexagonal algorithm is at least as efficient as the rectangular algorithm. For real rectangularly sampled images edge-noise removal requires the equivalent of an additional two passes of the algorithm. Counting passes, the average hexagonal computation requires only 80% of those for the rectangular

TABLE 1
A Comparison of the Number of Passes Required to Form
Skeletons by the Rectangular and Hexangular Algorithms

| Image | Rectangular algorithm passes | Hexagonal algorithm passes |
|---|---|---|
| Vertical rectangle | 20 | 20 |
| Horizontal rectangle | 23 | 16 |
| Triangle | 27 | 24 |
| Corner | 12 | 12 |
| Text | 19 | 15 |

computation. If the reduced time to compute the smaller templates (Figs. 35 and 36) is considered, the hexagonal computation will require only 63% of the time of the rectangular computation to calculate the average skeleton. The test images were obtained by subsampling a double-resolution rectangularly sampled image in such a way that each shape contained the same number of pixels whether sampled on the rectangular or hexagonal grid. If a regular hexagonal grid had been used, then 13.4% fewer pixels would have been required for each shape, and the time to calculate a skeleton would have been reduced to 55% of that to calculate it on the rectangular grid.

In conclusion, both schemes produced good-quality skeletons, although there were differences in skeletal attributes. The hexagonal scheme could compute the average skeleton in 55% of the time required to compute it with the rectangular scheme.

Further work including a comparison between two fully parallel thinning algorithms designed for images sampled on the square and hexagonal grids has been reported (Staunton, 1999, 2001). Using techniques from mathematical morphology, a hexagonal algorithm has been designed to closely match the operation of a well-known square-grid algorithm. Proofs of the connectivity and single-pixel width of the resulting converged hexagonal skeleton have been presented. Implementations of both algorithms were found to produce accurate skeletons, but the hexagonal algorithm could be implemented with only 50% of the logical operations required by the square. This new algorithm will produce a more accurately positioned skeleton than the one described previously in this section, but it will take longer to compute.

## G. Tomography

Binary tomography on the hexagonal grid has been researched (Matej et al., 1998). Gibbs priors that describe the local characteristics in images were used to simplify the reconstruction from the three natural projections through the grid. The three projections, each at 60° to the others, align with the grid but are a small number from which to reconstruct the image. By using the appropriate prior information, it was possible to limit the number of possible solutions to the reconstruction equations and efficiently process the image.

## V. Monochrome Image Processing

This section contains a review of gray-scale operators that have been designed, to work on hexagonally sampled images. As they are grid-dependent, they can be defined as low-level processes (Section I). Some operations can be computed more efficiently in the Fourier domain and thus hexagonal transforms have been developed; others are applied in the spatial domain. Where possible, comparisons have been made between these and similar operators designed for the rectangular grid.

### A. The Hexagonal Fourier Transform

A hexagonal Fourier transform and hexagonal fast Fourier transform (HFFT) have been developed (Mersereau, 1979; Mersereau and Speake, 1981; Dudgeon and Mersereau, 1984; Guessoum and Mersereau, 1986). It was found that the HFFT required 25% less storage of complex variables than the rectangular fast Fourier transform (RFFT) and that it computed more efficiently. The algorithm is based on the Rivard procedure (Rivard, 1977), rather than the decomposition of the 2D kernel into 1D FFT method. Decomposition to 1D FFTs is not possible in the hexagonal case. This alternative procedure is a direct extension of the 1D FFT algorithm to the 2D case, which can increase the computational efficiency of the RFFT by 25%. Mersereau has shown that his HFFT increased computational efficiency by an additional 25% in comparison to the Rivard RFFT.

### B. Geometric Transformations

Geometric transformations have been researched using a three-integer coordinate frame (Her, 1995) as described in Section III.A. The coordinate frame is easy to use and the symmetry of the grid has enabled the design of simple efficient operators. Operators have been designed for: rounding that finds the nearest integer grid point to a point calculated with real coordinates; translations and reflections; scalings and shearings; and rotations.

Pscudoinvariant transformations on the hexagonal grid have been considered (Sheridan *et al.,* 2000). The work was inspired by the arrangement of the cones in the primate eye. A spiral honeycomb geometry of pixels was defined, together with an algebra. Spiral addition corresponded to translation, and spiral multiplication corresponded to rotation of the image.

### C. Point Source Location

This task, also known as star tracking, involves the tracking of a moving point light source across the array. The image of the source is a blurred spot. The centroid of the spot is calculated to within subpixel accuracy to give the position of the source. Accuracy is improved if the sensor array has a high fill factor, that is, the sensor elements tile the image window as completely as possible. For a 100% fill factor, a hexagonal array of hexagonally shaped sensors has been shown (Cox, 1987) to out-perform a square array of square-shaped sensors. Detection error and sensitivity to noise is reduced, and computational load and data storage are reduced by 24%. For lower fill factors the advantages of a hexagonal array are less pronounced (Cox, 1989).

### D. Image-Processing Filters

#### 1. Linear Filters

A series of general-purpose hexagonal FIR and IIR filters have been developed (Mersereau, 1979; Mersereau and Speake, 1983) and compared to rectangular filters with similar frequency responses. The hexagonal filters were found to be superior in terms of computational efficiency, and as they could be designed with 12-fold symmetry, they had a more circular frequency response. These filters concern 2D signal processing, in general, as opposed to only image processing. Savings of up to 58% in memory and similar gains in computational efficiency were reported for hexagonal filters compared to their rectangular counterparts.

Considering filters for image processing in more detail, the regular hexagonal structure leads to easy spatial plane local operator design. The local area can be defined to include the central pixel and any number of concentric "shells" of pixels at increasing distances from the center. All the members of a particular shell can be assigned equal weighting factors in many local operator designs. For example, consider a 4-shell Gaussian filter operating on a hexagonal grid,

```
        n   m   n              f  e  d  e  f

    m   l   l   m              e  c  b  c  e

n   l   k   l   n              d  b  a  b  d

    m   l   l   m              e  c  b  c  e

        n   m   n              f  e  d  e  f
```

FIGURE 40. Four-shell hexagonal and six-shell square local operators.

where four weighting factors are initially calculated as shown in Figure 40, and the final algorithm will be of the form of Eq. (14):

$$P_H = k i_{1,1} + l \sum_{p=1}^{6} i_{2,p} + m \sum_{q=1}^{6} i_{3,q} + n \sum_{r=1}^{6} i_{4,r} \qquad (14)$$

where $k$, $l$, $m$ and $n$ are filter weights associated with the four shells, and $i$ denotes image points. Four multiplications and 19 additions are required for the computation of each output pixel.

In comparison, a similar filter on a square grid ($5 \times 5$) requires six different weighting factors and a correspondingly more complicated algorithm of the form of Eq. (15).

$$P_S = a i_{1,1} + b \sum_{p=1}^{4} i_{2,p} + c \sum_{p=1}^{4} i_{3,p} + d \sum_{p=1}^{4} i_{4,p} + e \sum_{p=1}^{4} i_{5,p} + f \sum_{p=1}^{4} i_{6,p} \qquad (15)$$

where $a$, $b$, $c$, $d$, $e$, and $f$ are filter weights associated with the six shells. Six multiplications and 25 additions are required for the computation of each output pixel.

Both filters are convolved with a similar image area, but, in general, 13.4% fewer points will be required for the hexagonal filter than for the square. However, in this case, the square-system operator kernel is separable, giving an alternative computation algorithm of the form of Eq. (16):

$$P_S = [a i_{1,1} + b(i_{2,1} + i_{2,3}) + d(i_{4,1} + i_{4,3}] \otimes [a i_{1,1} + b(i_{2,2} + i_{2,4})$$
$$+ d(i_{4,2} + i_{4,4}] \qquad (16)$$

Now, six multiplications and eight additions are required for the computation of each output pixel. The hexagonal operator requires only four multiplications, compared with six for both rectangular algorithms, but the number of additions is larger than that for the separable kernel rectangular method. Computational efficiency will be determined by the architecture of the computer arithmetic and logic unit, and depend upon whether the filter coefficients are integer or real numbers.

Wavelet techniques have been applied to hexagonally sampled 2D signals and images (He and Lai, 1997). The Battle-Lemarie's wavelets have been generalized to the bivariate box spline wavelets. Here, these were applied to the hexagonal sampling grid and hexagonal box spline wavelets, filters, and filter banks constructed.

### 2. Nonlinear Filters

This class of filters includes designs such as the median filter and gray-scale morphologic filters (Sternberg, 1986; Haralick et al., 1987). Hexagonal grid median filters should be more computationally efficient than their square-grid counterparts, because for the same area of support, 13.4% fewer values exist. This will significantly simplify the sorting procedure.

### E. Edge Detectors

Edges correspond to intensity discontinuities in the image. These discontinuities may correspond to the edges of an object, but unfortunately sometimes they do not. For example, the edge of a shadow is likely to be detected. Many algorithms have been researched, but here some of the simplest are compared. Differential operators model local edges by fitting the best plane over a convenient size of neighborhood. In square arrays two orthogonal operators are applied to a pixel and from the response of these, the magnitude $m$ of the gradient of the plane and the edge angle, $a$, can be calculated:

$$m = (th^2 + tv^2)^{1/2} \tag{17}$$

$$a = \arctan(tv/th) \tag{18}$$

where $tv$ and $th$ are the responses of operators designed to respond maximally to vertical and horizontal edges. Figure 41 shows Sobel operators designed to be convolved with a $3 \times 3$ area of the image. For edge detection, the response magnitude is compared with a threshold to determine if a significant edge exists.

The Sobel operator has a computational processing time advantage over some other operators as only integer arithmetic is required and the local area

$$\begin{array}{ccc} 1 & 2 & 1 \\ 0 & 0 & 0 \\ -1 & -2 & -1 \end{array} \qquad \begin{array}{ccc} 1 & 0 & -1 \\ 2 & 0 & -2 \\ 1 & 0 & -1 \end{array}$$

FIGURE 41. Sobel differential operators with $3 \times 3$ area.

in which it operates is relatively small. It has been shown by some researchers to be the optimum $3 \times 3$ operator (Davies, 1984; Staunton, 1997b).

## F. Hexagonal Edge-Detection Operators

Hexagonal operators have been researched (Staunton, 1989). The regular hexagonal data structure leads to easy local operator design. The central element of the local area is surrounded by shells of elements. Figure 42 shows a set of edge-detection operators exploiting only the inner shell of neighbors, and these are of a comparable order to the $3 \times 3$ operators in Figure 41. These hexagonal operators will respond maximally to edges at 60° angular intervals from the horizontal. The weighting functions of the shell elements are chosen as 1 or −1 to reflect the regular structure of the grid of sampling points. Davies' design principle (Davies, 1984) indicates "1" to be nearly optimal. Again only integer arithmetic is required for computation. If these masks are used as differential operators, the slope magnitude $m$ becomes relatively complicated compared with Eq. (17). The equation of $m$ is derived as follows.

The output of each of the three hexagonal operators, as shown in Figure 42, can be represented as a vector. An edge can be modeled by a plane, and the three vectors, $t_1, t_2, t_3$, lie within this plane. Assuming orthogonal $x$ and $y$ axes, $t_3$ is aligned with the $y$ axis, $t_1$ is at 60° to $t_3$, and $t_2$ at 60° to $t_1$. The resultant vector, $m$ can be found:

$$m = t_1 + t_2 + t_3 \qquad (19)$$

$$\begin{array}{ccc} & 0 & 1 \\ -1 & 0 & 1 \\ & -1 & 0 \end{array} \qquad \begin{array}{ccc} -1 & 0 & \\ -1 & 0 & 1 \\ & 0 & 1 \end{array} \qquad \begin{array}{ccc} & 1 & 1 \\ 0 & 0 & 0 \\ & -1 & -1 \end{array}$$

FIGURE 42. Hexagonal differential edge-detection operators.

Examination of Figure 42 indicates the simple relationship $t_3 = t_1 - t_2$, giving

$$m = \frac{\sqrt{3}}{2}(t_1 + t_2)\hat{x} + \frac{3}{2}(t_1 - t_2)\hat{y} \qquad (20)$$

The slope magnitude, $m$, is

$$m = \left[3(t_1^2 + t_2^2 - t_1 t_2)\right]^{1/2} \qquad (21)$$

The angle that $m$ makes with the $x$ axis is known as the edge angle $a$

$$a = \arctan\left(\sqrt{3}\frac{t_1 - t_2}{t_1 + t_2}\right) \qquad (22)$$

A comparison between the computational efficiency and accuracy of local edge-detection operators in the two systems has been made (Staunton, 1989). The hexagonal system detector was found to compute more efficiently than the square-system Sobel detector as the mask weights are fewer in number and are all unity. On a SISD computer the hexagonal program is computed in 55% of the time required by the Sobel program. The accuracy of the two detectors was found to be equivalent, with the hexagonal being more accurate with one type of sensor model, and the square more accurate with a second type.

### G. The Visual Appearance of Edges and Features

The visual appearance of monochrome images is illustrated here using hexagonal and rectangular sampled images of a sand core used for metal casting. The core contained three small surface scratch defects that can be seen in Figures 43 and 44. The illumination employed divided the image of each defect into a bright and a dark (shadow) segment. There is one large circular defect, a long thin defect, and a small defect with dimensions comparable with the pixel size. The core was 14 cm high. The large circular defect has a diameter of 6 mm, the long thin scratch has dimensions 30 mm by 2 mm, and the small circular defect has a diameter of 1 mm.

On comparing the square and hexagonal images in Figures 43 and 44, the defects can generally be seen more clearly in the hexagonal. The offsetting of pixels on alternate lines enables the eye to trace their outlines more readily at this resolution. The large circular defect appears more circular, and the light and dark segments are more easily discerned. The long thin defect is more easily discernible as a connected component. The object edges, which in these examples are near vertical, are easier to localize in the hexagonal image. Long repeating brightness step sequences are observed in the rectangular image, whereas, a small castellated effect is observed in the hexagonal. In an attempt

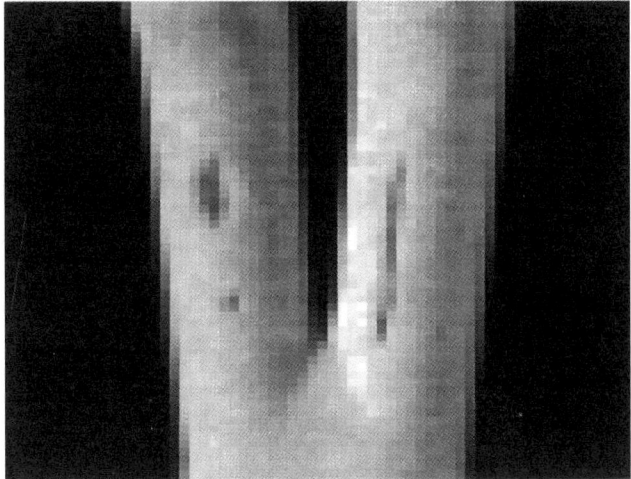

FIGURE 43. Rectangular sampled sand core image, 64 × 64 resolution.

to segment the defects from the remainder of the image, the two images were then edge detected using the optimum Sobel and corresponding hexagonal operators introduced in Section V(E). The threshold level was set manually so that the resulting edge images contained, where possible, connected edges around the defects, and so that the number of false detections was minimized.

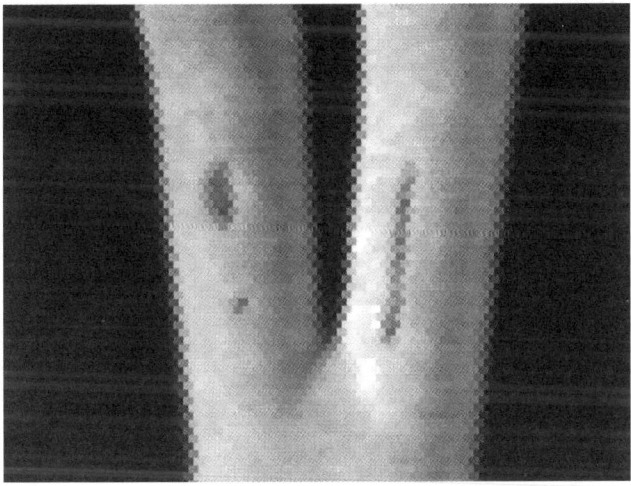

FIGURE 44. Hexagonal sampled sand core image, 64 × 64 resolution.

FIGURE 45. Rectangular sampled sand core image edge detected, 64 × 64 resolution.

Figure 45 shows the resulting square edge-detected image, and Figure 46 shows the resulting hexagonal image. The large circular defect appears to be square in overall shape in the rectangular image and there is a small disconnection in the outline. In the hexagonal image, it appears more circular, and the structure, such as the central dividing line between the light and dark segments, is more easily discerned. There is also a small gap in the outline.

The long thin defect has a break in its outline in the square image, whereas the outline is complete in the hexagonal. The equal width of this defect along its length is more discernible in the hexagonal image. The presence of the

FIGURE 46. Hexagonal sampled sand core image edge detected, 64 × 64 resolution.

FIGURE 47. Rectangular sampled sand core image, thinned, 64 × 64 resolution.

small defect is indicated by a small group of edge pixels in each image. There are also more detected false edge points in the square image. These are seen as unconnected black pixels in various parts of the image.

Figures 47 and 48 show the square and hexagonal edge-detected images after thinning. The same points as in the foregoing, concerning the defects,

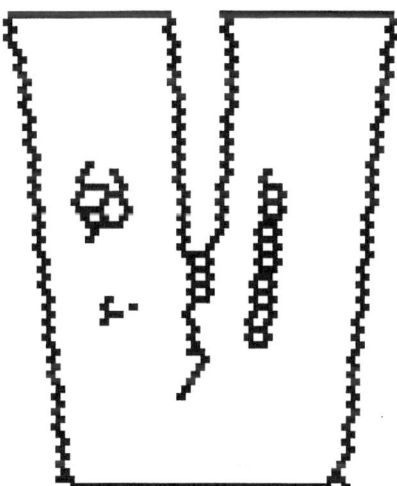

FIGURE 48. Hexagonal sampled sand core image, thinned, 64 × 64 resolution.

are still evident. The near vertical object edges appear as gradually increasing steps in the rectangular image, whereas in the hexagonal image a castellated effect is visible.

## 1. *Human Interpretation*

Interpretation of the images depends on the individual observer and the resolution of the image being viewed. At the low 64 × 64 resolution of the forementioned images, features are easily discerned in the hexagonal images, and their true shapes, whether circular or rectangular, can be more easily estimated. At the higher resolutions of 256 × 256 and 512 × 512, the aliasing effects at the object edges are less troubling to the eye and may be undiscernible at even higher resolutions. With the offsetting of pixels on alternate lines in the monochrome hexagonal images, the human eye may be able to estimate the boundaries between features more accurately as the pixel boundaries do not align to form long vertical features as in the square system. However, with the binary line images the pixel off setting may appear troublesome to the human eye. This has been reported by other researchers (Preston *et al.*, 1979). Machine interpretation will not depend on the visual appearance of the image, but on the efficiency of the higher-level processes. High-level processing will be easier if a detected edge contains fewer gaps.

## VI. CONCLUSIONS

This chapter has reviewed research on the sampling of images on a hexagonal grid, the processing of hexagonally sampled images by single- and multiprocessor computers, and the computation of image-processing operations on both binary and monochrome hexagonally sampled images. In the following, conclusions are drawn on each of these areas, and an attempt is made to answer the questions of when research should be conducted using hexagonally sampled images, and when it may be commercially advantageous to implement a hexagonally sampled image-processing system.

The hexagonal packing of sensors together with a hexagonal sensor shape is found in eyes. Evolution has favored the hexagon. Some manufactured sensors have hexagonal shapes, and others have circular or rectangular shapes. Each of these shapes has been shown to pack together efficiently on a hexagonal grid. A high fill factor or complete tiling of the area can lead to a high signal-to-noise ratio; however, for integrated sensor arrays, fill factors below 100% are necessary because communication circuits are required on the surface of the chip to transfer the image signals to the processor. Two-dimensional sampling theory was reviewed, with consideration being given to the aliasing of high-frequency components and the necessity to band limit analog signals before digitization to prevent this. If signals are circularly bandlimited, then their high-frequency information content is limited equally for any direction within the image. This is advantageous, as a feature detected when presented at one orientation to the sensor array can, in theory, be equally well detected when presented at any other orientation. If signals have been circularly bandlimited, then the hexagonal grid is more efficient than the square as 13.4% fewer sampling points will be required to give equal high-frequency information. This reduction in the number of sampling points for the hexagonally sampled images leads to reductions in image-storage requirements and faster subsequent processing.

A circular bandlimit has many advantages, and it was shown to be achievable for two CCD TV camera–frame grabber systems. The first discrete stage in such a system is the CCD sensor array, and if a circular bandlimit is to be achieved, then the lens and the active area of the sensor that integrates the brightness signal focused on it will be the main frequency-limiting components. The modulation transfer function (MTF) of the lens can be modeled most simply by the diffraction limit. More sophisticated models include aberration limits, but these are best evaluated using a specialist CAD system at the time the lens is designed. Simple 1D sensor models regard the sensor as a spatial window. Transforming this window to the Fourier domain results in a spectrum that is a sinc function of distance. The theoretical MTF of the system can be found by combining the lens and sensor MTFs, but this was found to overestimate the cutoff frequency. A knife-edge method for measuring the MTF of these discrete systems was outlined and applied to six TV camera–frame grabber systems. A circular bandlimit was found for two systems, and an elliptical bandlimit with a high vertical cutoff frequency was found for the others. Methods to reduce the vertical frequency response to make the bandlimiting more circular were discussed.

Once a hexagonal image has been acquired it can be processed using a conventional SISD or a multiprocessor computer. Hexagonal images can be processed by most computer architectures capable of processing square images. With some architectures the structure of the processor interconnections is fixed by the image-sampling grid. When processing hexagonal images, each processor will be connected to six neighbors, and when processing square images, each processor will be connected to four or eight neighbors. For some machines of this type it is possible to set up the connections to enable both types of sampled image to be processed. With architectures based on the sampling pattern, the processing task is divided between processors using spatial criteria. Other divisions are possible and can make it easier to use general-purpose multiprocessors such as sharedmemory, hypercube, or pipeline systems. It depends on the application how the task is best divided between processors. Communications need to be established between the processors, and within a 2D plane, general-purpose systems employing six-way (hexagonal) communications have been realized. Within a computer program a square image will map directly into a 2D array, and two-integer indexing is possible. For hexagonal images several indexing methods have been proposed, but the three-integer scheme appears to be the most efficient for general use.

Hexagonal pyramid systems are interesting to research first, because they can be used to model processing within the human visual cortex, and second, because the structure enables the efficient processing of low-medium-, and high-level operations on arrays of different processors at each level in the pyramid. An example of a pipeline processor that was capable of processing both square and hexagonal images was given. Small changes to each processor element were required to enable this dual role. For hexagonal-only processing, less data storage (13.4%), a lower clock rate for real-time operation (13.4%), fewer multipliers and adders, but twice as many convolution masks were required compared to square image processing. For recirculating pipelines faster processing was possible with hexagonal images.

The processing of binary hexagonal images was reviewed. It is preferred by some researchers to square processing due to its simple definition of connectivity, its large possible rotation group, and the 13.4% reduction in the number of sampling points. Advantages have been found for hexagonal images with distance measurement, distance functions, morphologic operators, and skeletonizing programs. Two similar skeletonizing algorithms, one for hexagonal and one for square images were compared. Both were designed according to the same criterion, and had been proved to produce good-quality skeletons. On a single-processor computer, the hexagonal program was found to calculate the skeleton in 55% of the time required by the square program.

The processing of monochrome hexagonal images was reviewed. Hexagonal FFT algorithms have been researched (Mersereau and Speake, 1981; Guessoum and Mersereau, 1986), and in a comparison with a similar square program, a hexagonal program was shown to require 25% less storage of complex variables and to exhibit a 25% increase in computational efficiency. Geometric transforms have been researched (Her, 1995) and shown to compute efficiently when a three-integer image indexing scheme was used. Hexagonal and square convolution filters have been compared (Mersereau, 1979; Mersereau and Speake, 1983). Due to the symmetry of the filter weight masks savings of up to 58% in memory and computations were demonstrated. The convolution mask weights tend to be arranged in equal value shells around a central value. Fewer shells are required to cover a particular area if hexagonal sampling is used. The details of the design of a hexagonal edge detector and its comparison with the square-system Sobel detector have been presented. Again the symmetry of the hexagonal convolution masks that leads to unit weight coefficients resulted in a detector with a similar accuracy to the Sobel detector, but that could be computed in 55% of the time required by the Sobel detector.

In Section V, a pair of resampled hexagonal and square-grid images was used for a visual comparison of edges and features. The hexagonal sampling enabled defects to be seen more clearly and their size better estimated. It is possible that the eye was better able to estimate boundaries more accurately when the pixels were offset in the hexagonal image. After edge detection and line thinning, the hexagonal edges were better connected, but the "zipper" effect caused by the offsetting of pixels in a binary brightness thin vertical line was not as pleasing to the observer as the single-pixel thick lines in the square-edge map.

To answer the question on when hexagonally sampled images should be used, the following conclusions can clarify the choice:

- The quality of circularly bandlimited images is similar between hexagonally and square-sampled images. This has been shown theoretically in terms of information content, and in practice by observation.
- Hexagonally sampled images can be processed by most types of computer.
- A circularly bandlimited hexagonal image requires 13.4% less storage than a square image.
- For image processes of a similar quality, a hexagonal process may compute in only 55% of the time required by the square process.

Hexagonal sampling and processing will always be important when modeling processes in human vision. For general research the position is less clear as vast libraries of software and a large choice of hardware is available to support the square scheme. This support is important if new ideas are to be tested and published quickly. Processing speed is important for real-time applications. At present, if a computer is not fast enough the researcher can rely on a faster one shortly becoming available. For this group of researchers switching to hexagonal processing may enable them to stay one jump ahead of the computer technology. The author has found that researching hexagonal processes at the same time as square processes can often lead to a deeper understanding of the problem.

Commercially, the higher processing speed and reduced storage requirement of hexagonally sampled images may be attractive. The printing of images and text on a hexagonal grid has already been done (Ulichney, 1987). Other self-contained products such as document scanners could well be produced at a lower cost if hexagonal processing was employed.

# REFERENCES

Annaratone, M., Arnold, E., Gross, T., Kung, H. T., Lam, M. S., Menzilcioglu, O., Sarocky, K., and Webb, J. A. (1986). Warp architecture and implementation. *Proc. IEEE 13th Int. Symposium on Computer Architecture,* 346–356.

Batchelor, B. G., Hill, D. A., and Hodgson, D. C. (1985). *Automated Visual Inspection,* Bedford, UK: IFS Publications Ltd.

Bell, S. B. M., Holroyd, F. C., and Mason, D. C. (1989). A digital geometry for hexagonal pixels. *Image and Vision Computing,* **7**(3), 194–204.

Black, G., and Linfoot, E. H. (1957). Spherical aberration and the information content of optical images. *Proc. Roy. Soc. A,* **239**, 522–540.

Boudin, J. P., Wang, D., Lecoq, J. P., and Xuan, N. P. (1998). Model for the charged coupled video camera and its application to image reconstruction. *Optical Engineering,* **37**(4), 1268–1274.

Bourbakis, N. G., and Mertoguno, J. S. (1996). Kydon: An autonomous multi-layer image-understanding system: Lower layers. *Engineering Applications of Artificial Intelligence,* **9**(1), 43–52.

Bowman, C. C., and Batchelor, B. G. (1987). Kiwivision a high speed architecture for machine vision. *Proc. SPIE,* **849**, 42–51.

Bowman, C. C. (1988). Getting the most from your pipelined processor. *Proc SPIE,* **1004**, 202–210.

Burt, P. J., Anderson, C. H., Sinniger, J. O., and van der Wal, G. (1986). A pipelined pyramid machine. In *Pyramidal Systems for Computer Vision,* V. Cantoni and S. Levialdi, eds., pp. 133–152, Berlin: Springer Verlag.

Burton, J., Miller, K., and Park, S. (1991). Fidelity metrics for hexagonally sampled digital imaging systems. *J. Imaging Technology,* **17**(6), 279–283.

Cantoni, V., and Levialdi, S. (1988). Multiprocessor computing for images. *Proc. IEEE,* **76**(8), 959–969.

Cox, J. A. (1987). Point source location using hexagonal detector arrays. *Optical Engineering,* **26**(1), 69–74.

Cox, J. A. (1989). Advantages of hexagonal detectors and variable focus for point source sensors. *Optical Engineering,* **28**(11), 1145–1150.

Crisman, J. D., and Webb, J. A. (1991). The warp machine on navlab. *IEEE Trans. PAMI,* **13**(5), 451–465.

Curcio, C. A., Sloan, K. R., Packer, O., Hendrickson, A. E., and Kalina, R. E. (1987). Distribution of cones in human and monkey retina: Individual variability and radical asymmetry. *Science,* **236**, 579–582.

Datacube Inc. (1989). *Maxvideo System,* Peabody, MA.

Davies, E. R. (1984). Circularity a new principle underlying the design of accurate edge orientation operators, *Image and Vision Computing,* **2**(3), 134–142.

Davies, E. R. (1990). *Machine Vision: Theory, Algorithms, Practicalities.* London: Academic Press.

Delbruck, T. (1993). Silicon retina with correlation-based, velocity-tuned pixels. *IEEE Trans. Neural Networks,* **4**(3), 529–541.

Deng, Y., Manjunath, B. S., Kenney, C., Moore, M. S., and Shin, H. (2001). An efficient color representation for image retrieval. *IEEE Trans. Image Process,* **10**(1), 140–147.

Deutsch, E. S. (1972). Thinning algorithms on rectangular hexagonal and triangular arrays. *Communications ACM,* **15**(9), 827–837.

Dolter, J. W., Ramanathan, P., and Shin, K. G. (1991). Performance analysis of virtual cut-through switching in HARTS: A hexagonal mesh multicomputer, *IEEE Trans. Computing,* **40**(6), 669–679.

Downton, A. C., Tregidgo, R. W. S., and Cuhadar, A. (1994). Top-down structured parallelization of embedded image-processing applications. *IEE Proc. Vision Image and Sign Processing,* **141**(6), 431–437.
Downton, A., and Crookes, D. (1998). Parallel architectures for image processing. *IEE Electronics Communication Engineering J.,* **10**(3), 139–151.
Dudgeon, D. E., and Mersereau, R. M. (1984). *Multidimensional Digital Signal Processing,* Upper Saddle River, N.J.: Prentice Hall Inc.
Duff, M. J. B. (1985). Real Applications on Clip4. In *Integrated Technology for Parallel Image Processing,* S. Levialdi, ed., London: Academic Press, pp. 153–165.
Dyer, C. R. (1989). Introduction to the special section on computer architectures and parallel algorithms, *IEEE Trans. PAMI,* **11**(3), 225–226.
Ekmecic, I., Tartalja, I., and Milutinovic, V. (1996). A survey of heterogeneous computing: Concepts and systems. *Proc. IEEE,* **84**(8), 1127–1143.
Fisher, A. L., Kung, H. T., Monier, L. M., Walker, H., and Dohi, Y. (1983). Design of the psc: a programmable systolic chip, *Proc. 3rd Caltech Conf. on VLSI,* pp. 287–302.
Flynn, M. J. (1966). Very high speed computing systems. *Proc. IEEE,* **54**(12), 1901–1909.
Fountain, T. J. (1987). *Processor Arrays Architecture and Applications.* London: Academic Press.
Gaskill, J. D. (1978). *Linear Systems, Fourier Transforms, and Optics.* New York: Wiley.
Golay, M. J. E. (1969). Hexagonal parallel patern transformations, *IEEE Trans. Computers,* **18**(8), 733–740.
Goldstein, M. D., and Nagler, M. (1987). Real time inspection of a large set of surface defects in metal parts. *Proc. SPIE,* **849,** 184–190.
Gonzalez, R. C., and Woods, R. E. (1992). *Digital Image Processing.* Reading, MA: Addison Wesley.
Guessoum, A., and Mersereau, R. M. (1986). Fast algorithms for the multidimensional discrete Fourier transform. *IEEE ASAP,* **34**(4), 937–943.
Guo, Z., and Hall, R. W. (1989). Parallel thinning algorithms: Parallel speed and connectivity preservation. *Communications ACM,* **32**(1), 124–131.
Handler, W. (1984). Multiprozessoren fur breite answendungsgebiete erlangen, general purpose array. GI NTG Fachtagung Architektur und Betrieb von Rechensystemen Informatik Fachbetrichte. Berlin: Springer-Verlag, pp. 195–208.
Hanzal, B. R., Joseph, J. D., Cox, J. A., and Schwanebeck, J. C. (1985). PtSi hexagonal detector focal plane arrays. *Proc. SPIE,* **570,** 163–171.
Haralick, R. M., Sternberg, S. R., and Zhuang, X. (1987). Image analysis using mathematical morphology, *IEEE Trans. PAMI,* **9**(4), 532–550.
Hartman, P., and Tanimoto, S. (1984). A hexagonal pyramid data structure for image processing, *IEEE Trans. SMC,* **14**(2), 247–256.
Helmholtz, H. L. F. (1911). *Handbuch der Physiologischen Optik.* Volume 2, Hamburg, Germany: Verlag von Leopold Voss.
Helmholtz, H. L. F. (1962). *Treatise on Physiological Optics.* Volume 2 (Translated by J. P. C. Southall), New York: Dover Publications.
He, W., and Lai, M. J. (1997). Digital filters associated with bivariate box spline wavelets. *J. Electron, Imaging,* **6**(4), 453–466.
Her, I. (1995). Geometric transformations on the hexagonal grid, *IEEE Trans. Image Processing,* **4**(9), 1213–1222.
Illingworth, J., and Kittler, J. (1988). A survey of the Hough transform. *Computer Vision Graphics and Image Processing,* **44,** 87–116.
Inmos Ltd. (1989). *The Transputer Databook,* 2nd ed., Bristol, UK.
Jang, B. K., and Chin, R. T. (1990). Analysis of thinning algorithms using mathematical morphology. *IEEE Trans. PAMI,* **12**(6), 541–551.

Kamgar-Parsi, B., and Kamgar-Parsi, B. (1989). Evaluation of quantization error in computer vision. *IEEE Trans. PAMI,* **11**(9), 929–940.

Kamgar-Parsi, B., and Kamgar-Parsi, B. (1992). Quantization error in hexagonal sensory configurations. *IEEE Trans. PAMI,* **14**(6), 665–671.

Kobayashi, H., Matsumoto, T., and Sanekata, J. (1995). Two dimensional spatio-temporal dynamics of analog image processing neural networks, *IEEE Trans. Neural Networks,* **6**(5), 1148–1164.

Kovac, M., and Ranganathan, N. (1995). Jaguar—a fully pipelined VLSI architecture for JPEG image compression standard. *Proc. IEEE,* **83**(2), 247–258.

Lam, L., Lee, S. W., and Suen, C. Y. (1992). Thinning methodologies, a comprehensive survey. *IEEE Trans. PAMI,* **14**(9), 869–885.

Lee, J. W., Yang, M. H., Kang, S. H., and Choe, Y. (1997). An efficient pipelined parallel architecture for blocking effect removal in HDTV, *IEEE Trans. Consumer Electronics,* **43**(2), 149–156.

Lee, J. W., Park, J. W., Yang, M. H., Kang, S. H., and Choe, Y. (1998). Efficient algorithm and architecture for post-processor in HDTV, *IEEE Trans. Consumer Electronics,* **44**(1), 16–26.

Lenoir, F., Bouzar, S., and Gauthier, M. (1989). Parallel architecture for mathematical morphology, *Proc. SPIE,* **1199,** 471–482.

Li, H., and Kender, J. R. (1988). Special issue on computer vision scanning the issue, *Proc. IEEE,* **76**(8), 859–862.

Li, H., and Maresca, M. (1989). Polymorphic torus architecture for computer vision, *IEEE Trans. PAMI,* **12**(3), 233–243.

Lin, T. P., and Hsieh, C. H. (1994). A modular and flexible architecture for real-time image template matching, *IEEE Trans. Circuits and Systems: I-Fundamental Theory and Applications,* **41**(6), 457–461.

Lougheed, R. M., and McCubbrey, D. L. (1980). The cytocomputer a practical pipelined image processor, *7th. Int. Symposium in Comput. Architecture,* pp. 271–277.

Lougheed, R. M. (1985). A high speed recirculating neighborhood processing architecture, *Proc. SPIE,* **534,** 22–33.

Luck, R. L. (1986). Using PIPE for inspection applications, *Proc. SPIE,* **730,** 12–19.

Luck, R. L. (1987). Implementing an image understanding system architecture using pipe, *Proc. SPIE,* **489,** 35–41.

Luczak, E., and Rosenfeld, A. (1976). Distance on a hexagonal grid, *IEEE Trans. Comput.,* **25,** 532–533.

Lundmark, A., Wadstromer, N., and Li, H. (2001). Hierarchical subsampling giving fractal regions, *IEEE Trans. Image Process.* **10**(1), 167–173.

Maresca, M., Lavin, M. A., and Hungwen, L. (1988). Parallel architectures for vision, *Proc. IEEE,* **76**(8), 970–981.

Matej, S., Herman, G. T., and Vardi, A. (1998). Binary tomography on the hexagonal grid using Gibbs priors, *Int. J. Imaging Syst. and Technol.* **9**(2–3), 126–131.

Matheron, G. (1975). *Random Sets and Integral Geometry,* New York: Wiley.

McCafferty, J. D., Fryer, R. J., Codutti, S., and Monai, G. (1987). Edge detection algorithm and its video rate implementation, *Image and Vision Computing,* **5**(2), 155–160.

McCormick, B. (1963). The Illinois pattern recognition computer—Illiac 3. *IEEE Trans. Electronic Computers,* **12**(6), 791–813.

McIlroy, C. D., Linggard, R., and Monteith, W. (1984). Hardware for real time image processing, *IEE Proc. Part E,* **131**(6), 223–229.

Mehnert, A. J. H., and Jackway, P. T. (1999). On computing the exact Euclidean distance transform on rectangular and hexagonal grids, *J. Math. Imaging and Vis.* **11**(3), 223–230.

Mersereau, R. M. (1979). The processing of hexagonally sampled two dimensional signals, *Proc. IEEE,* **67**(6), 930–949.

Mersereau, R. M., and Speake, T. C. (1981). A unified treatment of Cooley-Tukey algorithms for the evaluation of multidimensional DFT, *IEEE Trans. ASSP,* **29**(5), 1011–1018.

Mersereau, R. M., and Speake, T. C. (1983). The processing of periodically sampled multidimensional signals, *IEEE ASSP,* **31**(1), 188–194.

Meyer, F. (1988). Skeletons in digital spaces. In *Image Analysis and Mathematical Morphology, Volume 2: Theoretical Advances,* J. Serra, ed., London. Academic Press.

Minami, T., Kasai, R., Yamaauchi, H., Tashiro, Y., Takahashi, Y., and Data, S. (1991). A 300-mops video signal processor with a parallel architecture, *IEEE J. Solid-State Circuits,* **26**(12), 1868–1875.

Mitchell, J. W. (1993). The silver halide photographic emulsion grain, *J. Imaging Science and Technology,* **37**(4), 331–343.

Mylopoulos, J. P., and Pavlidis, T. (1971). On the topological properties of quantized spaces: II connectivity and order of connectivity, *J. Assoc. Comput. Machinery,* **18**(2), 247–254.

Neeser, W., Bocker, M., Buchholz, P., Fischer, P., Holl, P., Kemmer, J., Klein, P., Koch, H., Locker, M., Lutz, G., Matthay, H., Struder, L., Trimpl, M., Ulrici, J., and Wermes, N. (2000). The DEPFET pixel BIOSCOPE, *IEEE Trans. Nucl. Sci.* **47**(3), 1246–1250.

Nudd, G. R., Grinberg, J., Etchells, R. D., and Little, M. (1985). The application of three dimensional microelectronics to image analysis, In *Integrated Technology for Parallel Image Processing* (S. Leviadi, ed.), London. Academic Press, pp. 256–282.

Nudd, G. R., Atherton, T. J., Howarth, R. M., Clippingdate, S. C., Francis, N. D., Kerbyson, D. J., Packwood, R. A., Vaudin, G. J., and Walton, D. W. (1989). WPM: A multiple-simd architecture for image processing, *IEE 3rd Int. Conf. on Image Proc., Warwick, UK,* Publication No. 307, 161–165.

Nyquist, H. (1928). Certain topics in telegraph transmission theory, *Trans. AIEE,* **47**, 617–644.

Oakley, J. P., and Cunningham, M. J. (1990). A function space model for digital image sampling and its application to image reconstruction, *Computer Vision Graphics and Image Processing,* **49**, 171–197.

Olson, T. J., Taylor, J. R., and Lockwood, R. J. (1996). Programming a pipelined image-processor, *Computer Vision and Image Understanding,* **64**(3), 351–367.

Petersen, D. P., and Middleton, D. (1962). Sampling and reconstruction of wave number limited functions in $n$ dimensional euclidean spaces, *Information and Control,* **5**, 279–323.

Preston, K. (1971). Feature extraction by Golay hexagonal pattern transforms, *IEEE Trans. Computers,* **20**(9), 1007–1014.

Preston, K., Duff, M. J. B., Levialdi, S., Norgren, P. E., and Toriwaki, J. (1979). Basics of cellular logic with some applications in medical image processing, *Proc. IEEE,* **67**(5), 826–856.

Ray, S. F. (1988). *Applied Photographic Optics,* London. Focal Press.

Reichenbach, S. E., Park, S. K., and Narayanswamy, R. (1991). Characterizing digital image acquisition devices, *Optical Engineering,* **30**(2), 170–177.

Rivard, G. E. (1977). Direct fast Fourier transform of bivariate functions, *IEEE Trans. ASSP,* **25**, 250–252.

Rosenfeld, A., and Pfaltz, J. L. (1968). Distance functions on digital pictures, *Patt. Rec.* **1**, 33–61.

Rosenfeld, A. (1970). Connectivity in digital pictures, *J. Assoc. Comput. Machinery,* **17**(1), 146–160.

Rosenfeld, A., and Kak, A. C. (1982). *Digital Picture Processing.* Volume 1, New York. Academic Press.

Ruetz, P. A., and Broderson, R. W. (1986). A custom chip set for real time image processing, *Conf. ICASSP Tokyo,* 801–804.

Schroder, D. K. (1980). Extrinsic silicon focal plane arrays, In *Charge Coupled Devices* (D. F. Barbe, ed.), New York: Springer-Verlag.

Sensiper, M., Boreman, G. D., Ducharme, A. D., and Snyder, D. R. (1993). Modulation transfer function testing of detector arrays using narrow-band laser speckle, *Optical Engineering,* **32**(2), 395–400.

Serra, J. (1982). *Image Analysis and Mathematical Morphology.* London: Academic Press.

Serra, J. (1986). Introduction to Mathematical Morphology, *Computer Vision Graphics and Image Processing,* **35**, 283–305.

Serra, J. (1988). *Image Analysis and Mathematical Morphology. Volume 2, Theoretical Advances.* London: Academic Press.

Sharp, E. D. (1961). A triangular arrangement of planar-array elements that reduces the number needed, *IRE Trans. Antennas Propagat.,* **3**, 445–476.

Sheridan, P., Hintz., T., and Alexander, D. (2000). Pseudo-invariant image transformations on a hexagonal lattice, *Image Vis. Comput.* **18**(11), 907–917.

Sheu, M. H., Wang, J. F., Chen, A. N., Suen, A. N., Jeang, Y. L., and Lee, J. Y. (1992). A data-resuse architecture for gray-scale morphologic operations, *IEEE Trans. Circuits and Systems, II-Analog and Digital Signal Processing,* **39**(10), 753–756.

Shih, F. Y., King, C. P., and Pu, C. C. (1995). Pipeline architectures for recursive morphological operations, *IEEE Trans. Image Processing,* **4**(1), 11–18.

Smith, R. W. (1987). Computer processing of line images a survey, *Patt. Rec.,* **20**(1), 7–15.

Staunton, R. C. (1989). The design of hexagonal sampling structures for image digitization and their use with local operators, *Image and Vision Computing,* **7**(3), 162–166

Staunton, R. C., and Storey, N. (1990). A comparison between square and hexagonal sampling methods for pipeline image processing, *Proc. SPIE,* **1194**, 142–151.

Staunton, R. C. (1996a). An analysis of hexagonal thinning algorithms and skeletal shape representation, *Patt. Rec.,* **29**(7), 1131–1146.

Staunton, R. C. (1996b). Edge detector error estimation incorporating CCD camera limitations, *IEEE Norsig96 Signal Processing Conference,* Espoo, Finland, pp. 243–246.

Staunton, R. C. (1997a). Measuring the high frequency performance of digital image acquisition system, *IEE Electronics Letters,* **33**(17), 1448–1450.

Staunton, R. C. (1997b). Measuring image edge detector accuracy using realistically simulated edges, *IEE Electronics Letters,* **33**(24), 2031–2032

Staunton, R. C. (1998). Edge operator error estimation incorporating measurements of CCD TV camera transfer function, *IEE Proc. Vision, Image and Signal Processing,* **145**(3), 229–235.

Staunton, R. C. (1999). A one-pass parallel hexagonal thinning algorithm, *Proc. IPA99: IEE 7th Int. Conf. Image Processing and its Applications,* Vol II, 841–845. Manchester. UK.

Staunton, R. C. (2001). One-pass parallel hexagonal thinning algorithm, *IEE Proc. Vis. Image and Signal Processing.* **148**(1), 45–53.

Sternberg, S. R. (1986). Greyscale morphology, *Computer Vision, Graphics and Image Processing,* **35**, 333–355.

Storey, N., and Staunton, R. C. (1989). A pipeline processor employing hexagonal sampling for surface inspection, *3rd Int. Conf. on Image Processing and Its Applications,* IEE Conference Publication No. 307, 156–160.

Storey, N., and Staunton, R. C. (1990). An adaptive pipeline processor for real-time image processing, *Proc. SPIE,* **1197**, 238–246.

Tanimoto, S. L., Ligocki, T. J., and Ling, R. (1987). A prototype pyramid machine for hierarchical cellular logic. In *Parallel Computer Vision,* L. Uhr, ed., Boston: Academic Press, pp. 43–83.

Tzannes, A. P., and Mooney, J. M. (1995). Measurement of the modulation transfer function of infrared cameras, *Optical Engineering,* **34**(6), 1808–1817.

Ulichney, R. (1987). *Digital Halftoning,* Cambridge, Mass. MIT Press.

Valkenburg, R. J., and Bowman, C. C. (1988). Kiwivision II a hybrid pipelined multitransputer architecture for machine vision, *Proc. SPIE,* **1004**, 91–96.

Wandell, B. A. (1995). *Foundations of Vision,* Sunderland, Mass.: Sinauer Associates Inc.

Watson, A. B., and Ahumada, A. J. (1989). A hexagonal orthogonal oriented pyramid as a model of image representation in visual cortex, *IEEE Trans. BME,* **36**(1), 97–106.

Wehner, B. (1989). Parallel recirculating pipeline for signal and image processing, *Proc. SPIE,* **1058**, 27–33.

Whitehouse, D. J., and Phillips, M. J. (1985). Sampling in a two-dimensional plane, *J. Physics A, Math. Gen.,* **18**, 2465–2477.

Zandhuis, J. A., Pycock, D., Quigley, S. F., and Webb, P. W. (1997). Sub-pixel non-parametric PSF estimation for image enhancement, *IEE Proc. Vis. Image Signal Process.,* **144**(5), 285–292.

# Space-Variant Two-Dimensional Filtering of Noisy Images*

## ALBERTO DE SANTIS,[1] ALFREDO GERMANI,[2] AND LEOPOLDO JETTO[3]

[1] *Dipartimento di Informatica e Sistemistica, Università degli Studi "La Sapienza" di Roma 00184 Rome, Italy*
[2] *Dipartimento di Ingegneria Elettrica, Università dell'Aquila, 67100 Monteluco (L'Aquila) Italy and Istituto di Analisi dei Sistemi ed Informatica del CNR—Viale Manzoni 30, 00185 Rome, Italy*
[3] *Dipartimento di Elettronica e Automatica, Università di Ancona, 60131 Ancona, Italy*

|  |  |
|---|---|
| I. Introduction | 268 |
|    A. The State of the Art | 268 |
|    B. The Image Signal: Basic Assumptions | 270 |
| II. Kalman Filtering | 271 |
|    A. State-Space Representation | 271 |
|    B. The Estimation Algorithm | 274 |
|    C. The Steady-State Solution | 275 |
| III. The Image Model | 276 |
|    A. The Homogeneous Image Equation | 276 |
|    B. The Component Equations of the Sampled Image | 279 |
|    C. Modeling the State Noise | 280 |
|    D. The Constitutive Equation | 282 |
| IV. Image Restoration | 285 |
|    A. Space-Variant Realization of the Image | 285 |
|    B. The Edge Problem | 289 |
|    C. The Filtering Algorithm | 291 |
|    D. Deblurring | 292 |
| V. New Research Developments | 294 |
|    A. Line-Scan Filter Implementation | 294 |
|       1. A First Suboptimal Implementation | 295 |
|       2. The Semicausal Filter | 297 |
|    B. An Edge-Estimation Algorithm | 298 |
|    C. Polynomial Filtering for the Non-Gaussian Case | 300 |
| VI. Numerical Results | 303 |
| VII. Conclusions | 310 |
| VIII. Appendices | 310 |
|    A. Appendix A | 310 |
|    B. Appendix B | 311 |
|    C. Appendix C | 314 |
|    References | 316 |

*Portions reprinted, with permission, from Space-variant recursive restoration of noisy images, by A. De Santis, A. Germani and L. Jetto, *IEEE Transactions on Circuits and Systems-Part II: Analog and Digital Signal Processing* **41** (4): 249–261, April 1994, © 1994 IEEE.

## I. Introduction

### A. The State of the Art

The considerable attention that the image-restoration problem has been receiving in the literature has motivated the interest in extending optimal one-dimensional (1D) filtering procedures to two-dimensional (2D) data fields. In particular most authors investigated the applicability of Kalman filtering techniques to the restoration of images corrupted by additive noise; as a consequence many efforts have been made toward the synthesis of image models suitable to recursive restoration techniques.

In Nahi (1972), Nahi and Assefi (1972), Nahi and Franco (1973), Jain and Angel (1974), Powell and Silverman (1974), Woods and Radewan (1977), Murphy and Silverman (1974), Katayama (1980), Suresh and Shenoi (1981), Azimi-Sajadi and Khorasani (1990), a 2D image is transformed into 1D scalar or vector stochastic process using a line-by-line scan or a vector-scanning scheme. Other approaches that use a 2D model can be found in Habibi (1972), Attasi (1976), Jain (1977), Jain and Jain (1978), Katayama and Kosaka (1978, 1979), and Suresh and Shenoi (1979). All these papers are based on the common assumption that the image is the realization of a wide sense stationary random field. By this simplifying hypothesis an image model suitable to a state-space representation can be derived; nevertheless the corresponding space-invariant filters are insensitive to abrupt changes in the image signal and give restored images with reduced contrast and blurred edges. Actually, a real image is composed of an ensemble of several different regions and, in general, no correlation among them may be assumed save that of casually being elements of the same picture. Thus the stationarity assumption may fit for the statistics of each single region, but not for the whole image; consequently blurring and oversmoothing phenomena occur at edge locations.

Adaptive space-variant filters based on identification-estimation algorithms have been proposed (Keshavan and Srinath, 1977, 1978; Katayama, 1980; Kaufman *et al.*, 1983; Yum and Park, 1983; Wellstead and Caldas Pinto, 1985a, 1985b; Azimi-Sajadi and Bannour, 1991). These methods allow the parameters describing the image model to vary inside the image itself according to the local statistics. The specific problem of reducing the numerical complexity involved in the adaptive parameter estimation procedures for a 2D image model is considered in Zou *et al.* (1994).

Other space-variant filters have been proposed (Rajala and De Figueiredo, 1981; Woods *et al.*, 1987; Tekalp *et al.*, 1989; Jeng and Woods, 1988, 1991; Wu and Kundu, 1992). In Rajala and De Figueiredo (1981) the image is partitioned into disjoint subregions with similar spatial activity; the *masking function*, first introduced in Anderson and Netravali (1976), is then used to properly weigh the response of human eye to additive noise. In Woods *et al.* (1987) the image is modeled as a globally homogeneous random field with a local structure created by a 2D hidden Markov chain. This hidden chain controls the switching of the coefficients of a conditionally Gaussian autoregressive model to take into account the presence of image edges. The method proposed in Woods *et al.* (1987) is extended in Tekalp *et al.* (1989) to deconvolution type problems. In Jeng and Woods (1988) each image pixel is processed by a switched linear filter with the switching governed by a visibility function; in Jeng and Woods (1991) images are described as a compound of several submodels having different characteristics, and an underlying structure model governs the transition between these submodels. In Wu and Kundu (1992) the image is modeled as a nonstationary mean and stationary variance autoregressive Gaussian process, some modifications of the reduced update Kalman filter described in Woods and Radewan (1977) are also proposed to obtain a filtering algorithm with a reduced numerical complexity.

The method proposed in Biemond and Jerbrands (1979) deals with the edge problem using the output of a linear edge detector as an additional input to improve the step response of a space-invariant Kalman filter.

All these papers are based on a description of the image in terms of its statistical properties. The self-tuning methods attempt to draw this information starting from noisy data. Their main drawbacks are the computational cost, which in many cases may be unacceptable, and/or the increased structural complexity of the algorithm. Moreover, it seems difficult to obtain a fast switching of the image model parameters in correspondence to a sudden change in the image statistics, such as at edge points. The other methods assume that the information on the image model is available *a priori* or that it can be obtained from the noisy free image or by a sample of similar pictures. In many practical situations it is unrealistic to assume that these data are available.

Based on the results of Germani and Jetto (1988) and Bedini and Jetto (1991), the method proposed in De Santis *et al.* (1994) starts from a completely different point of view, based on very physical assumptions on the structure of the stochastic image model (see Section I.B).

As shown in De Santis *et al.* (1994), such assumptions allow one to construct a space-variant image model where the problem of image parameter identification is greatly simplified and where the presence of image edges is intrinsically taken into account, so that edge oversmoothing is automatically avoided. The resulting filtering algorithm is suitable for implementation as a strip processor (Woods and Radewan, 1977).

A different filter implementation was devised in Concetti and Jetto (1997) and in Jetto (1999a); it has a lower memory occupancy and avoids some numerical drawbacks typical of the strip processor instrumentation. In De Santis and Sinisgalli (1999), starting from the image model proposed in De Santis *et al.* (1994), a Bayesian edge-detection procedure was designed; it is an adaptive algorithm based on the local signal characteristics estimated at any pixel and does not requires any global thresholding. Finally, in Dalla Mora *et al.* (1999) a polynomial-filtering algorithm was proposed for image restoration in presence of non-Gaussian noise; this method relies on a polynomial extension of the Kalman filter (De Santis *et al.*, 1995; Carravetta *et al.*, 1996).

Other 2D recursive filtering algorithms paralleling the 1D Kalman filter and not requiring strip processing have been proposed in Habibi (1972), Panda and Kak (1976), Katayama and Kosaka (1979), Biemond and Gerbrands (1980). These filters are based on the quarter-plane system, first introduced in Habibi (1972), and their nonoptimality was proved in Strintzis (1976) and Murphy (1980). As shown in Barry *et al.* (1976) there is no optimal finite-dimensional causal filter for the quarter-plane system, whereas a finite-dimensional approximation to the optimal half-plane filter has been presented in Attasi (1976).

### B. *The Image Signal: Basic Assumptions*

The new approach first proposed in (De Santis *et al.*, 1994) allows one to define an image model incorporating an *a priori* structural information about edge locations. In general, such information can be more reliably obtained than a complete statistical description of the image process. The model is derived from the following assumptions:

1. **Smoothness assumption:** The image is modeled by the union of open disjoint subregions whose interior is regular enough to be well described by a 2D surface of class $C^{\bar{n}}$.
2. **Stochastic assumption:** All the derivatives of order $\bar{n}+1$ of the 2D signal are modeled by means of zero-mean independent Gaussian random fields.
3. **Inhomogeneity assumption:** The random fields representing the image process relative to different subregions are independent.

Hypotheses 1 and 2 are based on the consideration that most images are composed of open disjoint subregions whose interior is regular enough to be well described as a finite support restriction of a smooth two-dimensional Gaussian process. The boundary of each subregion is constituted by the image edges, which represent sharp discontinuities in the distribution of the gray level. Assumption 3 means that no correlation can be assumed among pixels belonging to different subregions.

Hypotheses 1 and 2 were exploited in Germani and Jetto (1988) to derive a space-invariant image model that does not take into account the edges'presence. To reduce the consequent blurring phenomenon, a heuristic restoration procedure was defined by forcing the filter with the output of an edge detector. In this case it is crucial the detector performance in estimating both edge's location and amplitude.

The inhomogeneity assumption introduced in this paper allows the edges to be directly taken into account by the image model. A stochastic image-generating process is so obtained describing the gray-level discontinuities by a space-varying model, where only the information on edge location is needed. Consequently, the optimal restoration procedure is guaranteed by the corresponding nonstationary Kalman filter. Blurring is intrinsically avoided because at any pixel the estimate is obtained by using the information carried by the neighboring pixels belonging to a convex set contained in the same subregion. The information on edge location can be obtained by an edge-detector operator whose main feature should be robustness with respect to noise (see, for example, Argile, 1971; From and Deutsch, 1975, From and Deutsch, 1978; Rosenfield and Kak, 1982; Lunscher and Beddoes, 1986; Nalva and Binford, 1986).

In the sequel we will use the following definition.

***Definition I.1***  Let $P_1$, $P_2$ be two points in a given subregion $R_i$. We say $P_1$, $P_2$ are adjacent if and only if they belong to a convex set contained in $R_i$.

***Remark I.1***  Note that $P_1$, $P_2$ are not adjacent if and only if either they belong to different subregions or their convex combination

$$\{P \in \mathbb{R}^2 : P = \alpha P_1 + (1-\alpha)P_2, \ \alpha \in [0, 1]\}$$

contains at least a boundary point of $R_i$.

## II. Kalman Filtering

In this section a brief account on the Kalman approach to the filtering of noisy signal is given. Although the theory is well established even for signals defined in the continuum (the independent variable can be indifferently time or space), we shall restrict the attention to sampled signals because all the data processing is nowadays performed by means of computers; moreover the mathematics involved is greatly simplified.

### A. State-Space Representation

In any practical situation data are collected by periodically sampling the quantities of interest obtaining, at any step, a vector $S_k \in \mathbb{R}^q$ of observations. Yet, the physical determination of $S_k$ is invariably corrupted by a random quantity representing the measurement noise $N_k^o \in \mathbb{R}^q$; its influence on the observed data $Y_k \in \mathbb{R}^q$ may vary at any step, so that we can write

$$Y_k = S_k + G_k N_k^o, \quad k = 1, 2, \ldots \quad (1)$$

with $G_k \in \mathbb{R}^{q \times q}$.

The sequence $\{S_k\}$ of the "true signal" samples may be a random sequence as well, owing to the randomness of the generating process.

A full statistical description of $\{Y_k\}$ would require the knowledge of the finite distributions $p(Y_{i_1}, \ldots, Y_{i_k})$ for any set of indexes $i_1, \ldots, i_k$. Such information can be efficiently obtained from the knowledge of a limited number of parameters once a model for the mechanism that generates the sequence $\{S_k\}$ is adopted. Then, in addition to the measurement equation (1), the following signal *state-space model* may be considered

$$X_{k+1} = A_k X_k + F_k N_k^s \quad (2)$$

$$S_{k+1} = C_{k+1} X_{k+1} \quad (3)$$

with $A_k \in \mathbb{R}^{n \times n}$, $F_k \in \mathbb{R}^{n \times p}$, and $C_k \in \mathbb{R}^{q \times n}$. Vector $X_k \in \mathbb{R}^n$ is the *system's memory*, because it accounts for the process past history up to the $k$th step; in this respect the state *initial condition* $X_0$ is correctly modeled as a random variable. Vector $N_k^s \in \mathbb{R}^p$ is a random excitation so that, according to (2.2), the new system outcome $X_{k+1}$ is computed as linear combination of the available information $X_k$ and the occurring input $N_k^s$. Then the signal sample $S_{k+1}$ is obtained as partial observation of $X_{k+1}$ through Eq. (3).

Equations (2) and (3) define a linear nonstationary model. This scheme is sufficiently general to describe a great deal of practical situations or can be considered as approximation of more sophisticated nonlinear processes.

Then we see that the statistics of sequences $\{X_k\}$, $\{S_k\}$, and $\{Y_k\}$ are completely determined by those of $X_0$, $\{N_k^s\}$, and $\{N_k^0\}$. Now, our intuition about the behavior of system's evolution suggests that the probabilistic structure induced by Eqs. (1), (2), and (3) should be such that the information at any step $k$ depended only on the past information and not on the future one. In other words we want to model the causality usually featured by the physical processes we deal with. Indeed, this would be accomplished by the causal form of relations (2) and (3) in the deterministic case. Causality is retained at statistical level providing some assumptions are made on the random variables involved. Thus, $\{N_k^s\}$ and $\{N_k^o\}$ are assumed to be independent zero mean white sequences, i.e., sequences of independent identically distributed (i.i.d.) random variables

$$\mathrm{E}\left[N_k^s N_h^{oT}\right] = 0, \forall h, k, \quad \mathrm{E}\left[N_k^s N_h^{sT}\right] = \delta_{k,h} I_s, \quad \mathrm{E}\left[N_k^o N_h^{oT}\right] = \delta_{k,h} I_o$$

where $I_s \in \mathbb{R}^{p \times p}$, $I_o \in \mathbb{R}^{q \times q}$, and $\delta_{k,h} = \begin{cases} 1 & k=h \\ 0 & k \neq h \end{cases}$ is the Kronecker delta.

The random variable $X_0$ is assumed independent of $\{N_k^s\}$ and $\{N_k^o\}$ as well, i.e.,

$$E[N_k^s X_0^T] = 0, \quad E[N_k^o X_0^T] = 0, \forall k$$

Then, according to Eqs. (1), (2), and (3) and the just-stated hypothesis, the sequences $\{X_k\}$, $\{S_k\}$, and $\{Y_k\}$ are *nonanticipative*, i.e., $X_k$, $S_k$ and $Y_k$ are independent of $X_{k+p}$, $S_{k+p}$, and $Y_{k+p}$ for any $k$ and $p \geq 1$. Moreover, $\{X_k\}$ features the so-called Markov property:

$$p(X_k|X_{k-1},\ldots,X_{k-h}) = p(X_k|X_{k-1}) \qquad (4)$$

with $p(u|v)$ denoting the probability density function of $u$ conditioned to the knowledge of $v$. Relation (4) states that, for any $k$, $X_{k-1}$ conveys all the past information. Property (4) need not to hold for the sequences $\{S_k\}$ and $\{Y_k\}$.

For signal-processing purposes we shall be concerned only with second-order theory, so that only mean and covariance do matter about sequences $\{X_k\}$, $\{S_k\}$, $\{Y_k\}$. Moreover, if the Gaussian assumption is made, it turns out that the second-order statistics suffice for a signal full description. Then, suppose that $X_0$ is a Gaussian random variable with mean $\bar{X}_0$ and covariance matrix $P_0$, and that $\{N_k^s\}$ and $\{N_k^o\}$ are Gaussian as well. From Eqs. (1), (2), and (3) we readily obtain that

$$X_{k+1} = \Phi_{k,0}X_0 + \sum_{j=0}^{k} \Phi_{k,j} F_j N_j^s \qquad (5)$$

$$S_{k+1} = C_{k+1}\Phi_{k,0}X_0 + \sum_{j=0}^{k} C_{k+1}\Phi_{k,j} F_j N_j^s \qquad (6)$$

$$Y_{k+1} = S_{k+1} + G_{k+1} N_{k+1}^o \qquad (7)$$

with $\Phi_{\ell,m} = A_\ell \cdot A_{\ell-1} \cdots A_m$ We see that $X_{k+1}, S_{k+1}, Y_{k+1}, k=0,1,\ldots,$ are linear transformations of the random variables $X_0, N_0^s, \ldots, N_k^s, N_{k+1}^o$, which, because of independence, are jointly Gaussian. Consequently, $X_k$, $S_k$, and $Y_k$ are Gaussian sequences. The mean evolution is obtained as follows:

$$E[X_{k+1}] = \Phi_{k,0}\bar{X}_0 = \bar{X}_{k+1} = A_k \bar{X}_k \qquad (8)$$

$$E[S_{k+1}] = C_{k+1}\bar{X}_{k+1} = \bar{S}_{k+1} \qquad (9)$$

$$E[Y_{k+1}] = \bar{S}_{k+1} = \bar{Y}_{k+1} \qquad (10)$$

which hold because the noise sequences have zero mean. For the covariance matrices

$$R^x_{k+1} = E[(X_{k+1} - \bar{X}_{k+1})(X_{k+1} - \bar{X}_{k+1})^T]$$

$$R^s_{k+1} = E[(S_{k+1} - \bar{S}_{k+1})(S_{k+1} - \bar{S}_{k+1})^T]$$

$$R^y_{k+1} = E[(Y_{k+1} - \bar{Y}_{k+1})(Y_{k+1} - \bar{Y}_{k+1})^T]$$

it readily follows that

$$R^x_{k+1} = \Phi_{k,0} P_0 \Phi^T_{k,0} + \sum_{j=0}^{k} \Phi_{k,j} F_j F^T_j \Phi^T_{k,j}$$

$$= A_k R^x_k A^T_k + F_k F^T_k \tag{11}$$

$$R^s_{k+1} = C_{k+1} R^x_{k+1} C^T_{k+1} \tag{12}$$

$$R^y_{k+1} = R^s_{k+1} + G_{k+1} G^T_{k+1} \tag{13}$$

In particular, Eqs. (8) and (11) provide recursive relations to compute the state mean and covariance matrix.

## B. The Estimation Algorithm

Our goal consists in estimating the value $\tilde{S}_{k|k}$ of $S_k$, $k = 1, 2, \ldots$, given the measurement sample $Y_1, Y_2, \ldots, Y_k$. This is the filtering problem; when a measurement sample $Y_1, Y_2, \ldots, Y_m$ is used, we talk about prediction if $m < k$ and smoothing if $m > k$. The estimate is then correctly intended as a measurable function of the available data, $\tilde{S}_{k|k} = \psi(Y_1, \ldots, Y_k)$. Such a function is usually selected according to an optimization criterion. For instance, the optimal estimate $\hat{S}_{k|k}$ should be unbiased, i.e., $E[\hat{S}_{k|k}] = E[S_k]$, and may be chosen in order to minimize the trace $E[(S_k - \hat{S}_{k|k})^T (S_k - \hat{S}_{k|k})]$ of the estimation error covariance matrix. A standard result in estimation theory states that such an optimal estimate coincide with the conditional mean expectation $E[S_k|Y_k, \ldots, Y_1]$ of the signal $S_k$ given the measurement sample $Y_k, \ldots, Y_1$. Now, from Eq. (3) we see that

$$\hat{S}_{k|k} = E[S_k|Y_k, \ldots, Y_1] = C_k E[X_k|Y_k, \ldots, Y_1] = C_k \hat{X}_{k|k} \tag{14}$$

The celebrated Kalman filter is a linear system that recursively provides the optimal state estimate $\hat{X}_{k|k}$ on real time with data acquisition. With the assumption that for any $k$ matrix $G_k G_k^T$ is full rank, the following scheme is obtained:

$$\hat{X}_{k|k} = A_k \hat{X}_{k-1|k-1} + K_k(Y_k - A_k \hat{X}_{k-1|k-1}), \quad \hat{X}_0 = \bar{X}_0 \quad (15)$$

$$K_k = R_k C_k^T \left( G_k G_k^T \right)^{-1} \quad (16)$$

$$R_k = \mathrm{E}[(X_k - \hat{X}_{k|k})(X_k - \hat{X}_{k|k})^T]$$
$$= \left( I + H_{k-1} C_k^T \left( G_k G_k^T \right)^{-1} C_k \right)^{-1} H_{k-1}, \quad R_0 = P_0 \quad (17)$$

$$H_{k-1} = \mathrm{E}[(X_k - A_k \hat{X}_{k-1|k-1})(X_k - A_k \hat{X}_{k-1|k-1})^T]$$
$$= A_k R_{k-1} A_k^T + F_k F_k^T \quad (18)$$

According to Eq. (15) the optimal estimate is composed of two terms. The first one represents the *one-step prediction*, i.e., the best estimate at the $k$th step (in the minimum variance sense) that we can obtain without processing further data than $Y_1, \ldots Y_{k-1}$. The second term is a *correction* based on the innovation carried by the $k$th measurement with respect to the information contained in the past measurements already accounted for in the prediction term. The $n \times q$ matrix $K_k$ is called *Kalman gain* and, according to Eqs. (16), (17) and (18), can be computed on the base of the signal model coefficients $A_k$, $F_k$, $C_k$, $G_k$ and the initial condition covariance matrix $P_0$. The choice $\hat{X}_0 = \bar{X}_0$ yields an unbiased estimate for any $k$. The matrix $R_k \in \mathbb{R}^{n \times n}$ is the state estimation error covariance and Eq. (17) is known as the (dynamical) *Riccati equation*. Matrix $H_{k-1} \in \mathbb{R}^{n \times n}$ is the state one-step prediction error covariance; moreover it can be shown that matrix $(I + H_{k-1} C_k^T (G_k G_k^T)^{-1} C_k$ is always invertible.

The remarkable feature of the Kalman filter is that it allows for a non-stationary data processing, as opposed to the frequency domain methods for filtering, which are ultimately a steady-state approach.

## C. The Steady-State Solution

When the signal model is stationary, it is interesting to study the asymptotic behavior of the Kalman filter, once Eqs. (15)–(18) are updated with constant system matrices $A$, $F$, $C$, and $G$. Here the notion of system stability, as well as the structural properties of system controllability and observability, play an essential role. We briefly recall them. A state $X \in \mathbb{R}^n$ is $A$-stable if $\|A^k X\|$ tends to zero as $k$ increases. We say that $X$ is $(C, A)$ *unobservable* if $C A^k X = 0$ for any $k \geq 0$, whereas it is $(A, F)$ *uncontrollable* if $F^T A^{T^k} X = 0$ for any $k \geq 0$. The states featuring either one of the stated properties form a linear subspace of the state space $\mathbb{R}^n$. Concerning the system behavior, the notions of stability, observability, and controllability can be considered together to give the properties usually referred as *detectability* and *stabilizability*. Then we say that a system is detectable if all the $(A, C)$ unobservable states are $A$-stable and that it is stabilizable if all the $(A, F)$ uncontrollable states are $A^T$-stable.

Besides the inherent interest, the steady-state investigation is needed because no Kalman filter can be optimal unless the initial condition covariance $P_0$ is known exactly. To overcome this difficulty consider Eqs. (17) and (18) in the stationary case

$$R_k = (I + H_{k-1}C^T(GG^T)^{-1}C)^{-1}H_{k-1}, \qquad H_{k-1} = AR_{k-1}A^T + FF^T \tag{19}$$

Then we want to check if the Riccati equation admits a steady-state solution $R_\infty = \lim_{k \to \infty} R_k$. The advantage would be twofold: first, if the limit exists it is unique regardless the way the sequence $R_k$ is started up according to Eq. (19), thus obtaining a filtering algorithm robust with respect to initial data knowledge inaccuracies; then, from a practical point of view, we would not need to update matrix $R_k$ for $k$ sufficiently large, savings computational resources (time and memory).

It can be shown that if the signal model is stabilizable and detectable, then the Riccati equation (19) admits a unique steady-state nonnegative definite and self-adjoint solution $R_\infty$. This matrix solves the so-called steady-state Riccati equation (SSRE)

$$R_\infty = (I + H_\infty C^T(GG^T)^{-1}C)^{-1}H_\infty, \qquad H_\infty = AR_\infty A^T + FF^T \tag{20}$$

Then the following suboptimal filter can be designed:

$$\tilde{X}_{k|k} = A\tilde{X}_{k-1|k-1} + K_\infty(Y_k - A\tilde{X}_{k-1|k-1}), \quad \tilde{X}_0 = \bar{X}_0 \tag{21}$$

$$K_\infty = R_\infty C^T(GG^T)^{-1} \tag{22}$$

Nevertheless, it can be shown that the previous algorithm is asymptotically optimal, that is the state estimation error covariance matrix $\tilde{R}_k = E[(X_k - \tilde{X}_{k|k})(X_k - \tilde{X}_{k|k})^T]$ approaches $R_\infty$ as $k$ increases.

## III. THE IMAGE MODEL

### A. The Homogeneous Image Equation

In this section we describe an image by means of the gray-level signal together with its partial derivatives with respect to the spatial coordinates, up to a certain order. The vector so obtained is assumed as the state vector of the image model.

Moreover, by assumptions 1 and 2, a stochastic relation between the states evaluated at two different points in the same subregion is obtained.

Let us indicate by $x(r, s)$ the value of the original monochromatic image at spatial coordinate $(r, s)$ inside a smooth subregion. The continuous variables $r$ and $s$ denote the vertical and horizontal position respectively. For simplicity, but without loss of generality, we assume $(r, s) \in [0, 1]^2$.

Because of the smoothness assumption, it is possible to define a state vector composed of the signal $x(r, s)$ and its partial derivatives with respect to $r$ and $s$

$$X(r, s) = \left[ \frac{\partial^n x(r, s)}{\partial r^{n-\alpha} \partial s^\alpha}, \quad n = 0, 1, \ldots, \bar{n}; \quad \alpha = 0, 1, \ldots, n \right]^T \quad (23)$$

If $\bar{n}$ is the maximum order of derivation used, the dimension of $X(r, s)$ is $N = [(\bar{n} + 1)(\bar{n} + 2)]/2$. The $k$th component of $X(r, s)$, denoted by $X_k(r, s)$, is given by

$$X_k(r, s) = \frac{\partial^n x(r, s)}{\partial r^{n-\alpha} \partial s^\alpha} \quad (24)$$

with

$$n = \left[ \frac{\sqrt{8k - 7} - 1}{2} \right], \quad \alpha = k - 1 - \frac{n(n+1)}{2}$$

where $[\cdot]$ stands for the integer part. Let

$$r = r(u) = r_0 + \gamma u$$
$$s = s(u) = s_0 + \beta u$$

denote a parametric representation in $u$ of a straight line passing through the point $(r_0, s_0)$ belonging to a homogeneous subregion $R_1 \subset [0, 1]^2$. As a direct consequence of the state vector definition (23), the following equation can be written for any $u \in [0, u_{\max}]$ such that the point $(r(u), s(u))$ is adjacent to $(r_0, s_0)$ according to Definition I.1:

$$\dot{X}(r(u), s(u)) = \left( \frac{\partial}{\partial r} X(r(u), s(u)) \right) \gamma + \left( \frac{\partial}{\partial s} X(r(u), s(u)) \right) \beta \quad (25)$$

the dot denoting the derivative with respect to $u$. Moreover, by direct computation we have

$$\frac{\partial}{\partial r} X(r(u), s(u)) = AX(r(u), s(u)) + BW_r(r(u), s(u)) \quad (26)$$

$$\frac{\partial}{\partial s} X(r(u), s(u)) = A'X(r(u), s(u)) + BW_s(r(u), s(u)) \quad (27)$$

where

- $A$ and $A'$ are $(N \times N)$ commuting matrices (Germani and Jetto, 1988) (see Section VIII. A), whose elements $a_{\ell,m}$ and $a'_{\ell,m}$ are such that

$$a_{\ell,m} = \begin{cases} 1, & \text{if } m = \ell + 1 + \left[\frac{(\sqrt{8\ell-7}-1)}{2}\right] \leq N \\ 0, & \text{otherwise} \end{cases} \qquad (28)$$

$$a'_{\ell,m} = \begin{cases} 1, & \text{if } m = \ell + 2 + \left[\frac{(\sqrt{8\ell-7}-1)}{2}\right] \leq N \\ 0, & \text{otherwise} \end{cases} \qquad (29)$$

- The vectors $W_r(r(u), s(u))$ and $W_s(r(u), s(u))$ have dimension $\bar{n} + 1$ and are given by

$$W_r(r, s) = \left[\frac{\partial^{\bar{n}+1} x(r, s)}{\partial r^{\bar{n}-\alpha+1} \partial s^{\alpha}}, \quad \alpha = 0, 1, \ldots, \bar{n}\right]^T \qquad (30)$$

$$W_s(r, s) = \left[\frac{\partial^{\bar{n}+1} x(r, s)}{\partial r^{\bar{n}-\alpha} \partial s^{\alpha+1}}, \quad \alpha = 0, 1, \ldots, \bar{n}\right]^T \qquad (31)$$

- The $(N \times (\bar{n} + 1))$ matrix $B$ has the form $B = [0^T \ I]^T$

The dimension of the null block and the identity matrix are $\{(N - (\bar{n} + 1)) \times (\bar{n} + 1)\}$ and $(\bar{n} + 1) \times (\bar{n} + 1)$, respectively.

Using Eqs. (26) and (27), Eq. (25) can be rewritten in the following form

$$\dot{X}(r(u), s(u)) = (\gamma A + \beta A') X(r(u), s(u)) + B[\gamma W_r(r(u), s(u)) + \beta W_s(r(u), s(u))] \qquad (32)$$

Formal integration of Eq. (32) with respect to $u$ between $u_0$ and $u_1$, allows us to derive a relation between the state vectors evaluated at two generic points $(r_0 + \gamma u_0, s_0 + \beta u_0)$ and $(r_0 + \gamma u_1, s_0 + \beta u_1)$. Exploiting the commutativity of $A$ and $A'$ we obtain

$$X(r_0 + \gamma u_1, s_0 + \beta u_1) = e^{(\gamma A + \beta A')(u_1 - u_0)} X(r_0 + \gamma u_0, s_0 + \beta u_0)$$
$$+ \int_{u_0}^{u_1} e^{(\gamma A + \beta A')(u_1 - \tau)} B[\gamma W_r(r_0 + \gamma \tau, s_0 + \beta \tau)$$
$$+ \beta W_s(r_0 + \gamma \tau, s_0 + \beta \tau)] d\tau \qquad (33)$$

By the stochastic assumption, $W_r(\cdot, \cdot)$, $W_s(\cdot, \cdot)$ are white Gaussian vector fields, so that the integral term in Eq. (33) is intended as a stochastic Wiener integral.

## B. The Component Equations of the Sampled Image

In this section we use the results of Section III. A in order to get a semicausal statistical model for the sampled image, which takes into account each possible edge configuration at any pixel.

We denote by

$$x_{i,j} = x(i\Delta_r, j\Delta_s), \qquad i, j = 1, \ldots, \bar{m}$$

the true value of the sampled image at the pixel with vertical coordinate $i\Delta_r$ and horizontal coordinate $j\Delta_s$, where $\Delta_r$ and $\Delta_s$ denote, respectively, the vertical and the horizontal spatial sampling steps. If the image is sampled with an equal number $\bar{m}$ of pixels on each row and on each column, the normalized value of $\Delta_r$ and $\Delta_s$ are both equal to $1/(\bar{m}-1)$.

We consider the situation where the image is observed under additive white Gaussian noise $v_{i,j} \sim N(0, \sigma_v^2)$:

$$y_{i,j} = x_{i,j} + v_{i,j} \tag{34}$$

We assume that the image state vector at pixel coordinates $(i, j)$ depends on the state vectors at neighboring pixels according to the scheme shown in Figure 1, where causality is assumed only for the first coordinate (semicausal model).

The coefficients $c_{i,j}^{(\ell)}(\ell = 1, \ldots, 5)$ may be 0 or 1; they are 1 if the corresponding pixels are adjacent and 0 otherwise. This implies that $2^5$ different configurations of the image model can be obtained, depending on the edge shape at pixel $(i, j)$.

The relations between the state vector $X_{i,j}$ at the pixel $(i, j)$ and the state evaluated at neighbouring pixels for which $c_{i,j}^{(\ell)} = 1$ can be obtained by applying (33), with a suitable choice of $\gamma$ and $\beta$. The following component equations

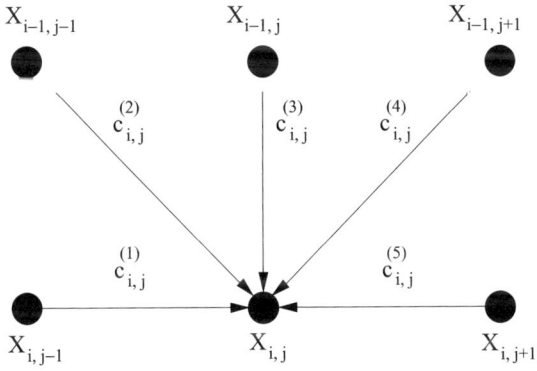

FIGURE 1. Spatial structure of the semicausal dependent model.

are derived

$$c_{i,j}^{(1)} X_{i,j} = c_{i,j}^{(1)} \left( e^{A' \Delta_s} X_{i,j-1} + W_{i,j}^{(1)} \right) \tag{35}$$

$$c_{i,j}^{(2)} X_{i,j} = c_{i,j}^{(2)} \left( e^{A' \Delta_s + A \Delta_r} X_{i-1,j-1} + W_{i,j}^{(2)} \right) \tag{36}$$

$$c_{i,j}^{(3)} X_{i,j} = c_{i,j}^{(3)} \left( e^{A \Delta_r} X_{i-1,j} + W_{i,j}^{(3)} \right) \tag{37}$$

$$c_{i,j}^{(4)} X_{i,j} = c_{i,j}^{(4)} \left( e^{-A' \Delta_s + A \Delta_r} X_{i-1,j+1} + W_{i,j}^{(4)} \right) \tag{38}$$

$$c_{i,j}^{(5)} X_{i,j} = c_{i,j}^{(5)} \left( e^{-A' \Delta_s} X_{i,j+1} + W_{i,j}^{(5)} \right) \tag{39}$$

where

$$W_{i,j}^{(1)} = \int_0^1 e^{A' \Delta_s (1-\tau)} \Delta_s B W_s(i \Delta_r, (j-1) \Delta_s + \Delta_s \tau) \, d\tau \tag{40}$$

$$W_{i,j}^{(2)} = \int_0^1 e^{(A \Delta_r + A' \Delta_s)(1-\tau)} B$$
$$\cdot [\Delta_r W_r((i-1) \Delta_r + \Delta_r \tau, (j-1) \Delta_s + \Delta_s \tau)$$
$$+ \Delta_s W_s((i-1) \Delta_r + \Delta_r \tau, (j-1) \Delta_s + \Delta_s \tau)] \, d\tau \tag{41}$$

$$W_{i,j}^{(3)} = \int_0^1 e^{A \Delta_r (1-\tau)} \Delta_r B W_r((i-1) \Delta_r + \tau \Delta_r, j \Delta_s) \, d\tau \tag{42}$$

$$W_{i,j}^{(4)} = \int_0^1 e^{(A \Delta_r - A' \Delta_s)(1-\tau)} B$$
$$\cdot [\Delta_r W_r((i-1) \Delta_r + \Delta_r \tau, (j+1) \Delta_s - \Delta_s \tau)$$
$$- \Delta_s W_s((i-1) \Delta_r + \Delta_r \tau, (j+1) \Delta_s - \Delta_s \tau)] \, d\tau \tag{43}$$

$$W_{i,j}^{(5)} = -\int_0^1 e^{-A' \Delta_s (1-\tau)} \Delta_s B W_s(i \Delta_r, (j+1) \Delta_s - \Delta_s \tau) \, d\tau \tag{44}$$

Note that only the component equations for which $c_{i,j}^{(\ell)} = 1$ give significant information. If some $c_{i,j}^{(\ell)} = 0$, only a trivial relation is obtained because (32) cannot be integrated along that direction.

A pixel is said to be *internal, boundary,* or *isolated* if, for it, the number $p_{i,j}$ of component equations that really hold is $p_{i,j} = 5$, $1 \leq p_{i,j} < 5$, $p_{i,j} = 0$, respectively.

## C. Modeling the State Noise

From the stochastic assumption and using Eqs. (40)–(44), it is possible to show that $W_{i,j}^{(\ell)}$ ($\ell = 1, \ldots, 5$) are zero-mean white Gaussian random fields with the following properties (Germani and Jetto, 1988):

$$E\left[W^{(1)}_{i,j} W^{(1)T}_{\ell,m}\right] = \delta_{i,j}\delta_{\ell,m} Q_s \qquad (45)$$

$$E\left[W^{(3)}_{i,j} W^{(3)T}_{\ell,m}\right] = \delta_{i,j}\delta_{\ell,m} Q_r \qquad (46)$$

$$E\left[W^{(1)}_{i,j} W^{(3)T}_{\ell,m}\right] = 0 \qquad (47)$$

with

$$Q_s = \int_0^1 e^{A'\Delta_s(1-\tau)} B \Psi_s B^T e^{A'^T \Delta_s(1-\tau)} d\tau$$

$$Q_r = \int_0^1 e^{A\Delta_r(1-\tau)} B \Psi_r B^T e^{A^T \Delta_r(1-\tau)} d\tau$$

where $\Psi_s$ and $\Psi_r$ are diagonal matrices such that

$$E\left[W_s(r,s) W_s^T(\bar{r},\bar{s})\right] = \Psi_s \delta(\|(r,s) - (\bar{r},\bar{s})\|)$$

$$E\left[W_r(r,s) W_r^T(\bar{r},\bar{s})\right] = \Psi_r \delta(\|(r,s) - (\bar{r},\bar{s})\|)$$

Estimates of $\Psi_s$ and $\Psi_r$ can be obtained as functions of the image spectrum (Germani and Jetto, 1988) (see Section VIII. B). Moreover, from (35)–(39) the following identities can be proven to hold

$$W^{(5)}_{i,j} = -e^{-A'\Delta_s} W^{(1)}_{i,j+1}, \qquad (48)$$

$$W^{(2)}_{i,j} = e^{A'\Delta_s} W^{(3)}_{i,j-1} + W^{(1)}_{i,j}, \qquad (49)$$

$$W^{(4)}_{i,j} = e^{-A'\Delta_s} \left(W^{(3)}_{i,j+1} - W^{(1)}_{i,j+1}\right). \qquad (50)$$

The previous identities, obtained by assuming $c^{(\ell)}_{i,j} = 1$, ($\ell = 1, \ldots, 5$), imply the following relations among the covariance matrices of the white Gaussian random fields $W^{(\ell)}_{i,j}$, ($\ell = 1, \ldots, 5$)

$$E\left[W^{(5)}_{i,j} W^{(5)T}_{l,m}\right] = \delta_{i,j}\delta_{l,m} e^{-A'\Delta_s} Q_s e^{-A'^T \Delta_s} \qquad (51)$$

$$E\left[W^{(2)}_{i,j} W^{(2)T}_{l,m}\right] = \delta_{i,j}\delta_{l,m} (e^{A'\Delta_s} Q_r e^{A'^T \Delta_s} + Q_s) \qquad (52)$$

$$E\left[W^{(4)}_{i,j} W^{(4)T}_{l,m}\right] = \delta_{i,j}\delta_{l,m} e^{-A'\Delta_s} (Q_s + Q_r) e^{-A'^T \Delta_s} \qquad (53)$$

The statistics of the infinite two-dimensional Gaussian process corresponding to a generic smooth subregion is completely defined by (45)–(47) and (51)–(53). As a consequence these equations define the statistics for any pixel of each subregion. Of course some of the right-hand sides of (48)–(50), might lose their physical meaning in the presence of an edge, but the statistical meaning is retained, because it is related to the infinite two-dimensional Gaussian random field. Therefore, relations (51)–(53) are considered true even if some $c^{(\ell)}_{i,j} = 0$.

## D. The Constitutive Equation

Assuming $p_{i,j} > 0$, we now exploit the component equations in order to derive a unique relation among $X_{i,j}$ and the state evaluated at its $p_{i,j}$ neighboring pixels. The case $p_{i,j} = 0$ (isolated pixels) will be considered separately. For convenience the following notation is introduced

$$H_1 := e^{A'\Delta_s}, \qquad H_2 := e^{A'\Delta_s + A\Delta_r}, \qquad H_3 := e^{A\Delta_r}$$
$$H_4 := e^{-A'\Delta_s + A\Delta_r}, \qquad H_5 := e^{-A'\Delta_s}$$

Defining:

$$C_{i,j} := \text{diag}\big[c_{i,j}^{(\ell)} I_N, \ell = 1, \ldots, 5\big]$$
$$\bar{C}_{i,j} := C_{i,j}[I_N, I_N, I_N, I_N, I_N]^T$$

$$\mathbf{Z}_{i,j} := \begin{bmatrix} H_1 X_{i,j-1} \\ H_2 X_{i-1,j-1} \\ H_3 X_{i-1,j} \\ H_4 X_{i-1,j+1} \\ H_5 X_{i,j+1} \end{bmatrix}, \qquad \mathbf{W}_{i,j} := \begin{bmatrix} W_{i,j}^{(1)} \\ W_{i,j}^{(2)} \\ W_{i,j}^{(3)} \\ W_{i,j}^{(4)} \\ W_{i,j}^{(5)} \end{bmatrix}$$

where $I_N$ is the $N \times N$ identity matrix, Eqs. (35)–(39) can be rewritten as

$$\bar{C}_{i,j} X_{i,j} = C_{i,j} \mathbf{Z}_{i,j} + C_{i,j} \mathbf{W}_{i,j} \tag{54}$$

Note that $\bar{C}_{i,j} X_{i,j}$ belongs to $\mathbb{R}^{5N}$, being composed of five $(N \times 1)$-dimensional blocks. We stress that only $p_{i,j}(1 \leq p_{i,j} \leq 5)$ of them are nonzero blocks corresponding to the $p_{i,j}$ component equations for which $c_{i,j}^{(\ell)} = 1$.

Now we define an operator $\Pi_{i,j} : \mathbb{R}^{5N} \to \mathbb{R}^{p_{i,j}N}$ selecting the nonzero block entries of $\bar{C}_{i,j} X_{i,j}$. It is given by a $p_{i,j} N \times 5N$ matrix whose generic block entry $\Pi_{i,j}(k, h)$, $k = 1, \ldots p_{i,j}$, $h = 1, \ldots, 5$, is

$$\Pi_{i,j}(k, h) = \begin{cases} c_{i,j}^{(\ell)} I_N, & \text{if } k = \sum_{\ell=1}^{h} c_{i,j}^{(\ell)} \\ O_{N \times N}, & \text{otherwise} \end{cases}$$

Note that definitions of $C_{i,j}$ and $\Pi_{i,j}$ imply that

$$\Pi_{i,j} C_{i,j} = \Pi_{i,j}$$

Hence, applying $\Pi_{i,j}$ to both sides of (54) one has

$$C_{i,j}^p X_{i,j} = \tilde{Z}_{i,j} + \tilde{W}_{i,j} \tag{55}$$

where $C_{i,j}^p := \Pi_{i,j} \bar{C}_{i,j}$, $\tilde{Z}_{i,j} := \Pi_{i,j} \mathbf{Z}_{i,j}$ and $\tilde{W}_{i,j} := \Pi_{i,j} \mathbf{W}_{i,j}$.

Equation (55) so obtained is a general compact form to express the $p_{i,j}$ component equations, which actually hold at pixel $(i, j)$. We want to combine such equations in order to obtain a unique relation with a minimum variance stochastic term. It is well established (see e.g. Billingsley, 1979, p. 403) that such an equation is of the type

$$X_{i,j} = \mathrm{E}[X_{i,j}/\tilde{Z}_{i,j}] + \hat{W}_{i,j} \tag{56}$$

where $\mathrm{E}[X_{i,j}/\tilde{Z}_{i,j}]$ denotes the conditional expectation of $X_{i,j}$ given $\tilde{Z}_{i,j}$, and $\hat{W}_{i,j}$ is the mentioned stochastic term. In the present linear Gaussian case, it is known that (Söderström and Stoica, 1989)

$$\mathrm{E}[X_{i,j}/\tilde{Z}_{i,j}] = \left(C_{i,j}^{p\ T}\tilde{\Psi}_{i,j}^{-1}C_{i,j}^{p}\right)^{-1}C_{i,j}^{p\ T}\tilde{\Psi}_{i,j}^{-1}\tilde{Z}_{i,j}$$

and

$$\hat{W}_{i,j} = \left(C_{i,j}^{p\ T}\tilde{\Psi}_{i,j}^{-1}C_{i,j}^{p}\right)^{-1}C_{i,j}^{p\ T}\tilde{\Psi}_{i,j}^{-1}\tilde{W}_{i,j}$$

where $\tilde{\Psi}_{i,j}$ is the $p_{i,j}N \times p_{i,j}N$ covariance matrix of $\tilde{W}_{i,j}$ given by

$$\tilde{\Psi}_{i,j} := \mathrm{E}[\tilde{W}_{i,j}\tilde{W}_{i,j}^{T}] = \Pi_{i,j}\Psi_w\Pi_{i,j}^{T}, \quad \text{with} \quad \Psi_w := \mathrm{E}[W_{i,j}W_{i,j}^{T}]$$

Using (45)–(47) and (51)–(53), it is found that

$$\Psi_w = \begin{bmatrix} Q_s & Q_s & 0 & 0 & 0 \\ Q_s & H_1Q_rH_1^T + Q_s & 0 & 0 & 0 \\ 0 & 0 & Q_r & 0 & 0 \\ 0 & 0 & 0 & H_1^{-1}(Q_s+Q_r)H_1^{-T} & H_1^{-1}Q_sH_1^{-T} \\ 0 & 0 & 0 & H_1^{-1}Q_sH_1^{-T} & H_1^{-1}Q_sH_1^{-T} \end{bmatrix} \tag{57}$$

Hence, (56) becomes

$$X_{i,j} = \left(C_{i,j}^{p\ T}\tilde{\Psi}_{i,j}^{-1}C_{i,j}^{p}\right)^{-1}C_{i,j}^{p}\tilde{\Psi}_{i,j}^{-1}(\tilde{Z}_{i,j} + \tilde{W}_{i,j}) \tag{58}$$

Now observe that, by definition of $C_{i,j}^{p}$, $\tilde{\Psi}_{i,j}$, $\tilde{Z}_{i,j}$ and $\tilde{W}_{i,j}$, Eq. (58) can be rewritten as

$$X_{i,j} = \left(\bar{C}_{i,j}^{T}\Pi_{i,j}^{T}\left(\Pi_{i,j}\Psi_w\Pi_{i,j}^{T}\right)^{-1}\Pi_{i,j}\bar{C}_{i,j}\right)^{-1}$$
$$\cdot \bar{C}_{i,j}^{T}\Pi_{i,j}^{T}\left(\Pi_{i,j}\Psi_w\Pi_{i,j}^{T}\right)^{-1}\Pi_{i,j}(Z_{i,j} + W_{i,j}) \tag{59}$$

and that, by (57) and definition of $\Pi_{i,j}$, the following general expression for $\mathcal{A}_{i,j} := \Pi_{i,j}^T (\Pi_{i,j} \Psi_w \Pi_{i,j}^T)^{-1} \Pi_{i,j}$ is found:

$$\mathcal{A}_{i,j} = \begin{bmatrix} A_{1(i,j)} & -A_{1,2(i,j)} & 0 & 0 & 0 \\ -A_{1,2(i,j)} & A_{2(i,j)} & 0 & 0 & 0 \\ 0 & 0 & A_{3(i,j)} & 0 & 0 \\ 0 & 0 & 0 & A_{4(i,j)} & -A_{4,5(i,j)} \\ 0 & 0 & 0 & -A_{4,5(i,j)} & A_{5(i,j)} \end{bmatrix} \quad (60)$$

where

$$A_{1(i,j)} := c_{i,j}^{(1)} \left( Q_s^{-1} + c_{i,j}^{(2)} (H_1 Q_r H_1^T)^{-1} \right)$$

$$A_{1,2(i,j)} := c_{i,j}^{(1)} c_{i,j}^{(2)} (H_1 Q_r H_1^T)^{-1}$$

$$A_{2(i,j)} := c_{i,j}^{(2)} \left( H_1 Q_r H_1^T + (1 - c_{i,j}^{(1)}) Q_s \right)^{-1}$$

$$A_{3(i,j)} := c_{i,j}^{(3)} Q_r^{-1} \quad (61)$$

$$A_{4(i,j)} := c_{i,j}^{(4)} H_1^T \left( (1 - c_{i,j}^{(5)}) Q_s + Q_r \right)^{-1} H_1$$

$$A_{4,5(i,j)} := -c_{i,j}^{(4)} c_{i,j}^{(5)} H_1^T Q_r^{-1} H_1$$

$$A_{5(i,j)} := c_{i,j}^{(5)} H_1^T \left( Q_s^{-1} + c_{i,j}^{(4)} Q_r^{-1} \right) H_1$$

as it can be directly verified for any possible configuration assumed by the general semicausal dependence scheme of Figure 1.

By definition of $\bar{C}_{i,j}$ one has that the $N \times N$ matrix

$$\left( \bar{C}_{i,j}^T \Pi_{i,j}^T (\Pi_{i,j} \Psi_w \Pi_{i,j}^T)^{-1} \Pi_{i,j} \bar{C}_{i,j} \right)^{-1}$$

is given by the sum of all the block entries of the matrix $\mathcal{A}_{i,j}$, whereas the $N \times 5N$ matrix

$$\left( \bar{C}_{i,j}^T \Pi_{i,j}^T (\Pi_{i,j} \Psi_w \Pi_{i,j}^T)^{-1} \Pi_{i,j} \right)$$

is given by the sum of the block rows of $\mathcal{A}_{i,j}$. Hence (59) results in

$$X_{i,j} = \left( A_{1(i,j)} - 2A_{1,2(i,j)} + A_{2(i,j)} + A_{3(i,j)} + A_{4(i,j)} - 2A_{4,5(i,j)} + A_{5(i,j)} \right)^{-1}$$
$$\cdot \left[ A_{1(i,j)} - A_{1,2(i,j)}; -A_{1,2(i,j)} + A_{2(i,j)}; A_{3(i,j)}; A_{4(i,j)} - A_{4,5(i,j)}; \right.$$
$$\left. - A_{4,5(i,j)} + A_{5(i,j)} \right] \cdot (\mathbf{Z}_{i,j} + \mathbf{W}_{i,j}). \quad (62)$$

Equation (62) is the unique relation with the minimum-variance stochastic term we were looking for. It does not hold for isolated pixels. In fact, in this case

Eq. (32) cannot be integrated along any one of the five directions of Figure 1. Hence $p_{i,j} = 0$ and the pseudoinverse cannot be defined. It seems natural to consider such pixels as initial conditions, namely, to assume $X_{i,j} = X_{i,j}^0$, with $X_{i,j}^0$ externally imposed. Thus Eq. (62) is modified in the following way:

$$X_{i,j} = \rho_{i,j}(A_{1(i,j)} - 2A_{1,2(i,j)} + A_{2(i,j)} + A_{3(i,j)} + A_{4(i,j)} - 2A_{4,5(i,j)}$$
$$+ A_{5(i,j)} + (1 - \rho_{i,j})I_N)^{-1} \cdot [A_{1(i,j)} - A_{1,2(i,j)}; -A_{1,2(i,j)}$$
$$+ A_{2(i,j)}; A_{3(i,j)}; A_{4(i,j)} - A_{4,5(i,j)}; -A_{4,5(i,j)} + A_{5(i,j)}]$$
$$\cdot (\mathbf{Z}_{i,j} + \mathbf{W}_{i,j}) + (1 - \rho_{i,j})X_{i,j}^0 \tag{63}$$

where $\rho_{i,j}$ is

$$\rho_{i,j} = \begin{cases} 1, & \text{if } p_{i,j} \geq 1 \\ 0, & \text{if } p_{i,j} = 0 \end{cases}$$

Equation (63) is referred to as the *constitutive equation* (CE) of the sampled image. For internal or boundary pixels it provides a relation between the state evaluated at spatial point $(i, j)$ and the state evaluated at neighbouring points. For isolated pixels Eq. (63) resets the state. The form of the CE is identical for each image pixel, but its actual expression depends on the spatial position $(i, j)$ through the coefficients $c_{i,j}^{(\ell)}$.

From Eq. (61) we note that

$$c_{i,j}^{(1)} = c_{i,j}^{(2)} = 1 \implies A_{1,2(i,j)} = A_{2(i,j)}$$
$$c_{i,j}^{(4)} = c_{i,j}^{(5)} = 1 \implies A_{4,5(i,j)} = A_{4(i,j)}$$

In this case the matrices $-A_{1,2(i,j)} + A_{2(i,j)}$ and $A_{4(i,j)} - A_{4,5(i,j)}$, which in Eq. (63) weigh the contribution of the diagonal directions (corresponding to $\ell = 2$ and $\ell = 4$, respectively), are null. This means that for every "internal" pixel, all the information necessary to estimate $X_{i,j}$ is contained in the horizontal and vertical directions. Diagonal directions are useful to estimate boundary pixels for which $c_{i,j}^{(1)}$ and/or $c_{i,j}^{(5)} = 0$.

## IV. IMAGE RESTORATION

### A. Space-Variant Realization of the Image

In this section we exploit the CE (63) to derive a state-space representation of the sampled image suitable for the Kalman filter implementation as a strip processor (Woods and Radewan, 1977). For this purpose consider the ensemble of pixels composed of $L$ columns and two contiguous rows (Fig. 2) where

$$j_2 := j_1 + 1, \quad j_3 = j_1 + 2, \ldots, \quad j_L - 1 := j_1 + L - 2, \quad j_L = j_1 + L - 1$$

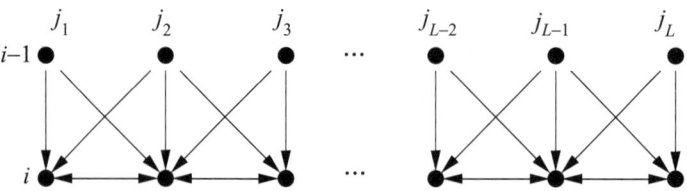

FIGURE 2. The considered ensemble of pixels.

We derive the desired image state-space representation by expressing in a suitable compact form the complete set of CE (63) that can be written for each pixel at coordinates $(i, j)$ with $j = j_1, \ldots, j_L$. By defining the matrices

$$F_{i,j} := \rho_{i,j}\big(A_{1(i,j)} - 2A_{1,2(i,j)} + A_{2(i,j)} + A_{3(i,j)} + A_{4(i,j)} - 2A_{4,5(i,j)}$$
$$+ A_{5(i,j)} + (1 - \rho_{i,j})I_N\big)^{-1}$$

$$F_{i,j}^{(1,2)} := F_{i,j}\big(A_{1(i,j)} - A_{1,2(i,j)}\big) \qquad F_{i,j}^{(2,1)} := F_{i,j}\big(-A_{1,2(i,j)} + A_{2(i,j)}\big)$$

$$F_{i,j}^{(3)} := F_{i,j} A_{3(i,j)}$$

$$F_{i,j}^{(4,5)} := F_{i,j}\big(A_{4(i,j)} - A_{4,5(i,j)}\big) \qquad F_{i,j}^{(5,4)} := F_{i,j}\big(-A_{4,5(i,j)} + A_{5(i,j)}\big)$$

the CE (63) can be rewritten in shorter notation as

$$X_{i,j} = \big[F_{i,j}^{(1,2)}; F_{i,j}^{(2,1)}; F_{i,j}^{(3)}; F_{i,j}^{(4,5)}; F_{i,j}^{(5)}\big](Z_{i,j} + W_{i,j}) + (1 - \rho_{i,j})X_{i,j}^0 \quad (64)$$

For each of the $(L - 2)$ pixels at coordinates $(i, j)$, $j = j_2, j_3, \ldots, j_L - 1$, the CE has the general form of Eq. (64). The extremal pixels $(i, j_1)$ and $(i, j_L)$ are considered as boundary pixels, so that the relative boundary conditions must be taken into account. In particular for the pixel at coordinates $(i, j_1)$, the relative CE is derived by putting $c_{i,j_1}^{(1)} = c_{i,j_1}^{(2)} = 0$. By definition of $F_{i,j}^{(1,2)}$ and $F_{i,j}^{(2,1)}$ and taking into account Eq. (61), Eq. (64) gives

$$X_{i,j_1} = \big[0; 0; F_{i,j_1}^{(3)}; F_{i,j_1}^{(4)}; F_{i,j_1}^{(5)}\big](Z_{i,j_1} + W_{i,j_1}) + (1 - \rho_{i,j_1})X_{i,j_1}^0 \quad (65)$$

Analogously, for the pixel at coordinates $(i, j_L)$, the relative CE is obtained by assuming $c_{i,j_L}^{(4)} = c_{i,j_L}^{(5)} = 0$. In this case Eq. (64) assumes the following form:

$$X_{i,j_L} = \big[F_{i,j_L}^{(1,2)}; F_{i,j_L}^{(2,1)}; F_{i,j_L}^{(3)}; 0; 0\big](Z_{i,j_L} + W_{i,j_L}) + (1 - \rho_{i,j_L})X_{i,j_L}^0 \quad (66)$$

To express in a suitable compact form the complete set of CE composed of the $L-2$ equations (64) and Eqs. (65) and (66), it is convenient to define the following vectors:

$$\mathbf{X}_i^{(j_1,L)} := \begin{bmatrix} X_{i,j_1} \\ \vdots \\ X_{i,j_L} \end{bmatrix}, \quad \mathbf{X}_i^{0(j_1,L)} := \begin{bmatrix} (1-\rho_{i,j_1})X_{i,j_1}^0 \\ \vdots \\ (1-\rho_{i,j_L})X_{i,j_L}^0 \end{bmatrix}$$

$$\Theta_{i-1}^{(j_1,L)}$$

$$:= \begin{bmatrix} F_{i,j_1}^{(3)}W_{i,j_1}^{(3)} + F_{i,j_1}^{(4,5)}W_{i,j_1}^{(4)} + F_{i,j_1}^{(5,4)}W_{i,j_1}^{(5)} \\ F_{i,j_2}^{(1,2)}W_{i,j_2}^{(1)} + F_{i,j_2}^{(2,1)}W_{i,j_2}^{(2)} + F_{i,j_2}^{(3)}W_{i,j_2}^{(3)} + F_{i,j_2}^{(4,5)}W_{i,j_2}^{(4)} + F_{i,j_2}^{(5,4)}W_{i,j_2}^{(5)} \\ \vdots \\ F_{i,j_L-1}^{(1,2)}W_{i,j_L-1}^{(1)} + F_{i,j_L-1}^{(2,1)}W_{i,j_L-1}^{(2)} + F_{i,j_L-1}^{(3)}W_{i,j_L-1}^{(3)} \\ + F_{i,j_L-1}^{(4,5)}W_{i,j_L-1}^{(4)} + F_{i,j_L-1}^{(5,4)}W_{i,j_L-1}^{(5)} \\ F_{i,j_L}^{(1,2)}W_{i,j_L}^{(1)} + F_{i,j_L}^{(2,1)}W_{i,j_L}^{(2)} + F_{i,j_L}^{(3)}W_{i,j_L}^{(3)} \end{bmatrix}$$

(67)

and matrices

$$\Phi_i^{(j_1,L)} := \begin{bmatrix} 0 & F_{i,j_1}^{(5,4)}H_5 & 0 & \cdots & 0 \\ F_{i,j_2}^{(1,2)}H_1 & 0 & F_{i,j_2}^{(5,4)}H_5 & 0 & 0 \\ 0 & \ddots & \ddots & \ddots & \vdots \\ \vdots & \ddots & F_{i,j_L-1}^{(1,2)}H_1 & 0 & F_{i,j_L-1}^{(5,4)}H_5 \\ 0 & \cdots & 0 & F_{i,j_L}^{(1,2)}H_1 & 0 \end{bmatrix},$$

$$\Gamma_i^{(j_1,L)} := \begin{bmatrix} F_{i,j_1}^{(3)}H_3 & F_{i,j_1}^{(4,5)}H_4 & 0 & \cdots & 0 \\ F_{i,j_2}^{(2,1)}H_2 & F_{i,j_2}^{(3)}H_3 & F_{i,j_2}^{(4,5)}H_4 & 0 & 0 \\ 0 & \ddots & \ddots & \ddots & \vdots \\ \vdots & \ddots & F_{i,j_L-1}^{(2,1)}H_2 & F_{i,j_L-1}^{(3)}H_3 & F_{i,j_L-1}^{(4,5)}H_4 \\ 0 & \cdots & 0 & F_{i,j_L}^{(2,1)}H_2 & F_{i,j_L}^{(3)}H_3 \end{bmatrix},$$

so that we can write

$$\mathbf{X}_i^{(j_1,L)} = \Phi_i^{(j_1,L)}X_i^{(j_1,L)} + \Gamma_i^{(j_1,L)}X_{i-1}^{(j_1,L)} + \Theta_{i-1}^{(j_1,L)} + \mathbf{X}_i^{0(j_1,L)} \qquad (68)$$

From (68), when $(I - \Phi_i^{(j_1,L)})$ is nonsingular, the following linear dynamical state equation of the image state-space representation is finally obtained:

$$\mathbf{X}_i^{(j_1,L)} = (I - \Phi_i^{(j_1,L)})^{-1} \Gamma_i^{(j_1,L)} \mathbf{X}_{i-1}^{(j_1,L)} + (I - \Phi_i^{(j_1,L)})^{-1} \Theta_{i-1}^{(j_1,L)}$$
$$+ (I - \Phi_i^{(j_1,L)})^{-1} \mathbf{X}_i^{0(j_1,L)} \qquad i = 1, 2, \ldots \qquad (69)$$

We assume to know the mean value $\bar{\mathbf{X}}_0^{(j_1,L)}$ and the covariance matrix $P_0$ of the initial condition

$$\bar{\mathbf{X}}_0^{(j_1,L)} := E[\mathbf{X}_0^{(j_1,L)}],$$
$$\mathbf{P}_0 := E[(\mathbf{X}_0^{(j_1,L)} - \bar{\mathbf{X}}_0^{(j_1,L)})(\mathbf{X}_0^{(j_1,L)} - \bar{\mathbf{X}}_0^{(j_1,L)})^T]$$

To define $\Phi_1^{(j,L)}$ and $\Gamma_1^{(j,L)}$ a reasonable a priori choice of the coefficients $c_{1,j}^{(\ell)}, j = j_1, \ldots j_L$, is

$$c_{1,j}^{(3)} = 1, \qquad j = j_1, \ldots j_L$$
$$c_{1,j}^{(2)} = c_{1,j}^{(4)} = 0, \qquad j = j_1, \ldots j_L$$
$$c_{1,j_1}^{(1)} = c_{1,j_L}^{(5)} = 0$$

whereas the coefficients

$$c_{1,j}^{(1)} \quad j = j_2, \ldots, j_L, \qquad c_{1,j}^{(5)}, \quad j = j_1, \ldots, j_L - 1$$

depend on the actual edge configuration of the first row.

When $(I - \Phi_i^{(j_1,L)})$ is singular, Eq. (69) cannot be derived. This particular case corresponds to the existence at $i$th row of at least one *isolated ensemble* of pixels, as shown in Figure 3. Here the pixels of coordinates

$$(i, j_p), \ldots, (i, j_r), \qquad j_p < j_r$$

are isolated from the other pixels of the same strip belonging both to the same row and to the previous one.

In this situation the entries of the state vector $\mathbf{X}_i^{(j_1,L)}$ corresponding to the isolated pixels need to be reinitialized according to the same procedure used to

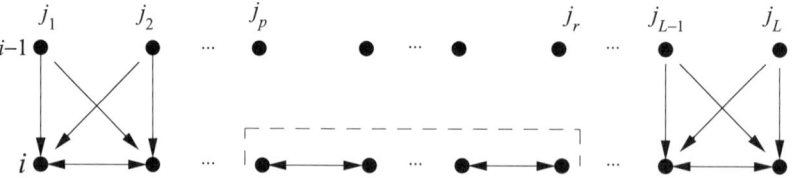

FIGURE 3. Edge configuration giving rise to an ensemble of isolated pixels.

define the initial conditions. By virtue of this reinitialization the matrix $\Phi_i^{(j_1,L)}$ is modified in such a way that $(I - \Phi_i^{(j_1,L)})$ is nonsingular.

An observation equation can be associated with (69) by writing Eq. (34) for each pixel at coordinates $(i, j)$, $j = j_1, \ldots, j_L$. To this aim define the following vectors:

$$Y_i^{(j_1,L)} := \begin{bmatrix} y_{i,j_1} \\ y_{i,j_2} \\ \vdots \\ y_{i,j_L} \end{bmatrix}, \quad V_i^{(j_1,L)} := \begin{bmatrix} v_{i,j_1} \\ v_{i,j_2} \\ \vdots \\ v_{i,j_L} \end{bmatrix}$$

and matrices

$$M := \begin{bmatrix} M' & 0 & \cdots & 0 \\ 0 & M' & \cdots & 0 \\ \vdots & \vdots & \cdots & \vdots \\ 0 & 0 & \cdots & M' \end{bmatrix}, \quad M' = [\underbrace{1\ 0\ \cdots\ 0}_{N\ \text{elements}}]$$

Taking into account that $M'X_{i,j} = x_{i,j}$, the complete set of scalar observation equations (34) can be written in the following compact form:

$$Y_i^{(j_1,L)} = MX_i^{(j_1,L)} + V_i^{(j_1,L)} \tag{70}$$

Because the state noise $\Theta_i^{(j_1,L)}$ is a white noise sequence (see Section VIII.C), the state-space representation given by Eqs. (69) and (70) have a form amenable to the Kalman filtering implementation as a strip processor. The resulting algorithm behaves like the classical Kalman filter if no isolated point is met in the image scanning. At isolated points, where no correlation with neighboring pixels exists, the dynamical model simply assigns the observed value, i.e.,

$$X_{i,j} = X_{i,j}^0 = \begin{bmatrix} y_{i,j} & \underbrace{0\ \cdots\ 0}_{N-1\ \text{elements}} \end{bmatrix}^T$$

### B. The Edge Problem

The proposed image model requires the preliminary identification of coefficients $c_{i,j}^{(\ell)}$ associated to each pixel; therefore, edge locations need to be estimated. This problem has been widely investigated in the literature and many methods have been proposed depending on the various definitions of edge and image models, either deterministic or stochastic.

The deterministic approach describes edge points as locations of suitable order discontinuities (Marr and Hildreth, 1980; Nalva and Binford, 1986; Torre and Poggio, 1986; Huertas and Medioni, 1986; Asada and Brady, 1986; Mallat and Hwang, 1992) so that the presence of a contour is associated to local extrema of the signal derivatives. For steplike edges a classical method for detecting such extrema consists in determining zero crossing points of the convolution of the data with a Laplacian of Gaussian masks (Huertas and Medioni, 1986). A later approach used multiscale wavelet analysis for identifying higher-order discontinuity points (Mallat and Hwang, 1992).

The stochastic approach is based on a probabilistic description of signals obtained either by defining the signal *a priori* distribution (Geman and Geman, 1984) or defining the signal generation model (Basseville and Benveniste, 1983). In the first case stochastic relaxation procedures are used to generate a sequence of images converging to the maximum a posteriori estimate of the true image. In the second case, sudden variations in the signal model parameters are detected by statistical hypothesis tests on the output innovation process.

We chose an edge-detection algorithm based on the gradient method (Rosenfield and Kak, 1982). The motivation for such a choice is twofold. The gradient method is relatively simple to implement and is based on the definition of edge points as locations where abrupt changes occur in the image gray-level distribution. Such definition well agrees with the image model here proposed. According to Rosenfield and Kak (1982), we estimated the gradient magnitude $M_{i,j}$ and direction $d_{i,j}$ as

$$M_{i,j} = \sqrt{f_{si,j}^2 + f_{ri,j}^2}, \qquad d_{i,j} = \tan^{-1}(f_{ri,j}/f_{si,j})$$

where $f_{ri,j}$ and $f_{si,j}$ represent the rate of change of the gray-level distribution along the vertical and horizontal direction, respectively. We thresholded $M_{i,j}$ with the mean value of the gradient computed over all the image pixels. As suggested in Rosenfield and Kak (1982), $f_{ri,j}$ and $f_{si,j}$ were estimated by means of noise-smoothing difference operators. Besides the differentiation procedure, these linear operators perform some smoothing action in order to reduce noise effects. From a practical point of view, the implementation of these operators requires a 2D discrete convolution of the noisy image with a matrix operator. As in Germani and Jetto (1988), to compute $f_{ri,j}$ we used the (6 × 3) edge detector operator, most sensitive to horizontal edges, given by

$$K_r = \frac{1}{14.5} \begin{bmatrix} -1 & -1 & -1 \\ -1.5 & -2 & -1.5 \\ -2 & -2.5 & -2 \\ 2 & 2.5 & 2 \\ 1.5 & 2 & 1.5 \\ 1 & 1 & 1 \end{bmatrix}$$

The matrix $K_s = K_r^T$ was used to estimate $f_{si,j}$. The entries have decreasing numerical values to weaken the influence on $f_{ri,j}$, $f_{si,j}$ of pixels lying on neighboring edges. The size (6 × 3) is a reasonable compromise in recognizing small objects while retaining a sufficient noise smoothing. Arguments about shape and dimensions of convolution operators are extensively discussed (Argile, 1971; From and Deutsch, 1975, 1978; Rosenfield and Kak, 1982). The response of the edge detectors was improved by a thinning procedure based on the nonmaximum suppression technique (Rosenfield and Kak, 1982). This facilitated the subsequent operation of determining the coefficients $c_{i,j}^{(\ell)}$, $\ell = 1, \ldots, 5$, for each image pixel, by exploiting the on-off information about edge locations.

### C. The Filtering Algorithm

Equations (69) and (70) provide us the sought image signal state-space representation to use in the Kalman filter design, according to the theory described in Section II. The particular filter implementation is named *strip processor* because Eq. (69) describes the gray-level spatial evolution for pixels lining up in a row of width $L$ so that, as the row index increases from 1 to $\bar{m}$, an image strip is scanned and processed.

To simplify the notation, let us rename same of the matrices appearing in Equation (69):

$$A_i^{(j_1,L)} := \left(I - \Phi_i^{(j_1,L)}\right)^{-1} \Gamma_i^{(j_1,L)} \tag{71}$$

$$F_i^{(j_1,L)} := \left(I - \Phi_i^{(j_1,L)}\right)^{-1} \tag{72}$$

Then, with $\hat{\mathbf{X}}_{k|k}^{(j_1,L)} = E[\mathbf{X}_k^{(j_1,L)} | Y_k^{(j_1,L)}, \ldots, Y_1^{(j_1,L)}]$, Eqs. from (15)–(18) become

$$\hat{\mathbf{X}}_{k|k}^{(j_1,L)} = A_k^{(j_1,L)} \hat{\mathbf{X}}_{k-1|k-1}^{(j_1,L)} + K_k^{(j_1,L)} \left(Y_k^{(j_1,L)} - A_k^{(j_1,L)} \hat{\mathbf{X}}_{k-1|k-1}^{(j_1,L)}\right)$$
$$+ F_k^{(j_1,L)} \hat{\mathbf{X}}_k^{0(j_1,L)}, \quad \hat{\mathbf{X}}_0^{(j_1,L)} = \mathbf{X}_0^{(j_1,L)} \tag{73}$$

$$K_k^{(j_1,L)} = \frac{1}{\sigma_v^2} R_k^{(j_1,L)} M^T \tag{74}$$

$$R_k^{(j_1,L)} = \left(I + \frac{1}{\sigma_v^2} H_{k-1}^{(j_1,L)} M^T M\right)^{-1} H_{k-1}^{(j_1,L)}, \quad R_0^{(j_1,L)} = P_0 \tag{75}$$

$$H_{k-1}^{(j_1,L)} = A_k^{(j_1,L)} R_{k-1}^{(j_1,L)} A_k^{(j_1,L)^T} + F_k^{(j_1,L)} F_k^{(j_1,L)^T} \tag{76}$$

Equations (74) and (75) are obtained considering that $E[V_k^{(j_1,L)} V_k^{(j_1,L)^T}] = \sigma_v^2 I$. The term $F_k^{(j_1,L)} \hat{X}_k^{0(j_1,L)}$ in Eq. (73) accounts for the singular cases discussed in Section IV.A; it resets the filter when isolated ensembles of pixels are met.

We finally note that a space variant filter is obtained, so that no steady-state argument can be applied. As a consequence, equations (74), (75), and (76) must be recursively solved on-line.

### D. Deblurring

In the image model devised through Sections II and III, only the degradation due to additive measurement noise was considered. The signal recording process usually introduces other deterministic kinds of perturbations, whose overall effect on the detected image is known as *blur* (Hwang, 1971). It mainly depends on the low-pass filter behavior of the measurement equipment, as well as on typical aberration of the optical components of the imaging system. Furthermore, the relative motion between source and sensor results in a defocused signal. The deblurring problem has been widely considered in the literature (see, e.g., Cannon, 1976; Tekalp *et al.*, 1986; Tekalp and Kaufman, 1988; Lagendijk *et al.*, 1990). The purpose of this section is to give an outline of how the image-modeling previously described can be extended to deal with this problem.

Referring directly to the discretized model, blurring is commonly described by a convolution of the original signal $x_{i,j}$, $(i, j) \in [1, \bar{m}]^2$, with a 2D linear space invariant system generally characterized by a *point-spread function* (PSF) $h_{k,l}$ with a rectangular support. Then, the noisy blurred image is given by

$$y_{i,j} = \sum_{k=-L_1}^{L_1} \sum_{l=-L_2}^{L_2} h_{k,l} \cdot x_{i-k, j-l} \qquad (77)$$

The blurred image can be considered as a unique smooth domain, so that any information about edge locations is lost. Models (69) and (70) can be well adapted to deal with this case by removing the inhomogeneity assumption. This implies $c_{i,j}^{(\ell)} = 1$ and $\varrho_{i,j} = 1$ for every $\ell$ and $(i, j)$. As a consequence system matrices $\Phi_i^{(j_1,L)}$ and $\Gamma_i^{(j_1,L)}$ in Eq. (69) become constant arrays $\Phi$ and $\Gamma$, depending only on the strip width $L$; moreover $X_0^{(j_1,L)} = \{0\}$ and the state noise vector sequence $\{\Theta_{i-1}^{(j_1,L)}\}$ becomes a stationary one with covariance matrix $Q$. To obtain matrices $\Phi$, $\Gamma$ and $Q$ substitute the stationary version of Eqs: (35)–(39) ($c_{i,j}^{(\ell)} = 1, \ell = 1, \ldots 5$) in formulas defining $\Phi_i^{(j_1,L)}$, $\Gamma_i^{(j_1,L)} \Gamma_i^{(j_1,L)}$ in Section IV, and $Q_{i-1}^{(j_1,L)}$ in Section VIII.C.

Defining the vectors and matrices

$$\underline{0} = [\underbrace{0, \ldots, 0}_{\bar{n}+1 \text{ elements}}], \quad \underline{h}_{k,l} = [\underbrace{h_{k,l}, 0, \ldots, 0}_{\bar{n}+1 \text{ elements}}],$$

$$M_k(\underline{h}) = \begin{bmatrix} \underline{h}_{k,0} & \cdots & \cdots & \underline{h}_{k,-L_2} & \underline{0} & \cdots & \cdots & \cdots & \underline{0} \\ \vdots & \ddots & \ddots & \ddots & \ddots & \ddots & \ddots & \ddots & \vdots \\ \underline{h}_{k,L_2} & \cdots & \underline{h}_{k,0} & \cdots & \underline{h}_{k,-L_2} & \underline{0} & \cdots & \cdots & \underline{0} \\ \vdots & \ddots & \ddots & \ddots & \ddots & \ddots & \ddots & \ddots & \vdots \\ \underline{0} & \cdots & \cdots & \underline{0} & \underline{h}_{k,L_2} & \cdots & \underline{h}_{k,0} & \cdots & \underline{h}_{k,-L_2} \\ \vdots & \ddots & \ddots & \ddots & \ddots & \ddots & \ddots & \ddots & \vdots \\ \underline{0} & \cdots & \cdots & \cdots & \underline{0} & \underline{h}_{k,L_2} & \cdots & \cdots & \underline{h}_{k,0} \end{bmatrix}$$

$$M(\underline{h}) = [M_{-L_1}(\underline{h}), \cdots, M_{L_1}(\underline{h})]$$

$$\underline{X}_i^{(j_1,L)} = \begin{bmatrix} X_{i+L_1}^{(j_1,L)} \\ \vdots \\ X_i^{(j_1,L)} \\ \vdots \\ X_{i-L_1}^{(j_1,L)} \end{bmatrix}, \quad \mathbf{B} = \begin{bmatrix} (I - \Phi)^{-1} \\ 0 \\ \vdots \\ 0 \end{bmatrix} \} 2L_1 + 1 \text{ blocks}$$

$$\mathbf{A} = \begin{bmatrix} (I - \Phi)^{-1}\Gamma & 0 & \cdots & \cdots & 0 \\ I & 0 & \cdots & \cdots & 0 \\ 0 & I & 0 & \cdots & 0 \\ \vdots & \ddots & \ddots & \ddots & \vdots \\ 0 & \cdots & 0 & I & 0 \end{bmatrix}$$

it can be shown that the following state-space representation of the blurred image is obtained:

$$\underline{X}_i^{(j_1,L)} = \mathbf{A}\underline{X}_{i-1}^{(j_1,L)} + \mathbf{B}\Theta_{i-1}^{(j_1,L)} \tag{78}$$

$$Y_i^{(j_1,L)} = \mathbf{M}(\underline{h})\underline{X}_i^{(j_1,L)} + V_i^{(j_1,L)} \tag{79}$$

Equations (78) and (79) define the sought model suitable for Kalman filter design as strip processor for blurred images. Moreover, a space-invariant scheme is obtained so that a steady-state implementation can be adopted.

## V. NEW RESEARCH DEVELOPMENTS

In this section, some algorithm updates and new research developments are reported. In particular, a filter implementation is considered, where the prediction estimate is first obtained for all the pixels in a row in one step; then, through a line-scan process, the filter estimate is computed recursively for any pixel at time.

The component equations (35)–(39) were also exploited in designing an optimal edge-detection procedure: at any pixel, the hypothesis that there is no edge crossing such a point is tested against the hypothesis that the considered pixel is indeed an edge point. This is obtained by the ratio test between the estimated likelihoods of two models, obtained from the component equations, corresponding to the two alternative hypotheses.

Finally, we will show that the proposed image model can be well adapted to those practical situations where the noises involved are not Gaussian. A new kind of filter is designed that is a polynomial approximation of the Kalman filter: it is well known indeed that in the non-Gaussian case the optimal estimate is no longer a linear transformation of the available measurement, so that an approximation is needed.

### A. Line-Scan Filter Implementation

In Concetti and Jetto (1997) and in Jetto (1999a), the problem of the filtering algorithm implementation requiring a lower memory allocation was studied. In the following we report some details of the proposed solution. Let us recall that if an edge occurs between pixels $(i, j)$ and $(i - 1, j)$ and/or between pixels $(i, j)$ and $(i, j - 1)$, Eq. (32) can not be integrated along the corresponding horizontal and/or vertical direction because, as a consequence of the inhomogeneity assumption, no relation exists between $X_{i,j}$ and $X_{i-1,j}$ and/or between $X_{i,j}$ and $X_{i-1,j-1}$. Equations (35) and (37) are then modified as follows:

$$X_{i,j} = H_1\big(c_{i,j}^{(1)} X_{i,j-1} + \big(1 - c_{i,j}^{(1)}\big) X_{i,j-1}^{0(s)}\big) + c_{i,j}^{(1)} W_{i,j}^{(1)} + \big(1 - c_{i,j}^{(1)}\big) W_{i,j}^{0(1)} \quad (80)$$

$$X_{i,j} = H_3\big(c_{i,j}^{(3)} X_{i-1,j} + \big(1 - c_{i,j}^{(3)}\big) X_{i-1,j}^{0(3)}\big) + c_{i,j}^{(3)} W_{i,j}^{(3)} + \big(1 - c_{i,j}^{(3)}\big) W_{i,j}^{0(3)} \quad (81)$$

where $X_{i,j-1}^{0(1)}$ and $W_{i,j}^{0(1)}$ are the initial state and the initial value of $\{W_{i,j}^{(1)}\}$, respectively, corresponding to each edge crossed during the horizontal scanning along a line and $X_{i-1,j}^{0(3)}$ and $W_{i,j}^{0(3)}$ are the initial state and the initial value of $\{W_{i,j}^{(3)}\}$, respectively, corresponding to each edge crossed during the vertical scanning along a column.

Equations (80) and (81) work like Eqs. (35) and (37), respectively, as long as $c_{i,j}^{(1)}$ and $c_{i,j}^{(3)}$ are one, namely, inside each smooth subregion. If for some pixel $(i, j)$ one has $c_{i,j}^{(1)} = 0$ (and/or $c_{i,j}^{(3)} = 0$), this means that an edge occurs between pixels $(i, j)$ and $(i, j-1)$ (and/or between pixels $(i, j)$ and $(i-1, j)$), so that a transition occurs between two contiguous smooth subregions; hence, in the light of the inhomogeneity assumption, no relation exists between $X_{i,j}$ and $X_{i,j-1}$ (and/or between $X_{i,j}$ and $X_{i-1,j}$). A state resetting is then performed in Eqs. (80) and/or (81) by expressing $X_{i,j}$ as a function of the initial conditions $X_{i,j-1}^{0(1)}$, $W_{i,j}^{0(1)}$ (and/or $X_{i-1,j}^{0(3)}$, $W_{i,j}^{0(3)}$) relative to the stochastic process describing the image inside the new smooth subregion.

Taking into account that the only observed component of the state vector is the image signal, the measure equation (34) can be associated with Eqs. (80) and (81):

$$y_{i,j} = C X_{i,j} + v_{i,j} \tag{82}$$

where $C$ is the $1 \times N$ row vector $[1, 0, \cdots, 0]$ and $v_{i,j}$ is a discrete white Gaussian noise $\sim \mathcal{N}(0, \sigma_v^2)$ uncorrelated with both $\{W_{i,j}^{(1)}\}$ and $\{W_{i,j}^{(3)}\}$. A system composed of Eqs. (80), (81), and (82) constitutes the particularization of the image model presented in Section IV, which will be used for a new filter implementation.

### 1. A First Suboptimal Implementation

Suboptimality stems from computing the signal estimate at pixel $(i, j)$ by using the measurements

$$\{y_{l,m}\}, \qquad \{l \leq i-1, m \leq \bar{m}\} \cup \{l = i, m \leq j\}$$

Given pixel $(i, j)$ on the $i$th row, let $Y_{i,j}$ and $Y_{i,j}^*$ be the sets of all the observations $y_{i,l}$ relative to pixels lying on the $i$th row that are not separated from pixel $(i, j)$ by an edge and with the additional requirement $l \leq j$ as for $Y_{i,j}$. Let $E[X_{i,j}/Y_{i,j-1}] = \hat{X}_{i,j}^{(s)-}$ and $E[X_{i,j}/Y_{i,j}] = \hat{X}_{i,j}^{(s)+}$, $1 \leq j \leq \bar{m}$, be the predicted and filtered estimates, respectively, of $X_{i,j}$, $1 \leq j \leq \bar{m}$, obtained by applying, on the $i$th row, the 1D Kalman filter to the image 1D submodel given by Eqs. (80) and (82); moreover, denote by $P_{i,j}^{(s)-}$ and $P_{i,j}^{(s)+}$ the error covariance matrices of $\hat{X}_{i,j}^{(s)-}$ and $\hat{X}_{i,j}^{(s)+}$, respectively. According to (80) and (82) one has

$$\hat{X}_{i,j}^{(s)+} = \hat{X}_{i,j}^{(s)-} + K_{i,j}^{(s)}\left(y_{i,j} - C\hat{X}_{i,j}^{(s)-}\right) \tag{83}$$

$$\hat{X}_{i,j}^{(s)-} = H_1\left(c_{i,j}^{(1)}\hat{X}_{i,j-1}^{(s)+} + \left(1 - c_{i,j}^{(1)}\right)\hat{X}_{i,j-1}^{0(1)}\right) \tag{84}$$

$$K_{i,j}^{(s)} = P_{i,j}^{(s)-}C^T\left(CP_{i,j}^{(s)-}C^T + \sigma_v^2\right)^{-1} \tag{85}$$

$$P_{i,j}^{(s)-} = H_1\big(P_{i,j-1}^{(s)+} + \big(1 - c_{i,j}^{(1)}\big) P_{i,j-1}^{0(1)}\big) H_1^T + c_{i,j}^{(1)} Q_s + \big(1 - c_{i,j}^{(1)}\big) Q_{i,j}^{0(1)} \tag{86}$$

$$P_{i,j}^{(s)+} = \big(I - K_{i,j}^{(s)} C\big) P_{i,j}^{(s)-} \tag{87}$$

where $\hat{X}_{i,j-1}^{0(1)}$ is the estimate of $X_{i,j-1}^{0(1)}$, $P_{i,j}^{0(1)}$ is the error covariance matrix of $\hat{X}_{i,j-1}^{0(1)}$, and $Q_{i,j}^{0(1)}$ is the covariance matrix of $W_{i,j}^{0(1)}$.

Assuming that $\hat{X}_{i-1,j}$ and $P_{i-1,j}$, $1 \leq j \leq \bar{m}$, have already been computed, their optimal predictions to the $i$th row can be obtained by the prediction equations of the 1D Kalman filter deriving from Eq. (81), obtaining

$$\hat{X}_{i,j}^{(r)-} = H_3\big(c_{i,j}^{(3)} \hat{X}_{i-1,j} + \big(1 - c_{i,j}^{(3)}\big) \hat{X}_{i-1,j}^{0(3)}\big) \tag{88}$$

$$P_{i,j}^{(r)-} = H_3\big(c_{i,j}^{(3)} P_{i-1,j} + \big(1 - c_{i,j}^{(3)}\big) P_{i-1,j}^{0(3)}\big) H_3^T + c_{i,j}^{(3)} Q_r + \big(1 - c_{i,j}^{(3)}\big) Q_{i,j}^{0(3)} \tag{89}$$

where $\hat{X}_{i-1,j}^{0(3)}$ is the estimate of $X_{i-1,j}^{0(3)}$, $P_{i-1,j}^{0(3)}$ is the error covariance matrix of $\hat{X}_{i-1,j}^{0(3)}$, and $Q_{i,j}^{0(3)}$ is the covariance matrix of $W_{i,j}^{0(3)}$. The final 2D estimates $\hat{X}_{i,j}$ and $P_{i,j}$ can be obtained through an optimal combination of $\hat{X}_{i,j}^{(s)+}$, $P_{i,j}^{(s)+}$, $\hat{X}_{i,j}^{(r)-}$ and $P_{i,j}^{(r)-}$. To this purpose let $(i, j_{i,\ell})$, $j_{i,\ell} = j_{i,1}, \ldots, j_{i,\bar{\ell}}$ with $j_{i,1} = 0$, $i = 1, \ldots, \bar{m}$, be the coordinates of discontinuity points on the $i$th row, namely, the coordinates of those pixels for which $\hat{X}_{i,j}^{0(1)}$ is defined, and denote by $\hat{E}_{i,j_{i,\ell}}^{0(1)} = X_{i,j_{i,\ell}}^{0(s)} - \hat{X}_{i,j_{i,\ell}}^{0(1)}$ the corresponding initial estimation error. Then we assume that each initial estimation error $\hat{E}_{i,j_{i,\ell}}^{0(1)}$, $j_{i,\ell} = j_{i,1}, \ldots, j_{i,\bar{\ell}}$, is uncorrelated with $\hat{E}_{i-1,j}$ and with $W_{i,j}^{(1)}$, $j = 1, \ldots, \bar{m}$.

This assumption is not restrictive; it simply means that at each edge point the available estimate $\hat{X}_{i,j_{\ell}}^{0(1)}$ of the initial value $X_{i,j_{\ell}}^{0(1)}$ is independent of the way the final estimates on the previous row have been computed and of the stochastic terms of state equation (81). The assumption allows us to show that each estimation error $\hat{E}_{i,j}^{(s)+} = X_{i,j} - \hat{X}_{i,j}^{(s)+}$ is uncorrelated with $\hat{E}_{i,j}^{(r)-} = X_{i,j} - \hat{X}_{i,j}^{(r)-}$, $j = 1, \ldots \bar{m}$. See Jetto (1999) for a proof. This, in turn, implies that the estimates $\hat{X}_{i,j}^{(s)+}$ and $\hat{X}_{i,j}^{(r)-}$ and the relative error covariance matrices $P_{i,j}^{(s)+}$ and $P_{i,j}^{(r)-}$, respectively, can be optimally combined to obtain $\hat{X}_{i,j}$ and $P_{i,j}$ according to

$$\hat{X}_{i,j} = P_{i,j}\big[\big(P_{i,j}^{(r)-}\big)^{-1} \hat{X}_{i,j}^{(r)-} + \big(P_{i,j}^{(s)+}\big)^{-1} \hat{X}_{i,j}^{(s)+}\big] \tag{90}$$

$$P_{i,j} = \big[\big(P_{i,j}^{(r)-}\big)^{-1} + \big(P_{i,j}^{(s)+}\big)^{-1}\big]^{-1} \tag{91}$$

The proposed causal filtering algorithm is given by Eqs. (83)–(91) applied to each row; it is referred to as the causal space-variant filter (CSVF).

## 2. The Semicausal Filter

Denote by $\hat{X}_{i,j}^{(s)*}$ the smoothed estimates $E[X_{i,j}/Y_{i,*}^*]$ on the generic $i$th row and by $P_{i,j}^{(s)*}$ the relative error covariance matrices corresponding to the image 1D submodel (80) and (82). The semicausal filter is obtained by replacing the filtered estimates $\hat{X}_{i,j}^{(s)+}$ with $\hat{X}_{i,j}^{(s)*}$ and $P_{i,j}^{(s)+}$ with $P_{i,j}^{(s)*}$. The smoothed estimates $X_{i,j}^{(s)*}$ and the relative error covariance matrices $P_{i,j}^{(s)*}$, $1 \le j \le \bar{m}$, can be obtained through the 1D fixed-interval smoother equations associated with Equations (83)–(91):

$$\hat{X}_{i,j}^{(s)*} = \hat{X}_{i,j}^{(s)+} + A_{i,j}\big(\hat{X}_{i,j+1}^{(s)*} - \hat{X}_{i,j+1}^{(s)-}\big) \tag{92}$$

$$P_{i,j}^{(s)*} = P_{i,j}^{(s)+} + A_{i,j}\big(P_{i,j+1}^{(s)*} - P_{i,j+1}^{(s)-}\big)A_{i,j}^T \tag{93}$$

$$A_{i,j} = c_{i,j+1}^{(1)} P_{i,j}^{(s)+} H_1^T \big(P_{i,j+1}^{(s)-}\big)^{-1} \tag{94}$$

Denote by $\hat{X}_{i,j}^*$ the final 2D semicausal estimates obtained by also exploiting Eq. (81) and by $P_{i,j}^*$ the relative error covariance matrices; as for the CSVF, assume that $\hat{X}_{i-1,j}^*$ and $P_{i-1,j}^*$, $1 \le j \le \bar{m}$, have already been computed, so Eqs. (90) and (91) are then replaced by

$$\hat{X}_{i,j}^{(r*)-} = H_3\big(c_{i,j}^{(3)} \hat{X}_{i-1,j}^* + \big(1 - c_{i,j}^{(3)}\big)\hat{X}_{i-1,j}^{0(3)}\big) \tag{95}$$

$$P_{i,j}^{(r*)-} = H_3\big(c_{i,j}^{(3)} P_{i-1,j}^* + \big(1 - c_{i,j}^{(3)}\big) P_{i-1,j}^{0(3)}\big)H_3^T + c_{i,j}^{(3)} Q_r + \big(1 - c_{i,j}^{(3)}\big) Q_{i,j}^{0(3)} \tag{96}$$

Analogously to the CSVF, the final semicausal estimates $\hat{X}_{i,j}^*$ and $P_{i,j}^*$ can be obtained through an optimal combination of $\hat{X}_{i,j}^{(r*)-}$ and $P_{i,j}^{(r*)-}$ with $\hat{X}_{i,j}^{(s)*}$ and $P_{i,j}^{(s)*}$. To this purpose, defining the final semicausal estimation error as $\hat{E}_{i,j}^* = X_{i,j} - \hat{X}_{i,j}^*$, this time we assume that each initial estimation error $\hat{E}_{i,j_{i,\ell}}^{0(s)}$, $j_{i,\ell} = j_{i,1}, \ldots, j_{i,\bar{\ell}}$, is uncorrelated with $\hat{E}_{i-1,j}^*$ and with $W_{i,j}^{(3)}$ (or $W_{i,j}^{0(3)}$), $j = 1, \ldots, \bar{m}$. This allows us to show that also $\hat{E}_{i,j}^{(s)*} = X_{i,j} - \hat{X}_{i,j}^{(s)*}$ and $\hat{E}_{i,j}^{(r*)-} = x_{i,j} - \hat{X}_{i,j}^{(r*)-}$, $j = 1, \ldots, \bar{m}$, are uncorrelated, and again we refer to Jetto (1999a) for the proof. Hence, the final semicausal estimates $\hat{X}_{i,j}^*$ and $P_{i,j}^*$ can be obtained with formulas analogous to (92) and (93):

$$\hat{X}_{i,j}^* = P_{i,j}^*\big[\big(P_{i,j}^{(r*)-}\big)^{-1}\hat{X}_{i,j}^{(r*)-} + \big(P_{i,j}^{(s)*}\big)^{-1}\hat{X}_{i,j}^{(s)*}\big] \tag{97}$$

$$P_{i,j}^* = \big[\big(P_{i,j}^{(r*)-}\big)^{-1} + \big(P_{i,j}^{(s)*}\big)^{-1}\big]^{-1} \tag{98}$$

The proposed semicausal filtering algorithm is given by equations (83)–(89), (90)–(96), (97), (98) applied to each row; it is referred to as the semicausal space-variant filter (SCSVF).

Jetto (1999a) showed that the state estimates provided by the CSVF and SCSVF are really minimum variance estimates. The interested reader can also check the given reference for a suitable numerical assessment. The approach of this section can be extended to 3D and $N$D filtering problems, as shown in Jetto (1999b).

## B. An Edge-Estimation Algorithm

The component equations (35)–(39), along with the measurement equation (34), can be reliably used to design an optimal procedure to mark a pixel as an edge point (see De Santis and Sinisgalli (1999) for a full account on this subject). So, for any pixel $(i, j)$ consider the following neighborhood $\Omega_p$ of size $p$:

$$\Omega_p = \{(h, k) : |h - i| \leq p, |k - j| \leq p\}$$

For any point $(h, k) \in \Omega_p$, the component equations (35)–(39) allow us to express the state vector $X_{h,k}$ as a linear combination of the state vector $X_{i,j}$ plus noise, i.e.,

$$X_{h,k} = A_{h-i,k-j} X_{i,j} + W_{h-i,k-j} \tag{99}$$

and therefore, from (34),

$$y_{h,k} = M'(A_{h-i,k-j} X_{i,j} + W_{h-i,k-j}) + v_{h,k} \tag{100}$$

Matrix $A_{h-i,k-j}$ is a suitable linear combination of matrices $H_1 - H_5$, and the stochastic term $W_{h-i,k-j}$ depends linearly on the noise vectors $W_{h,k}^{(l)}$, $l = 1-5$ for $(h, k) \in \Omega_p$. Then, ordering the pixels in $\Omega_p$, for instance columnwise, the aggregate of equations (99) and (100) yield the following linear stochastic model:

$$Y_{i,j} = L X_{i,j} + Z_{i,j} \tag{101}$$

with $Y_{i,j} = [y_{i-p, j-p} \cdots y_{i,j} \cdots y_{i+p, j+p}]^T \in \mathbb{R}^{(2p+1)\cdot(2p+1)}$, and matrices $L$ and $Z_{i,j}$ are obtained accordingly.

The statistics of the stochastic term $Z_{i,j}$ in (101) is fully specified according to the hypotheses developed in Sections III and IV. To obtain a Bayesian model we need to specify the statistics of the unknown $X_{i,j}$, along with the statistics of the model unknown parameters. For the latter flat priors will be assumed as usual, whereas the former is assumed to be a Gaussian vector with mean value $m_0 = [\mu_0 \ 0 \cdots 0]^T$ and covariance matrix $R_0 = \sigma_0^2 I$; the choice for $m_0$ stems from the assumed homogeneity of the gray level in any smooth subregion, so that signal gradients may be supposed to be zero mean random variables.

Model (101) is fine if no edge crosses set $\Omega_p$ and will be referred to as *model 1*. The statistical model correspondent to (101) is the joint probability density function $p(Y_{i,j}, X_{i,j}, \theta_1)$, where vector $\theta_1$ collects all the unknown model (random) parameters.

If pixel $(i, j)$ is an edge point, the edge splits set $\Omega_p$ in two disjoint subsets $\Omega_1$ and $\Omega_2$. In this case we may assume that the gray-level mean value in the two subsets differs for a constant value $\Delta$. Model (101) is then adapted as follows:

$$Y_{i,j} = LX_{i,j} + C\Delta + Z_{i,j} \qquad (102)$$

where, according to the chosen pixel ordering, $C$ is a vector with entries equal to one in correspondence of pixels belonging, for instance, to subset $\Omega_2$, and zero otherwise. Relation (102) defines *model 2* and is characterized by the joint density $p(Y_{i,j}, X_{i,j}, \theta_2)$, where $\theta_2 = \theta_1 \cup \Delta$.

For any pixel $(i, j)$ the edge-identification procedure is performed as follows:

**Step 1** Identify model 1 by solving the problem

$$(\hat{X}_{i,j}^{(1)}, \hat{\theta}^{(1)}) = \arg\max_{X_{i,j},\theta_1} p(Y_{i,j}, X_{i,j}, \theta_1)$$

**Step 2** Define subsets $\Omega_1$ and $\Omega_2$ according to the following rule (see De Santis and Sinisgalli, 1999):

$$\Omega_1 = \{(h, k) \in \Omega_p : \eta(i, j) < 0\}, \qquad \Omega_2 = \{(h, k) \in \Omega_p : \eta(i, j) \geq 0\}$$

where

$$\eta(i, j) = \sum_{\ell=1}^{\tilde{n}} \frac{1}{\ell!} \sum_{m=0}^{\ell} \binom{\ell}{m} \left[ \frac{\partial^\ell \hat{x}_{i,j}^{(1)}}{\partial r^{\ell-m} \partial s^m} \right] [(h-i)\Delta_r]^{\ell-m} [(k-j)\Delta_s]^m$$

the gradients being the entries of $\hat{X}_{i,j}^{(1)}$ obtained at step 1;

**Step 3** identify model 2 by solving the following problem

$$(\hat{X}_{i,j}^{(2)}, \hat{\theta}^{(2)}) = \arg\max_{X_{i,j},\theta_2} p(Y_{i,j}, X_{i,j}, \theta_2);$$

**Step 4** compare the identified model 1 and 2 by means of the generalized likelihood ratio test

$$\frac{p(Y_{i,j} | \hat{X}_{i,j}^{(2)}, \hat{\theta}^{(2)})}{p(Y_{i,j} | \hat{X}_{i,j}^{(1)}, \hat{\theta}^{(1)})} \geq c_\varepsilon$$

where the conditional densities of the data with respect to the identified quantities are used, and constant $c_\varepsilon$ is determined so to obtain the *most powerful test of level $\varepsilon$%* (see De Santis and Sinisgalli, 1999). Pixel $(i, j)$ is an edge point if the test is affirmative.

Function $\eta(i, j)$ is the projection on the 2D plane of the level contour equation $x(h, k) = x(i, j)$ and gives a good approximation of the edge segment possibly passing through pixel $(i, j)$, providing $\Omega_p$ is sufficiently small; this is indeed verified since images are nowadays high resolution data.

As a general remark we can say the procedure is quite robust with respect to the additive noise: on decreasing signal-to-noise ratio values, spurious detections hardly occur while retining good performances in terms of correct detections, as opposite to methods where a global thresholding is enforced. The method is adaptive since the decision rule to mark or not a pixel as an edge point relies on the local signal statistics identified at any point. Finally we stress that the method depends on the gradients estimated direction and not directly on the gradient estimated norm: this is indeed known to be sensitive to the scene illumination, while the direction is more reliably identified from noisy data also in case of low constrast images.

### C. Polynomial Filtering for the Non-Gaussian Case

A first significant example of this kind of data is the measurement noise simply introduced by approximation error derived from digitalization of the real image. Such error has obviously a flat distribution, which can be far from being Gaussian. Other non-Gaussian problems arise in digital communication when the noise interference includes noise components that are essentially non-Gaussian (this is a common situation below 100 MHz). Neglecting these components is a major source of errors in communication system design.

Certainly the classical filtering theory can be used even in this case. This leads to the definition of just *linear* optimal estimators, such as the 2D Kalman filtering developed in the previous sections. Indeed, such an algorithm can be improved by using the higher-order informations given by the moments of the error distribution by using the theory of polynomial filtering of a 1D process (De Santis *et al.*, Carravetta *et al.*, 1996). The key idea consists in finding a signal estimate that is more accurate than the simple linear one, yet retaining the features of easy computability and recursivity. This goal is achieved by projecting the conditional expectation on the Hilbert space generated by the polynomial transformations of the output measurements. The proposed approach requires the definition of an *extended system,* in which both the system state and output are defined as the aggregate of the original vectors with their Kronecker powers up to a desired order.

The derivation of the filter extension is quite involved and makes use of the Kronecker algebra. In the following, the main points of optimal estimate polynomial approximation will be illustrated, referring to Dalla Mora et al., (1999) for full details and due numerical assessment. First recall that given two matrices $M$ and $N$ of size $r \times s$ and $p \times q$, respectively, the Kronecker product $M \otimes N$ is defined as the $(r \cdot p) \times (s \cdot q)$ matrix

$$M \otimes N = \begin{bmatrix} m_{1,1}N & \cdots & m_{1,s}N \\ \vdots & \ddots & \vdots \\ m_{r,1}N & \cdots & m_{r,s}N \end{bmatrix}$$

where $m_{i,j}$ is a generic element of M. This kind of product is not commutative.

Consider again Eqs. (69) and (70) defining the image state-space representation; of course, the measurement noise is assumed non-Gaussian, with finite moments up to a suitable order. By using the Kronecker product we can construct the following augmented state and output vectors

$$\mathcal{X}(i) = \begin{bmatrix} \mathcal{X}(i) \\ \mathcal{X}^{[2]}(i) \\ \vdots \\ \mathcal{X}^{[\nu]}(i) \end{bmatrix}, \quad \mathcal{X}(i) = \mathbf{X}_i^{j_1,L}; \quad \mathcal{Y}(i) = \begin{bmatrix} \eta(i) \\ \eta^{[2]}(i) \\ \vdots \\ \eta^{[\nu]}(i) \end{bmatrix}, \quad \eta(i) = \mathbf{Y}_i^{j_1,L}$$

(103)

Rearranging Eqs. (69) and (70) according to definitions (103), the following system is obtained (see Dalla Mora et al., 1999):

$$\mathcal{X}(i+1) = \mathcal{A}(i)\mathcal{X}(i) + \mathcal{U}(i) + \mathcal{F}(i), \quad \mathcal{X}(0) = \bar{\mathcal{X}} \quad (104)$$
$$\mathcal{Y}(k) = \mathcal{C}(i)\mathcal{X}(i) + \mathcal{V}(i) + \mathcal{G}(i) \quad (105)$$

The following statements are true:

- Vectors $\mathcal{U}(i)$ and $\mathcal{V}(i)$ are deterministic and their entries are linear transformations of the mean value of the Kronecker powers up to the order $\nu$ of the state and output noise vectors, respectively.
- $\{\mathcal{F}(i)\}$ and $\{\mathcal{G}(i)\}$ are zero-mean white sequences with covariance matrices $\{\mathcal{Q}(i)\}$, $\{\mathcal{R}(i)\}$, respectively; they satisfy the following mutual relations:

$$E[\mathcal{F}(k)\mathcal{G}^T(j)] = 0, \quad k \neq j$$
$$E[\mathcal{F}(i)\mathcal{G}^T(i)] = \mathcal{T}(i)$$

- The stochastic sequences $\{\mathcal{F}(i)\}$ and $\{\mathcal{G}(i)\}$ are second-order asymptotically stationary processes, provided that matrix $(I - \Phi_i^{(j_1,L)})^{-1}$ is asymptotically stable.

Using the obtained results, taking in account the deterministic and the stochastic inputs and also that noise-sequences we have constructed are correlated at the same instant, the filter equations are (Goodwin and Payne, 1977)

$$\hat{\mathcal{X}}(i) = \hat{\mathcal{X}}(i|i-1) + \mathcal{K}(i)(\mathcal{Y}(i) - \mathcal{C}(i)\hat{\mathcal{X}}(i|i-1) - \mathcal{V}(k)) \quad (106)$$

$$\mathcal{Z}(i) = \mathcal{T}(i)(\mathcal{C}(i)\mathcal{P}(i|i-1)\mathcal{C}^T(i) + \mathcal{R}(i))^{-1} \quad (107)$$

$$\hat{\mathcal{X}}(i+1|i) = (\mathcal{A}(i) - \mathcal{C}(i)\mathcal{K}(i) + \mathcal{Z}(i))\mathcal{C}(i))\hat{\mathcal{X}}(i|i-1)$$
$$+ (\mathcal{A}(i)\mathcal{K}(i) + \mathcal{Z}(i))(\mathcal{Y}(i) - \mathcal{V}(i)) + \mathcal{U}(i) \quad (108)$$

$$\mathcal{P}(i+1|i) = \mathcal{A}(i)\mathcal{P}(i)\mathcal{A}^T(i) + \mathcal{Q}(i) - \mathcal{Z}(i)\mathcal{T}^T(i) - \mathcal{A}(i)\mathcal{K}(i)\mathcal{T}^T(i)$$
$$- \mathcal{T}(i)\mathcal{K}^T(i)\mathcal{A}^T(i) \quad (109)$$

$$\mathcal{P}(i) = (I - \mathcal{K}(i)\mathcal{C}(i))\mathcal{P}(i|i-1) \quad (110)$$

$$\mathcal{K}(i) = \mathcal{P}(i|i-1)\mathcal{C}^T(i)(\mathcal{C}(i)\mathcal{P}(i|i-1)\mathcal{C}^T(i) + \mathcal{R}(i))^{-1} \quad (111)$$

where $\mathcal{K}(i)$ is the filter gain and $\mathcal{P}(i)$, $\mathcal{P}(i|i-1)$ are the filter and prediction covariances, respectively. If the matrix $\mathcal{C}(i)\mathcal{P}(i|i-1)\mathcal{C}(i)^T + \mathcal{R}(i)$ is singular, we can use the Moore-Penrose pseudoinverse. The initial conditions for (106) and (110), respectively, are

$$\hat{\mathcal{X}}(0|-1) = E[\bar{\mathcal{X}}],$$

$$\mathcal{P}(0|-1) = E[(\bar{\mathcal{X}} - E[\bar{\mathcal{X}}])(\bar{\mathcal{X}} - E[\bar{\mathcal{X}}])^T)$$

The optimal linear estimate of the augmented state process $\mathcal{X}(i)$ with respect to the augmented observations $\mathcal{Y}(i)$ agrees with its optimal polynomial estimate with respect to the original observations $\eta(i) = \mathbf{Y}_i^{j_1, L}$, in the sense of taking into account all the powers, up to the $v$th order, of $\eta(i)$, $i = 0, \ldots, k$. We thus obtain the optimal polynomial estimate of the system (69) and (70). The optimal linear estimate of the original state $\mathcal{X}(i) = \mathbf{X}_i^{j_1, L}$ with respect to the same set of augmented observations is easily determined by extracting the first $NL$ components in the vector $\hat{\mathcal{X}}(i)$).

A reduced-order implementation can also be found in Dalla Mora *et al.* (1999) in order to decrease both the computation time and the memory occupancy.

## VI. NUMERICAL RESULTS

Two 256 × 256 pixels 8-bit images were used to test the proposed restoration method. The first one is a simulated image consisting of concentric rhombi with constant gray level in each homogeneous subregion (see Fig. 4). The second one is a picture of a common outdoor scene, shown in Figure 5.

The synthesized image has been chosen because it contains sharp edges, while the real image has been chosen to evaluate the filter performance on real data. For each of the original test images two different noisy versions were generated with an SNR (signal variance/noise variance) equal to 4 and 16. See Figures 6a and 7a and 6b and 7b. This experimental situation was considered to test the method capability in restoring noisy images of heavily different characteristics.

Once the parameters $c_{i,j}^{(\ell)}(\ell = 1, \ldots, 5)$ have been obtained, according to the procedure described in Section IV.B, the Kalman filter was implemented as a strip processor (Woods and Radewan, 1977) according to the image

FIGURE 4. "Rhombi" original image.

FIGURE 5. "Outdoor picture" original image.

representation of Eqs. (69) and (70). The images were partitioned into strips 15 pixels wide; these strips were overlapped and only the 11 middle pixels were retained as final estimates to avoid strip edge effects. For each strip, the Kalman filter estimate equation was initialized by assuming

$$\bar{\mathbf{X}}_0^{(j_1,L)} = \left[ y_{1,j_1} \quad \underbrace{0 \cdots 0}_{N-1 \text{ elements}} \quad \cdots \quad y_{1,j_L} \quad \underbrace{0 \cdots 0}_{N-1 \text{ elements}} \right]^T$$

The Riccati equation was implemented starting from an initial value of the $(LN) \times (LN)$ error covariance matrix $\mathbf{P}_0$ given by

$$\mathbf{P}_0 = \begin{bmatrix} P_0^{(1)} & 0 & \cdots & \cdots & \cdots & 0 \\ 0 & P_0^{(2)} & 0 & \cdots & \cdots & 0 \\ 0 & 0 & \ddots & \cdots & \ddots & 0 \\ 0 & \cdots & \cdots & \cdots & \cdots & P_0^{(L)} \end{bmatrix}$$

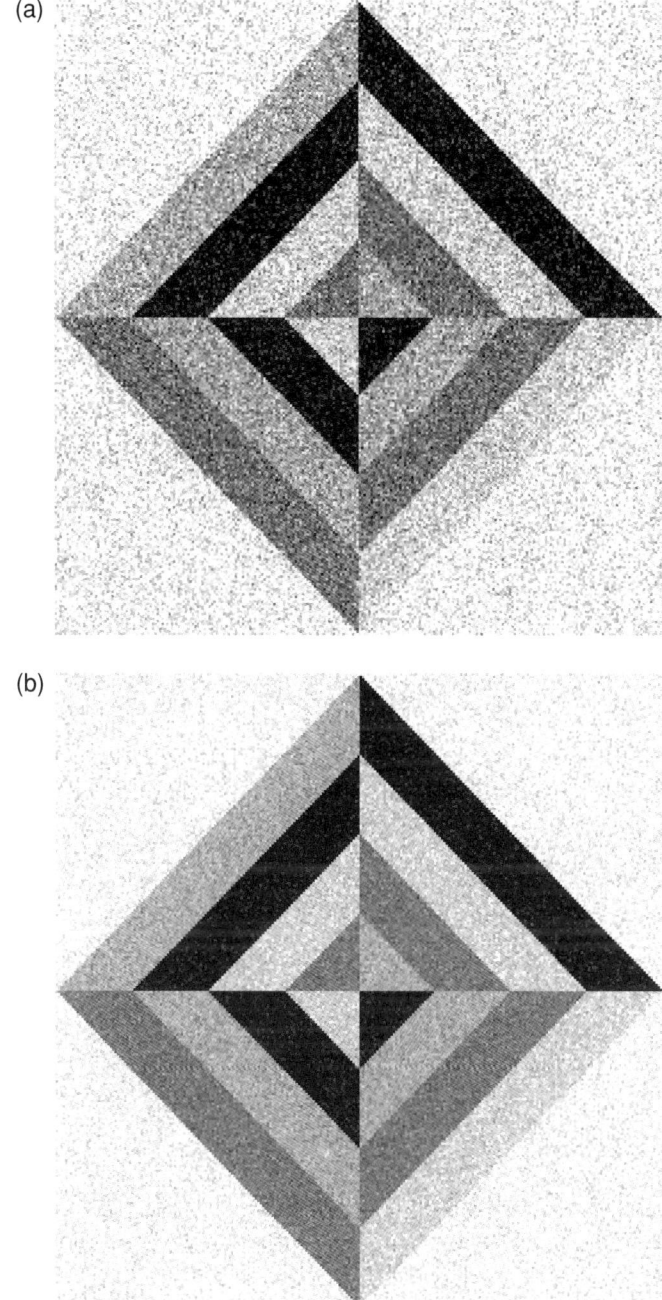

FIGURE 6. Noisy versions of "rhombi" corresponding to SNR = 4 (6a) and SNR = 16 (6b).

(a)

(b)

FIGURE 7. Noisy versions of "outdoor picture" corresponding to SNR = 4 (7a) and SNR = 16 (7b).

where $P_0^{(k)}$, $k = 1, \ldots, L$ are $N \times N$ matrices defined as

$$P_0^{(k)} = \begin{bmatrix} \sigma_v^2 & 0 \cdots & \cdots & \cdots & 0 \\ 0 & 0 & 0 & \cdots & 0 \\ 0 & 0 & \ddots & \cdots & 0 \\ 0 & \cdots & \cdots & 0 & 0 \end{bmatrix}$$

The image of concentric rhombi was processed with a model order corresponding to the choice $\bar{n} = 0$ because it can be considered a piecewise constant image, whereas the value $\bar{n} = 1$ seemed to be more appropriate for processing a shaded image such as in Figure 5.

To measure the improvement in SNR introduced by the filter, the following performance parameter $\eta$, expressed in decibels (dB), was defined as

$$\eta = 10 \log_{10} \frac{\sum_i \sum_j (y_{i,j} - x_{i,j})^2}{\sum_i \sum_j (\hat{x}_{i,j} - x_{i,j})^2}$$

where $y_{i,j}$ is the noisy signal observed at pixel $(i, j)$, $x_{i,j}$ is the corresponding true signal value, and $\hat{x}_{i,j}$ is the Kalman estimate of $x_{i,j}$.

Filtered images are reported in Figures 8a and 8b and Figures 9a and 9b. The values of $\eta$ for SNR = 4 (SNR = 16) were 5.78 and 3.55 (4.35 and 2.01) for Figures 4 and 5, respectively. Figures 8 and 9 reveal an effective reduction of the observation noise; edges are clearly demarcated and the original image contrast is well preserved. As final comments to the numerical simulation, it is worth underlining the following. Our filtering algorithm requires only the on-off information on edge location, so no edge amplitude estimate is needed. This simplifies the edge-detection procedure and fixes the two following limiting situations in the filter performance:

1. No edge is detected; the filter behaves according to the corresponding space invariant structure.
2. All image pixels are marked as edge points (all $\rho_{i,j}$'s are zero); the noisy image is reproduced.

This means that even in the theoretically worst possible cases, unrealistic images are not produced. Moreover, we mention that numerical experiments performed by varying entries and size of $K_r$ and $K_s$, produced filtered images very similar to those reported here. Hence the overall filtering algorithm can be considered robust enough with respect to the edge detection procedure.

The presented numerical results show improved filter performances with respect to the other existing methods similarly based on the information drawn from real data (noisy picture). Furthermore, they are even comparable with the best ones obtained by using noise-free image statistics, information which is not always available in practice.

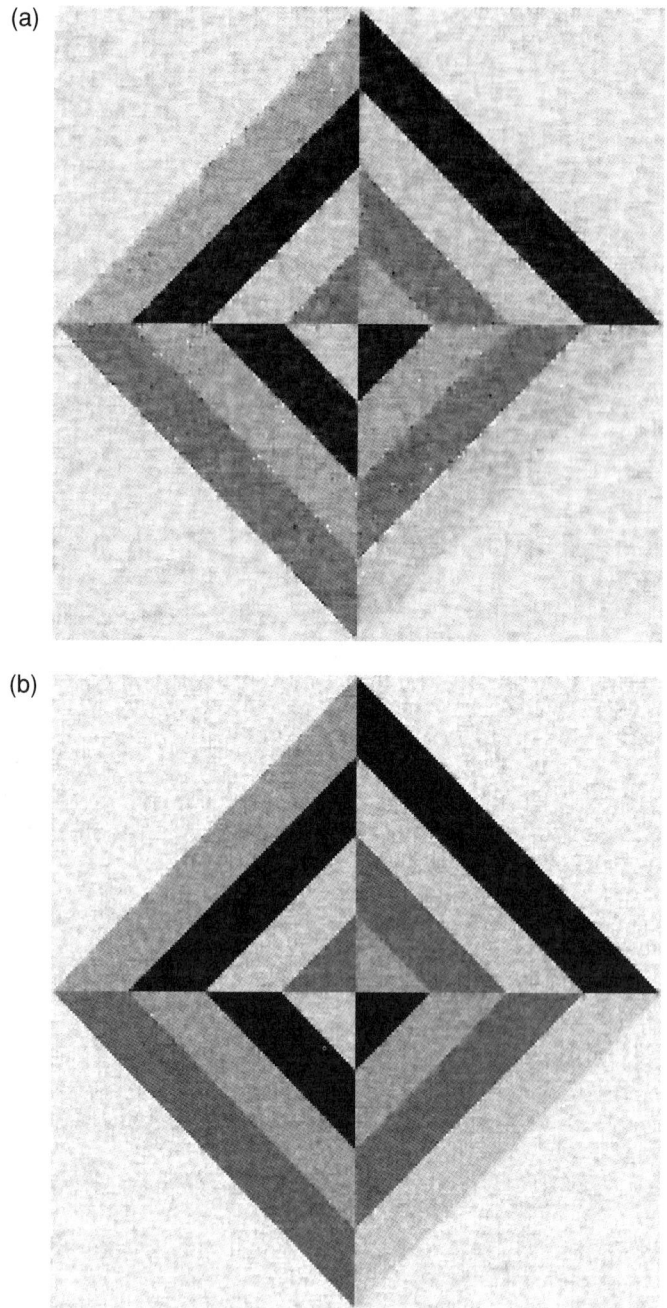

FIGURE 8. "Rhombi" filtered images corresponding to SNR = 4 (8a) and SNR = 16 (8b).

TWO-DIMENSIONAL FILTERING OF NOISY IMAGES 309

(a)

(b)

FIGURE 9. "Outdoor picture" filtered images corresponding to SNR = 4 (9a) and SNR = 16 (9b).

## VII. Conclusions

The necessity of processing real images calls for algorithms where a rapid switching of the filter characteristics are allowed. The method proposed in this work seems to be a simple and efficient way to meet this requirement. It should be emphasized that the main feature of the proposed approach is the analytical construction of the image model. Starting from the smoothness, stochastic, and inhomogeneity assumptions, a nonstationary state-space representation is obtained without the necessity of onerous identification procedures, or the *a priori* knowledge of the image autocorrelation function. The obtained model is space varying according to the presence of image edges. In this way the edge-defocusing phenomenon is greatly reduced.

The adaptive behavior of the proposed restoration method is obtained by including the information on edge locations into the image model. In this way, the filter transitions in correspondence of edge locations are not the result of heuristic procedures, but are justified on a theoretical basis because they are strictly related to the image model. It is stressed that the choice, here adopted, of detecting edges by the gradient method is just one among many existing possibilities. Other reliable edge detectors for noisy environments proposed in the literature can be used (see, e.g., references in Section IV.B and Section V.B). These characteristics make the method amenable to be applied to a large class of images; moreover, the experimental results presented in the previous section confirmed the merit of the approach by showing that high filter performances are really attainable (Section VI).

## VIII. Appendices

### A. Appendix A

In order to show that $A$ and $A'$ commute, we need to prove that

$$\sum_{m=1}^{N} a_{\ell,m} a'_{m,r} = \sum_{m=1}^{N} a'_{\ell,m} a_{m,r}, \qquad \ell, r = 1, 2, \ldots, N \qquad (112)$$

According to Eqs. (28) and (29) we obtain

$$\sum_{m=1}^{N} a_{\ell,m} a'_{m,r} = a'_{\bar{\ell},r} \qquad (113)$$

$$\sum_{m=1}^{N} a'_{\ell,m} a_{m,r} = a_{\bar{\ell}+1,r} \qquad (114)$$

where

$$\bar{\ell} = \ell + 1 + \left[\frac{\sqrt{8\ell - 7} - 1}{2}\right] \quad (115)$$

Then, from (29) and (113), it follows that all the entries of each row of the product matrix $AA'$ are zero except for the $k$th entry which is equal to 1, with $k$ given by

$$k = \ell + 2 + \left[\frac{\sqrt{8\ell - 7} - 1}{2}\right] \leq N \quad (116)$$

In the same way, from equations (28) and (114), each row of $A'A$ has null entries but the $h$th one, which is equal to 1, with $h$ given by

$$h = \ell + 1 + \left[\frac{\sqrt{8(\ell + 1) - 7} - 1}{2}\right] \leq N \quad (117)$$

According to (115), it is easy to verify that (116) and (117) are equivalent, thus proving (112).

## B. Appendix B

Here we propose a method to obtain feasible estimates of $\Psi_r$ and $\Psi_s$ under the hypothesis of a finite limited and isotropic spectrum in each subregion. Because both these matrices can be estimated in a fully analogous manner, the calculations are developed only for one of them, say, $\Psi_s$.

If, for convenience, we indicate the vector composed of the last $\bar{n} + 1$ derivatives of order $\bar{n}$ with $X^{(\bar{n})}(r, s)$, the last $\bar{n} + 1$ equations of system (27) can be rewritten as

$$\frac{\partial}{\partial s} X^{(\bar{n})}(r, s) = W_s(r, s) \quad (118)$$

By integrating this equation with respect to $r$ between two adjacent pixels of coordinates $(i + 1, j)$ and $(i + 1, j + 1)$, we only obtain the last $\bar{n} + 1$ equations of system (35) ($c^{(1)}_{i+1,j+1} = 1$):

$$X^{(\bar{n})}_{i+1,j+1} = X^{(\bar{n})}_{i+1,j} + W^{(1)\bar{n}}_{i+1,j+1} \quad (119)$$

The vector $W^{(1)\bar{n}}_{i+1,j+1}$ is composed of the last $\bar{n} + 1$ elements of $W^{(1)}_{i+1,J+1}$ and, under the hypothesis (45), it is a discrete white noise $\sim N(0, Q^{(\bar{n})}_s))$. Taking into account that from (45), (118), and (119) we obtain $Q^{(\bar{n})}_s = \Delta_s \Psi_s$, an estimate of $\Psi_s$ can be obtained through a feasible estimate of $Q_s^{(\bar{n})}$. To this purpose let us rewrite system (35) as

$$\left[\frac{\partial^n x(r,s)}{\partial r^{n-\alpha}\partial s^\alpha}\right]_{i+1,j+1} = \left[\frac{\partial^n x(r,s)}{\partial r^{n-\alpha}\partial s^\alpha}\right]_{i+1,j}$$

$$+ \sum_{l=1}^{\bar{n}-n}\left[\frac{\partial}{\partial s^l}\left(\frac{\partial^n x(r,s)}{\partial r^{n-\alpha}\partial s^\alpha}\right)\right]_{i+1,j}\frac{\Delta_s^l}{l!} + W_{k_{i+1,j+1}}^{(1)},$$

(120)

$$n = 0, 1\ldots, \bar{n}, \qquad \alpha = 0, 1, \ldots, n, \qquad k = 1, 2\ldots, N$$

where $W_{k_{i+1,j+1}}^{(1)}$ is the $k$th component of $W_{i+1,j+1}^{(1)}$.

Equation (120) is formally identical to the Taylor series expansion for the signal and its derivatives. Our idea is to get an estimate of $Q_s^{(\bar{n})}$ through a feasible estimate of the Taylor remainder. Let us indicate by $G(\omega_r,\omega_s)$ the spectrum of the image as a function of the spatial frequencies $\omega_r$ and $\omega_s$ and let us assume the following hypotheses: there exist $\bar{\omega}_s$ and $\bar{\omega}_s$ such that

$$G(\omega_s,\omega_r) = 0 \quad \text{if } \omega_s \geq \bar{\omega}_s \quad \text{and/or} \quad \omega_r \geq \bar{\omega}_r \qquad (121)$$

$$|G(\omega_s,\omega_r)| \leq \bar{K}, \quad \forall(\omega_s,\omega_r) \in [0,\bar{\omega}_s] \times [0,\bar{\omega}_r] \qquad (122)$$

By consequence of (121), the signal is of class $C^\infty$ (Papoulis, 1977). Using the two-dimensional Fourier transform, we can now rewrite Eq. (120) in the following way:

$$\left[\frac{\partial^n x(r,s)}{\partial r^{n-\alpha}\partial s^\alpha}\right]_{i+1,j+1} = \left[\frac{\partial^n x(r,s)}{\partial r^{n-\alpha}\partial s^\alpha}\right]_{i+1,j}$$

$$+ \sum_{l=1}^{\bar{n}-n} F^{-1}[(j\omega_s)^{l+\alpha}(j\omega_r)^{n-\alpha}G(\omega_s,\omega_r)]\bigg|_{i+1,j}$$

$$\times \frac{\Delta_s^l}{l!} + W_{k_i+1,j+1}^{(1)} \qquad (123)$$

where

$$F^{-1}[(j\omega_s)^{l+\alpha}(j\omega_r)^{n-\alpha}G(\omega_s,\omega_r)]|_{i+1,j}$$

$$= \frac{1}{(2\pi)^2}\int_{-\infty}^{\infty}\int_{-\infty}^{\infty}(j\omega_s)^{l+\alpha}(j\omega_r)^{n-\alpha}G(\omega_s,\omega_r)e^{(j\omega_s s+j\omega_r r)}d\omega_s d\omega_r\bigg|_{i+1,j}$$

(124)

From Eq. (123) it follows that

$$\left|W^{(1)}_{k_{i+1},j+1}\right| \leq \sum_{l=\bar{n}-n+1}^{\infty} |F^{-1}(j\omega_s)^{l+\alpha}(j\omega_r)^{n-\alpha} G(\omega_s, \omega_r)]|_{i+1,j} \frac{\Delta_s^l}{l!} \quad (125)$$

Moreover, using (121) and (122) we have

$$|F^{-1}(j\omega_s)^{l+\alpha}(j\omega_r)^{n-\alpha} G(\omega_s, \omega_r)]|_{i+1,j}$$

$$\leq \frac{1}{(2\pi)^2} \int_0^\infty \int_0^\infty 4|(j\omega_s)^{l+\alpha}(j\omega_r)^{n-\alpha} G(\omega_s, \omega_r)| d\omega_s d\omega_r \bigg|_{i+1,j}$$

$$\leq \frac{\bar{k}\bar{\omega}_s^{l+\alpha+1} \bar{\omega}_r^{n-\alpha+1}}{\pi^2 (l+\alpha+1)(n-\alpha+1)} \leq \frac{\bar{k}\bar{\omega}^{l+n+2}}{\pi^2 (l+n+2)} \quad (126)$$

where $\bar{\omega} = \max[\bar{\omega}_s, \bar{\omega}_r]$ and the condition $\alpha \leq n$ has been used.

Substituting (126) into (125) we obtain

$$\left|W^{(1)}_{k_{i+1},j+1}\right| \leq \frac{\bar{k}}{\pi^2} \sum_{l=\bar{n}-n+1}^{\infty} \frac{\bar{\omega}^{l+n+2}}{(l+n+2)!} \frac{\Delta_s^l}{l!}$$

$$= \frac{\bar{k}}{\pi^2} \sum_{j=0}^{\infty} \frac{\bar{\omega}^{\bar{n}+j+3} \Delta_s^{\bar{n}-n+1+j}}{(\bar{n}+j+3)(\bar{n}-n+1+j)!} \leq \frac{\bar{k}}{\pi^2} \bar{\omega}^{\bar{n}+3} \Delta^{\bar{n}-n+1}$$

$$\times \sum_{j=0}^{\infty} \frac{(\bar{\omega} \Delta_s)^j}{(\bar{n}-n+j+2)} \leq \frac{\bar{k}}{\pi^2} \frac{\bar{\omega}^{\bar{n}+3} \Delta_s^{\bar{n}-n+1}}{(\bar{n}-n+2)!} e^{\bar{\omega}\Delta_s} =: q_n \quad (127)$$

Inequality (127) represents an upper bound for all the $n+1$ components of $W^{(1)}_{i+1,j+1}$ that corresponds to the same order $n$ of derivation for $x(r,s)$.

Hence, for $n = \bar{n}$ we have that all the elements of $W^{(1)\bar{n}}_{i+1,j+1}$ satisfy the following inequality:

$$\left|W^{(1)}_{k_{i+1},j+1}\right| \leq \frac{\bar{k}\bar{\omega}^{\bar{n}+3} \Delta_s}{2\pi^2} e^{\bar{\omega}\Delta_s} =: q_{\bar{n}}, \quad \text{for } k = \frac{\bar{n}(\bar{n}+1)}{2} + 1, \ldots, N \quad (128)$$

which states an upper bound for the remainder relative to the Taylor series expansion of order 0 for the derivatives of order $\bar{n}$. It seems reasonable to estimate $Q_s^{(\bar{n})}$ as

$$Q_s^{(\bar{n})} = \frac{q_{\bar{n}}^2}{3} I \quad (129)$$

where $q_{\bar{n}}^2/3$ is the variance of a random variable uniformly distributed between

$[-q_{\bar{n}}, q_{\bar{n}}]$. Therefore,

$$\Psi_s = \frac{q_{\bar{n}}^2}{3\Delta_s} I \tag{130}$$

In a fully analogous manner the following estimate for $\Psi_r$ is found:

$$\Psi_r = \frac{q_{\bar{n}}'^2}{3\Delta_r} I \tag{131}$$

where

$$q_{\bar{n}}' = \frac{\bar{k}\bar{\omega}^{\bar{n}+3}\Delta_r}{2\pi^2} e^{\bar{\omega}\Delta_r}$$

## C. Appendix C

In this section we show that the noise sequence $\Theta_i^{(j_1,L)}$ satisfies the whiteness condition

$$E[\Theta_i^{(j_1,L)} \Theta_k^{(j_1,L)^T}] = \delta_{i,k} Q_i^{(j_1,L)} \tag{132}$$

where $\delta_{i,k}$ is the Kronecker delta, and we compute the covariance matrix $Q_i^{(j_1,L)}$. Whiteness of $\Theta_i^{(j_1,L)}$ is a straightforward consequence of (45)–(47) and (51)–(53).

For convenience denote by $\theta_{i,j}$, $j = j_1, \ldots, j_L$, the entries of the vector $\Theta_{i-1}^{(j_1,L)}$ defined in (67). Then the covariance matrix $Q_{i-1}^{(j_1,L)}$ can be computed as follows.

Taking into account (45)–(47) and (51)–(53) we have, for $m, k = j_2, \ldots, j_L - 1$,

$$\begin{aligned}
E[\theta_{i,m}\theta_{i,k}^T] &= \left(F_{i,m}^{(1,2)} Q_s F_{i,k}^{(1,2)^T} + F_{i,m}^{(2,1)}(H_1 Q_r H_1^T + Q_s) F_{i,k}^{(2,1)^T} + F_{i,m}^{(3)} Q_r F_{i,k}^{(3)^T} \right. \\
&\quad \left. + F_{i,m}^{(4,5)} H_5 (Q_r + Q_s) H_5^T F_{i,k}^{(4,5)^T} + F_{i,m}^{(5,4)} H_5 Q_s H_5^T F_{i,k}^{(5,4)^T} \right) \delta_{m,k} \\
&\quad + \left( F_{i,m}^{(2,1)} H_1 Q_r F_{i,k}^{(3)^T} + F_{i,m}^{(3)} Q_r H_5^T F_{i,k}^{(4,5)^T} \right. \\
&\quad - F_{i,m}^{(1,2)} Q_s H_5^T F_{i,k}^{(4,5)^T} - F_{i,m}^{(1,2)} Q_s H_5^T F_{i,k}^{(5,4)^T} \\
&\quad \left. - F_{i,m}^{(2,1)} Q_s H_5^T F_{i,k}^{(4,5)^T} - F_{i,m}^{(2,1)} Q_s H_5^T F_{i,k}^{(5,4)^T} \right) \delta_{m,k+1} \\
&\quad + F_{i,m}^{(2,1)} H_1 Q_r H_5^T F_{i,k}^{(4,5)^T} \delta_{m,k+2}
\end{aligned} \tag{133}$$

If we put

$$S_{1(m,k)} := F^{(3)}_{i,m} Q_r F^{(3)^T}_{i,k} + F^{(4,5)}_{i,m} H_5 (Q_r + Q_s) H_5^T F^{(4,5)^T}_{i,k} + F^{(5,4)}_{i,m} H_5 Q_s H_5^T F^{(5,4)^T}_{i,k} \tag{134}$$

$$S_{2(m,k)} := F^{(1,2)}_{i,m} Q_s F^{(1,2)^T}_{i,k} + F^{(2,1)}_{i,m} (H_1 Q_r H_1^T + Q_s) F^{(2,1)^T}_{i,k} + F^{(3)}_{i,m} Q_r F^{(3)^T}_{i,k} \tag{135}$$

$$\begin{aligned} T_{(m,k)} := & F^{(2,1)}_{i,m} H_1 Q_r F^{(3)^T}_{i,k} + F^{(3)}_{i,m} Q_r H_5^T F^{(4,5)^T}_{i,k} \\ & - F^{(1,2)}_{i,m} Q_s H_5^T F^{(4,5)^T}_{i,k} - F^{(1,2)}_{i,m} Q_s H_5^T F^{(5,4)^T}_{i,k} \\ & - F^{(2,1)}_{i,m} Q_s H_5^T F^{(4,5)^T}_{i,k} - F^{(2,1)}_{i,m} Q_s H_5^T F^{(5,4)^T}_{i,k} \end{aligned} \tag{136}$$

$$U_{m,k} := F^{(2,1)}_{i,m} H_1 Q_r H_5^T F^{(4,5)^T}_{i,k} \tag{137}$$

Eq. (133) can be rewritten as

$$\begin{aligned} \mathrm{E}[\theta_{i,m} \theta_{i,k}^T] = & \left(S_{1(m,k)} + S_{2(m,k)} - F^{(3)}_{i,m} Q_r F^{(3)^T}_{i,k}\right) \delta_{m,k} \\ & + T_{(m,k)} \delta_{m,k+1} + U_{(m,k)} \delta_{m,k+2}. \end{aligned} \tag{138}$$

By arguing as for (133) and using the notation of (134)–(137) it is easily verified that

$$\mathrm{E}[\theta_{i,m} \theta_{i,j_1}^T] = S_{1(m,j_1)} \delta_{m,j_1} + T_{(m,j_1)} \delta_{m,j_2} + U_{(m,j_1)} \delta_{m,j_3}, \quad m = j_1, \ldots, j_L,$$

$$\mathrm{E}[\theta_{i,j_L} \theta_{i,k}^T] = S_{2(j_L,j_L)} \delta_{j_L,k} + T_{(j_L,k)} \delta_{j_L,k+1} + U_{(j_L,k)} \delta_{j_L,k+2}, \quad k = j_1, \ldots, j_L.$$

By setting

$$S_{(j_m,j_m)} := S_{1(j_m,j_m)} + S_{2(j_m,j_m)} - F^{(3)}_{i,j_m} Q_r F^{(3)^T}_{i,j_m}$$

the matrix $Q^{(j_1,L)}_{i-1}$ has the following five-band structure:

$$Q^{(j_1,L)}_{i-1} = \begin{bmatrix} S_{1(j_1,j_1)} & T^T_{(j_2,j_1)} & U^T_{(j_3,j_1)} & 0 & 0 & \cdots & 0 \\ T_{(j_2,j_1)} & S_{(j_2,j_2)} & T^T_{(j_2,j_3)} & U^T_{(j_2,j_4)} & 0 & \cdots & 0 \\ U_{(j_3,j_1)} & T_{(j_3,j_2)} & S_{(j_3,j_3)} & T^T_{(j_3,j_4)} & U^T_{(j_3,j_5)} & 0 & 0 \\ 0 & \ddots & \ddots & \ddots & \ddots & \ddots & 0 \\ 0 & \cdots & U_{(j_L-3,j_L-4)} & T_{(j_L-2,j_L-3)} & S_{(j_L-2,j_L-2)} & T^T_{(j_L-2,j_L-3)} & U^T_{(j_L-3,j_L-2)} \\ 0 & \cdots & 0 & U_{(j_L-1,j_L-3)} & T_{(j_L-1,j_L-2)} & S_{(j_L-1,j_L-1)} & T^T_{(j_L,j_L-1)} \\ 0 & \cdots & 0 & 0 & U_{(j_L,j_L-2)} & T_{(j_L,j_L-1)} & S_{2(j_L,j_L)} \end{bmatrix}$$

## References

Anderson, G. L., and Netravali, A. N. (1976). Image restoration based on a subjective criterion, *IEEE Trans. Syst. Man. Cybern.* **6,** 845–853.

Argile, E. (1971). Techniques for edge detection, *Proc. IEEE* **59,** 285–286.

Asada, H., and Brady, M. (1986). The curvature primal sketch, *IEEE Trans. on Pattern Anal. and Machine Intell.* **8,** 2–14.

Attasi, S. (1976). Modeling and recursive estimation for double indexed sequences, in *System Identification: Advances and Case Studies,* R. K. Mehra and D. G. Lainiotis, eds., New York: Academic Press.

Azimi-Sadjadi, M. A., and Bannour, S. (1991). Two-dimensional recursive parameter identification for adaptive Kalman filtering, *IEEE Trans. on Circ. Syst.* **38,** 1077–1081.

Azimi-Sadjadi, M. R., and Khorasani, K. (1990). Reduced order strip Kalman filtering using singular perturbation method, *IEEE Trans. on Circ. Syst.* **37,** 284–290.

Barry, P. E., Gran, R., and Waters, C. R. (1976). 'Two-dimensional filtering—a state space approach', *Proc. of Conf. Decision and Control,* 613–618.

Basseville, M., and Benveniste, A. (1983). Design and comparative study of some sequential jump detection algorithms for digital signals, *IEEE Trans. Acous. Speech Sign. Process.* **31,** 521–534.

Bedini, M. A., and Jetto, L. (1991). Realization and performance evaluation of a class of image models for recursive restoration problems, *Int. J. Syst. Science* **22,** 2499–2519.

Biemond, J., and Gerbrands, J. J. (1979). An edge-preserving recursive noise-smoothing algorithm for image data, *IEEE Trans. System Man Cybernetics* **9,** 622–627.

Biemond, J., and Gerbrands, J. J. (1980). Comparison of some two-dimensional recursive point-to-point estimators based on a DPCM image model, *IEEE Trans. System Man Cybernetics* **10,** 929–936.

Billingsley, P. (1979). *Probability and Measure,* New York: Wiley.

Cannon, M. (1976). Blind deconvolution of spatially invariant image blurs with phase, *IEEE Trans. Acous. Speech Sign. Proces.* **24,** 58–63.

Carravetta, F., Germani, A., and Raimondi, M. (1996). Polynomial filtering for linear discrete non-Gaussian systems, *SIAM Jou. Contr. Opt.* **34,** 1666–1690.

Concetti, A., and Jetto, L. (1997). Two-dimensional recursive filtering algorithm with edge preserving properties and reduced numerical complexity, *IEEE Trans. Circ. Syst. Part II: Analog and Digital Signal Processing* **44**(7), 587–591.

Dalla Mora, E., Germani, A., and Nardecchia, A. (1999). 2D Filtering for images corrupted by non-Gaussian noise, *Proceedings of the 38th Conference on Decision and Control* **4,** 4167–4172.

De Santis, A., Germani, A., and Jetto, L. (1994). Space-variant recursive restoration of noisy images, *IEEE Trans. Circ. Syst. Part II: Analog and Digital Signal Processing* **41,** 249–261.

De Santis, A., Germani, A., and Raimondi, M. (1995). Optimal quadratic filtering of linear discrete-time non-Gaussian systems, *IEEE Trans. Automatic Contr.* **40**(7), 1274–1278.

De Santis, A., and Sinisgalli, C. (1999). A Bayesian approach to edge detection in noisy images, *IEEE Trans. Circ. Syst. Part I: Fundamental Theory and Applications* **46**(6), 686–699.

From, J. R., and Deutsch, E. S. (1975). On the quantitative evaluation of edge detection schemes and their comparison with human performance, *IEEE Trans. Computers* **24,** 616–628.

From, J. R., and Deutsch, E. S. (1978). A quantitative study on the orientation bias of some edge detection schemes, *IEEE Trans. Computers* **27,** 205–213.

Geman, S., and Geman, D. (1984). Stochastic relaxation, Gibbs distributions and the Bayesian restoration of images, *IEEE Trans. on Pattern Anal. and Machine Intell.* **6,** 721–741.

Germani, A., and Jetto, L. (1988). Image modeling and restoration: A new approach, *Circuits Syst. Sign. Process.* **7,** 427–457.

Goodwin, G. C., and Payne, R. L. (1977). Dynamical system identification: Experiment design and data analysis, in *Mathematical Science Engineering,* R. Bellman, ed., New York: Academic Press.

Habibi, A. (1972). Two-dimensional Bayesian estimate of images, *Proc. IEEE* **60,** 878–883.

Huertas, A., and Medioni, G. (1986). Detection of intensity changes with subpixel accuracy using Laplacian-Gaussian masks, *IEEE Trans. on Pattern Anal. and Machine Intell.* **8,** 651–664.

Hwang, T. S. (1971). Image processing, *Proc. IEEE* **59,** 1586–1609.

Jain, A. K. (1977). Partial differential equations and finite-difference methods in image processing. Part I: Image representation, *J. Optim. Theory Appl.* **23,** 470–476.

Jain, A. K., and Angel, E. (1974). Image restoration, modeling and reduction of dimensionality, *Trans. Comput.* **23,** 470–476.

Jain, A. K., and Jain, J. R. (1978). Partial differential equations and finite difference methods in image processing. Part II: image restoration, *IEEE Trans. Autom. Control* **23,** 817–833.

Jeng, F. C., and Woods, J. W. (1988). Inhomogeneous Gaussian image models for estimation and restoration, *IEEE Trans. on Acous. Speech and Sign. Process.* **36,** 1305–1312.

Jeng, F. C., and Woods, J. W. (1991). Compound Gauss-Markov random fields for image estimation, *IEEE Trans. Signal Process.* **10,** 225–254.

Jetto, L. (1999a). On the optimality of a new class of 2D recursive filters, *Kybernetica* **35**(6), 777–792.

Jetto, L. (1999b). Stochastic modelling and 3-D minimum variance recursive estimation of image sequences, *Multidimensional Systems and Signal Processing* **39,** 683–697.

Katayama, T. (1980). Estimation of images modeled by a two-dimensional separable autoregressive process, *IEEE Trans. Autom. Control* **26,** 1199–1201.

Katayama, T., and Kosaka, M. (1978). Smoothing algorithms for two-dimensional image processing, *IEEE Trans. Systems Man Cybernet* **8,** 62–66.

Katayama, T., and Kosaka, M. (1979). Recursive filtering algorithm for a two-dimensional system, *IEEE Trans. Autom. Control* **24,** 130–132.

Kaufman, H., Woods, J. W., Dravida, S., and Tekalp, A. M. (1983). Estimation and identification of two-dimensional images, *IEEE Trans. Autom. Control* **28,** 745–756.

Keshavan, H. R., and Srinath, M. D. (1977). Sequential estimation technique for enhancement of noisy images, *IEEE Trans. on Computers* **26,** 971–987.

Keshavan, H. R., and Srinath, M. D. (1978). Enhancement of noisy images using an interpolative model in two dimensions, *IEEE Trans. Syst. Man Cybern.* **8,** 247–259.

Lagendijk, R. L., Biemond, J., and Boekee, D. E. (1990). Identification and restoration of noisy blurred images using the expectation-maximization algorithm, *IEEE Trans. Acous. Speech Sign. Proces.* **38,** 1180–1191.

Lunscher, H. J., and Beddoes, M. P. (1986). Optimal edge detector design I: parameter selection and noise effects, *IEEE Trans. Pattern Analysis and Mach. Intell.* **8,** 164–177.

Mallat, S., and Hwang, W. L. (1992). Singularity detection and processing with wavelets, *IEEE Trans. on Inf. Theory* **38,** 617–643.

Marr, D., and Hildreth, E. (1980). Theory of edge detection, *Proc. Royal Society of London.* **207,** 187–217.

Murphy, M. S., and Silverman, L. M. (1978). Image model representation and line-by-line recursive restoration, *IEEE Trans. Autom. Control* **23,** 809–816.

Murphy, M. S. (1980). Comments on recursive filtering algorithm for a two-dimensional system, *IEEE Trans. Autom. Control* **25,** 336–338.

Nahi, N. E. (1972). Role of recursive estimation in statistical image enhancement, *Proc. IEEE* **60,** 872–877.

Nahi, N. E., and Assefi, T. (1972). Bayesian recursive image estimation, *IEEE Trans. Comput.* **21,** 734–738.
Nahi, N. E., and Franco, C. A. (1973). Recursive image enhancement—vector processing, *IEEE Trans. Comm.* **21,** 305–311.
Nalva, V. S., and Binford, T. O. (1986). On detecting edges, *IEEE on Pattern Analysis and Mach. Intell.* **8,** 699–714.
Panda, D. P., and Kak, A. C. (1976). Recursive filtering of pictures, *Tech. Rep. TR–EE–76,* School of Electr. Engin., Purdue University, Lafayette, also in Rosenfield, A., ad Kak, A. C., (1976). *Digital Picture Processing,* New York: Academic-Press.
Papoulis, A. (1977). *Signal Analysis,* New York: McGraw-Hill.
Powell, S. R., and Silverman, L. M. (1974). Modelling of two-dimensional covariance function with application to image enhancement, *IEEE Trans. Autom. Control* **19,** 8–13.
Rajala, S. A., and De Figueiredo, R. J. P. (1981). Adaptive nonlinear restoration by a modified Kalman filtering approach, *IEEE Trans. Acoust. Speech and Sign. Process.* **29,** 1033–1042.
Rosenfield, A., and Kak, A. C. (1982). Digital picture processing, 2, New York: Academic Press.
Söderström, T., and Stoica, P. (1989). *System Identification,* London: Prentice Hall.
Strintzis, M. G. (1976). Comments on two-dimensional Bayesian estimate of images, *Proc. IEEE* **64,** 1255–1257.
Suresh, B. R., and Shenoi, B. A. (1979). The state-space realization of a certain class of two-dimensional systems with applications to image restoration, *Computer Graphics and Image Processing* **11,** 101–110.
Suresh, B. R., and Shenoi, B. A. (1981). New results in two-dimensional Kalman filtering with applications to image restoration, *IEEE Trans. on Circ. Syst.* **28,** 307–319.
Tekalp, A. M., Kaufman, H., and Woods, J. W. (1986). Identification of image and blur parameters for the restoration of noncausal blurs, *IEEE Trans. Acous. Speech Sign. Process.* **34,** 963–971.
Tekalp, A. M., and Kaufman, H. (1988). On statistical identification of a class of linear space-invariant image blurs using non-minimum phase ARMA models, *IEEE Trans. Acous. Speech Sign. Process.* **36,** 1360–1363.
Tekalp, A. M., Kaufman, H., and Woods, J. W. (1989). Edge-adaptive Kalman filtering for image restoration with ringing suppression, *IEEE Trans. Acoust. Speech and Sign. Process* **29,** 892–899.
Torre, V., and Poggio, T. A. (1986). On edge detection, *IEEE Trans. on Pattern Analysis, and Mach. Intell.* **8,** 147–163.
Wellstead, P. E., and Caldas Pinto, J. R. (1985a). Self-tuning filters and predictors for two-dimensional systems. Part I: Algorithms, *Int. J. Contr.,* **42,** 457–478.
Wellstead, P. E., and Caldas Pinto, J. R. (1985b). Self-tuning filters and predictors for two-dimensional systems. Part II: Smoothing applications, *Int. J. Contr.* **42,** 479–496.
Woods, J. W., and Radewan, C. H. (1977). Kalman filtering in two dimensions, *IEEE Trans. Inform. Theory* **23,** 809–816.
Woods, J. W., Dravida, S., and Mediavilla, R. (1987). Image estimation using doubly stochastic Gaussian random field models, *IEEE Trans. Pattern Analysis and Mach. Intell.* **9,** 245–253.
Wu, W. R., and Kundu, A. (1992). Image estimation using fast modified reduced update Kalman filter, *IEEE Trans. Sign. Process* **40,** 915–926.
Yum, Y. H., and Park, S. B. (1983). Optimum recursive filtering of noisy two-dimensional data with sequential parameter identification, *IEEE Trans. Pattern Anal. Mach. Intell.* **5,** 337–344.
Zou, C. T., Plotkin, E. I., and Swamy, M. N. S. (1994). 2D fast Kalman algorithms for adaptive estimation of nonhomogeneous Gaussian Markov random field model, *IEEE Trans. Circ. Syst.* **41,** 678–692.

# Index

## A

Affinity, 165
Algebraic properties, basic, 5–6
  for fuzzy morphology, 35–37
  for soft morphology, 19–20
Analog neural network, 224
Antialias filtering, 206
Antiextensivity-extensivity, 6
  fuzzy soft, 36
  vector soft, 19–20
Arithmetic unit, 47
Array of registers, 47

## B

Bandlimit shape, 211–217
Barnes-Wall lattice, 85
Binary image processing, 195
Binary linear code, 93, 195
  comparison of hexagonal and rectangular skeletonization programs, 240–246
  connectivity, 237–238
  distance functions, 238–239
  line thinning and skeleton of an object, 239–240
  measurement of distance, 238
  morphological operators, 239
  tomography, 246–247
Binary mathematical morphology, 3–5
  soft, 14–15

Blocking effect, 56
Blurring, 292–293

## C

Cartan matrices, 80–81, 117–118
Causality, 129
Causal space-variant filter (CSVF), 296
Charge coupled devices (CCDs), TV camera, 209–211
Cheng Kung University, 233
Chung Cheng Institute modular system, 233
Clip4 array, 223–224
Closing-opening scale-space, multiscale, 146–153
Coarse-grain arrays, 225
Coarse-to-fine tracking, 126
Color image processing, vector morphology for, 9–13
  soft, 16–23
Color impulse noise model, 23
Compression ratio, 56, 69
Connectivity, hexagonal, 195, 237–238
Constitutive equation (CE), 282–285
Construction A, 95–96
Construction B, 96–97
Construction C, 97
Continuity property, 126, 141–143
Convex structuring functions, 136–137
Covering radius, 84
Cutoff frequency, 210, 211
Cytocomputer, 231–232
Cyto-HSS, 232

## D

Datacube, 233
DCT, 56
Deblurring, 292–293
Deep holes, 84
Digital image acquisition, 205–211
Digitization, 205
Dilation, 4–5, 6
  binary soft, 14–15
  fuzzy, 13–14, 27–28
  multiscale, 139–145
  vector soft, 19
Dimensionality, 164–168
Dirac delta functions, 106
Discrete wavelet transform (DWT), 58–59, 61
Distance functions, 238–239
Distortion function, 65–66
  minimization of, 69–70
Distributivity, 5
  fuzzy soft, 35–36
Dual gray-scale reconstruction, 173
Duality theorem, 5
  fuzzy soft, 35
  vector soft, 19
Dual lattice, 75–76
  closest point of, 87

## E

Edge detection, 289–291
Edge detectors
  hexagonal, 251–252
  Sobel, 194, 250–251
Edge effects, 132
Edge estimation algorithm, 298–300
Edge spread function (ESF), 213
Elliptic poweroid structuring function, 138
ElorOptronics Ltd., 233

Entropy coding, 105–111
Equal-slope algorithm, 60
Erosion, 2, 3–4
  binary soft, 14–15
  fuzzy, 13–14, 27–28
  multiscale, 139–145
  vector soft, 19

## F

Fast Fourier transform (FFT), 196
Fine-grain arrays, 222–224
Fingerprints
  definitions, 153–154
  equivalence of, 154–156
  reduced, 156–157
  reduced, computation of, 157–161
  of a signal, 127–128
Fourier transform, 56
  hexagonal, 247
Frame-grabber digitizer, 212
Fuzzy fitting, 27, 29
Fuzzy soft mathematical morphology
  algebraic properties for, 35–37
  compatibility with soft morphology, 34
  definitions, 27–33
  erosion and dilation, 13–14, 27–28
  role of, 2

## G

Gamma function, 68
Gaussian scale-space, 125–128
Generalized Gaussian function (GGF), 67–68, 73–74
Geometric transformations, 247–248
GLOPR, 233–234
Glue vectors, 76, 87
Gradient watershed region, 176–179
Gram matrix, 75, 76, 81

Granulometries, 134
Gray-scale mathematical morphology
  flat structuring elements and, 6–7
  flat structuring elements and soft, 15–16
  gray-scale structuring elements and, 7–9
  gray-scale structuring elements and soft, 16
  role of, 2
Green's function, 128–129

# H

Hamming distance, 94
Hamming weight, 94
HARTS, 225
Heat equation, 129
Hexagonal edge detectors, 251–252
Hexagonal fast Fourier transform (HFFT), 247
Hexagonal Fourier transform, 247
Hexagonal image processing, binary, 195
  comparison of hexagonal and rectangular skeletonization programs, 240–246
  connectivity, 237–238
  distance functions, 238–239
  line thinning and skeleton of an object, 239–240
  measurement of distance, 238
  morphological operators, 239
  tomography, 246–247
Hexagonal image processing, processor architectures, 194
  coarse-grain arrays, 225
  fine-grain arrays, 222–224
  hexagonal pipelines, 233–237
  parallel, 221–222
  pipelined, 229–233
  pyramid, 225–229
  single-instruction, single-datum (SISD) computers, 194, 218–221
  two- and multidimensional arrays, 222–225
Hexagonal sampling of images
  conclusions, 256–259
  digital image acquisition, 205–211
  hexagon shaped sensor elements, 198
  measurement of 2D modulation transfer function and bandlimit shape, 211–217
  noise and quantization error, 205
  packing of retinal sensory elements, 196–198
  sampling grids, 192
  two-dimensional theory, 199–204
  use of term, 193
Histogram technique, 47–49
Homogeneous image equation, 276–278
Homothety, 165
Homotopy modification of gradient functions, 173–176
Hough transform, 194
Huffman codebook, 57, 114
Human eye, packing of retinal sensory elements, 196–198
Human perception factor, 66

# I

Idempotency, 6, 132
  fuzzy soft, 36–37
  vector soft, 19
IDSP, 233
Illiac III (Illinois pattern recognition computer), 224

Image compression system
  components of, 56
  compression ratio, 56, 69
Image gathering, 205
Image restoration
  deblurring, 292–293
  edge problem, 289–291
  filtering algorithm, 291–292
  state-space representation, 285–289
Image widow-management module, 44
Increasing operations, 5
  fuzzy soft, 35
  vector soft, 19
Inhomogeneity assumption, 270, 271
Interline transfer (ILT) device, 209

## J

Jaguar, 233

## K

Kalman filtering, 268
  correction, 275
  detectability, 276
  estimation algorithm, 274–275
  one-step prediction, 275
  stabilizability, 276
  state-space model, 271–274
  steady-state solution, 275–276
Kalman gain, 275
Kiwivision, 233
Knife-edge technique, 212
Kydon, 224

## L

Lagrange multiplier technique, 69
Laminated latices, 82–85
Laplacian distribution, 68

Laplacian-of-Gaussian filter, 129
Lattice points, counting
  construction A, 95–96
  construction B, 96–97
  construction C, 97
  estimation of points within a sphere, 89–91
  number of points on a sphere, 91–93
  relationship between lattices and codes, 93–100
Lattice quantization (LQ)
  advantages of, 70
  codebook for, 71
  conclusions, 116–117
  distortion measure and quantization regions, 71–73
  entropy coding, 105–111
  experimental results, 111–116
  optimal quantizer for wavelet coefficients, 73–75
  scaling algorithm, 97–100
Lattices
  Barnes-Wall, 85
  defined, 75
  dual, 75–76
  laminated, 82–85
  selecting, for quantization, 100–105
Lattices, quantization algorithms for, 85
  closest point of a dual lattice, 87
  $D_n$ lattice and its dual, 88–89
  laminated $\Lambda_{16}$ lattice, 89
  utility functions, 86–87
  $Z_n$, 88
Lattices, root
  Cartan matrices, 80–81, 117–118
  construction of, 81–82
  root systems, 76–80
Lie algebras, 76, 77
Linear filters, 248–250

Line-scan filters, 294
  causal space-variant filter (CSVF), 295–296
  semicausal space-variant filter (CSVF), 297–298
Line segmentation, 240
Line tracking, 239

# M

Majority gate algorithm
  architecture for decomposition of soft structuring elements, 44–47
  description of, 39–40
  systolic array implementation for soft filtering, 40–44
Marker function, 173
Marr-Hildreth edge detector, 126
Mathematical morphology
  basic algebraic properties, 5–6
  binary, 3–5
  role of, 1–2
  vector, for color image processing, 9–13
Mathematical morphology, fuzzy soft
  algebraic properties for, 35–37
  compatibility with soft morphology, 34
  definitions, 27–33
  erosion and dilation, 13–14, 27–28
  role of, 2
Mathematical morphology, gray-scale
  flat structuring elements and, 6–7
  flat structuring elements and soft, 15–16
  gray-scale structuring elements and, 7–9
  gray-scale structuring elements and soft, 16
  role of, 2
Mathematical morphology, soft
  algebraic properties for, 19–20
  application, 2
  binary, 14–15
  compatibility with fuzzy morphology, 34
  conclusions, 52
  gray-scale, with flat structuring elements, 15–16
  gray-scale, with gray-scale structuring elements, 16
  majority gate algorithm, 39–47
  role of, 2
  threshold decomposition, 38–39
  vector, for color image processing, 16–23
MATLAB, 61
Maximum principle, 129
Mean-squared error (MSE)
  distortion measure, 73, 100
mergesort, 37
Modulation transfer function (MTF), 210
  measurement of 2D, 211–217
Monochrome image processing, 195–196
  edge detectors, 250–251
  edge detectors, hexagonal, 251–252
  geometric transformations, 247–248
  hexagonal Fourier transform, 247
  point source location/star tracking, 248
  visual appearance of edges and features, 252–256
Monochrome image processing filters
  linear, 248–250
  nonlinear, 250

Monotone property, 126, 144–146
  for gradient watershed region, 177–179
  for multiscale closing-opening, 151–153
  for watershed transform, 172
Moore-Penrose pseudoinverse, 302
Multidimensional arrays, 222–225
Multiple-instruction, multiple-datum (MIMD) computers, 222
Multiscale morphology
  definition and development of mathematical, 131–133
  scale-dependent, 134–138
  semigroup and general properties of structuring function, 138
Multiscale scale-space, closing-opening
  defined, 132, 147
  monotone property for, 151–153
  fingerprint, 153
  properties of, 147–151
  zero-crossings, 146–147
Multiscale scale-space, dilation-erosion
  continuity and order properties, 141–143
  defined, 131–132, 139
  fingerprint, 153
  monotone property, 144–146
  signal extrema, 143–145
  smoothing example, 139–141

# N

Neighborhood of a point, 160
New Jersey Institute of Technology, 233
Noise, 205

Nonlinear filters, 250
Nyquist's sampling theorem, 197

# O

Optimal bit allocation, 65–70
Order index, 14, 20
Order statistic modules (OSMs), 47

# P

Packing radius, 84
Parallel processors, 221–222
Patches, 149–151
PIPE, 232
Pipelined processors, 229–233
  hexagonal, 233–237
Pipeline Processor Farm, 233
Point source location/star tracking, 248
Polynomial filtering, 300–302
Poweroid structuring function, 137–138, 168–170
Prefiltering, 30
PREP, 233
Probability distribution function (PDF), 65–68, 73–75
Processor architectures, 194
  coarse-grain arrays, 225
  fine-grain arrays, 222–224
  hexagonal pipelines, 233–237
  parallel, 221–222
  pipelined, 229–233
  pyramid, 225–229
  single-instruction, single-datum (SISD) computers, 194, 218–221
  two- and multidimensional arrays, 222–225
PSC circuit, 224
Pyramid processors, 225–229

## Q

Quadratic form, 75
Quadratic structuring function (QSF), 138, 170
Quadrature pyramid, 225
Quantization
  *See also* Lattice quantization
  vector, 56, 70
  error, 205
Quantization of wavelet coefficients, 57
  discrete wavelet transform (DWT), 58–59
  distortion function, 65–66
  distortion function, minimization of, 69–70
  fundamentals of, 59–60
  information distribution across coefficient matrix, 60–65
  optimal bit allocation, 65–70
  statistical model of wavelet coefficients, 66–68
quicksort, 37

## R

Rectangular fast Fourier transform (RFFT), 247
Rectangular skeletonization programs, comparison of hexagonal and, 240–246
Reed-Muller code of length, 85, 89, 93
Riccati equation, 275
  steady-state, 276
Root lattices. *See* Lattices, root
Root systems, 76–80

## S

Scaled structuring function, 135

Scale-space(s)
  *See also* Multiscale morphology; Multiscale scale-space
  computer code, 184–186
  development of, 124–125, 126
  filtering, 125
  future work, 180–181
  Gaussian, 125–128
  gradient watershed region, 176–179
  homotopy modification of gradient functions, 173–176
  image, 125–126
  limitations, 180
  mathematical results, proof of, 181–184
  for regions, 170–179
  related work and extensions, 128–131
  summary, 179
  watershed transform, 171–173
Scale-space, fingerprints
  definitions, 153–154
  equivalence of, 154–156
  reduced, 156–157
  reduced, computation of, 157–161
  of a signal, 127–128
Scale-space, structuring functions, 132
  convex, 136–137
  dimensionality, 164–168
  patches of, 149–151
  powcroid, 137–138, 168–170
  quadratic, 138, 170
  scaled, 135
  semigroup and general properties of, 138, 161–162
  threshold set of, 135
  umbra and shape, 162–164
Scaling algorithm, 97–100
Scaling operation, 95

# INDEX

Semicausal space-variant filter (SCSVF), 297
Semigroup property, 138, 161–162
Shannon lower bound, 72
Shrinking algorithm, 160
Signal-to-noise ratio (SNR), 61, 63–65, 198
Silicon retina with correlation-based, velocity-tuned pixels, 224
Single-instruction, multiple-datum (SIMD) computers, 222
Single-instruction, single-datum (SISD) computers, 194, 218–221
Six-connectedness, 238
Skeletonization programs, comparison of hexagonal and rectangular, 240–246
Smoothness assumption, 270–271
Sobel edge detector, 194, 250–251
Soft mathematical morphology
  algebraic properties for, 19–20
  application, 2
  binary, 14–15
  compatibility with fuzzy morphology, 34
  conclusions, 52
  gray-scale, with flat structuring elements, 15–16
  gray-scale, with gray-scale structuring elements, 16
  majority gate algorithm, 39–47
  role of, 2
  threshold decomposition, 38–39
  vector, for color image processing, 16–23
Space-variant filters
  causal (CSVF), 295–296
  component equations, 279–280
  conclusions, 310
  constitutive equation, 282–285
  deblurring, 292–293
  edge problem, 289–291
  filtering algorithm, 291–292
  homogeneous image equation, 276–278
  image restoration, 285–293
  image signal, basic assumptions, 270–271
  Kalman filtering, 271–276
  noise, modeling of, 280–281
  numerical results, 303–309
  research on, 268–270
  semicausal (CSVF), 297–298
  state-space representation, 285–289
Standard Gaussian distribution, 68
Star tracking, 248
State-space model, 271–274
Steady-state Riccati equation (SSRE), 276
Stochastic assumption, 270–271
Strip processor, 291
Structuring element decomposition, soft morphology, 23–27
  architecture for, 44–47
Structuring element-management module, 44
Structuring elements, 2, 6
Structuring elements, flat
  gray-scale morphology with, 6–7
  gray-scale soft morphology with, 15–16
Structuring elements, gray-scale
  gray-scale morphology with, 7–9
  gray-scale soft morphology with, 16
Structuring functions, 132
  convex, 136–137
  dimensionality, 164–168
  patches of, 149–151
  poweroid, 137–138, 168–170

quadratic, 138, 170
scaled, 135
semigroup and general properties of, 138, 161–162
threshold set of, 135
umbra and shape, 162–164
Successive approximation technique, 47
Symmetrical hexagonal coordinate frame, 219
Systolic array implementation for soft morphological filtering, 40–44

## T

Texas Instruments Pipelines, 233
Theta functions, 92–93
Thinning, 239, 240
Threshold decomposition, 38–39
Thresholding, 239
Threshold set, 135
TITAN, 232
Tomography, 246–247
Translation invariance, 5
   fuzzy soft, 35
   vector soft, 19
Two-dimensional sampling theory, 199–204

## U

Uncertainty principle, 126
Uniform vector quantizer, 72
University of Belfast pipeline, 232
University of California pipeline, 232
University of Strathclyde pipeline, 232
University of Warwick pipeline, 234

## V

Vector morphology for color image processing, 9–13
   soft, 16–23
Vector quantization (VQ), 56, 70
Vector standard morphological operation implementation, 50–52

## W

Watershed transform, 171–173
   gradient watershed region, 176–179
Wavelet coefficients
   *See also* Quantization of wavelet coefficients
   high-frequency, 67–68
   low-frequency, 68
   statistical model of, 66–68
Wavelet transform, applications, 56
Weight distribution of C, 94
Weighted order statistics. *See* Majority gate algorithm
Weight enumerator of C, 94

## Y

Yonsei University real-time system, 233

## Z

Zero-crossings, 146–147

ISBN 0-12-014761-0